T0343061

Palgrave Studies in the History of Science and Technology

Designed to bridge the gap between the history of science and the history of technology, this series publishes the best new work by promising and accomplished authors in both areas. In particular, it offers historical perspectives on issues of current and ongoing concern, provides international and global perspectives on scientific issues, and encourages productive communication between historians and practicing scientists.

More information about this series at
http://www.palgrave.com/gp/series/14581

John M. Logsdon

Ronald Reagan and the Space Frontier

John M. Logsdon
Space Policy Institute
The George Washington University
Washington, DC, USA

Palgrave Studies in the History of Science and Technology
ISBN 978-3-319-98961-7 ISBN 978-3-319-98962-4 (eBook)
https://doi.org/10.1007/978-3-319-98962-4

Library of Congress Control Number: 2018953334

Cover image courtesy of Danita Delimont / Alamy Stock Photo
Cover design by Tom Howey
Author photo © Robert Levy Photography/rhlphoto.com

This Palgrave Macmillan imprint is published by the registered company Springer Nature Switzerland AG
The registered company address is: Gewerbestrasse 11, 6330 Cham, Switzerland

For Roslyn, again and always

Preface and Acknowledgments

This study of space policy during the presidency of Ronald Reagan is the third in my studies of the space policy decisions of U.S. presidents, following *John F. Kennedy and the Race to the Moon* (2010) and *After Apollo? Richard Nixon and the American Space Program* (2015). It has been by far the most difficult of the three to bring to completion.

There are a number of reasons for that observation. Both the Kennedy and the Nixon books covered decisions made during a three-year period, and each basically had a single narrative thread. In Kennedy's case, that thread was first deciding to enter a space race with the Soviet Union focused on send Americans to the lunar surface and safely back to Earth "before this decade is out," and then taking the steps to turn that decision into a warlike but peaceful mobilization of the human, financial, and technological resources required for success. In Nixon's case, it was rejecting the Apollo-like post-Apollo program proposed in the immediate aftermath of the *Apollo 11* lunar landing and spending the next two years debating what the post-Apollo program should be. The outcome was a program to be carried out at a lower priority and funding level, centered on developing the space shuttle, a means for providing highly capable but regular and lower-cost access to orbit.

This study covers the totality of Ronald Reagan's eight years as president. It does not deal with Reagan's Strategic Defense Initiative ("Star Wars") or other national security space issues; its focus is civilian and commercial space policy during the Reagan administration. There is no single narrative that runs throughout the eight years Reagan was in the White House, except perhaps his consistent enthusiasm regarding how the space effort embodied the best qualities of American exceptionalism and leadership. The Reagan administration had to deal with the realities of the space shuttle, both once it began flying in 1981 and after the tragic *Challenger* accident in 1986. The president decided to approve the "next logical step" in space development, a multiyear, multibillion dollar space station program. Administration interest in the economic payoffs from space activity led to an overly enthusiastic effort to commercialize various aspects of that activity. The administration made international partnerships a key element of its approach to space policy, first with "friends and allies" in the space station and then tentatively with the Soviet Union under Mikhail Gorbachev's leadership. Despite continuing suggestions of the need for setting a destination for human space flight, either a return to the Moon or initial journeys to Mars, the administration went no further than in its final year that setting as a long-range goal expanding "human presence and activity beyond Earth orbit." Telling each of these separate stories adequately has required a lengthy manuscript.

Most Reagan administration space decisions emerged from a contentious and slow-moving interagency process, involving a wide range of executive agencies and diverse elements of the White House staff. Ronald Reagan as president almost never made "top down" decisions, at least with respect to the space program. He much preferred to have options for his choice emerge from this process. Thus, chronicling space policy development during the eight years of the Reagan administration also required delving deeply into the details of interagency arguments on a wide variety of issues.

This proved to be a challenge. As I began research on this book, I discovered that many of the relevant documents, especially at the Ronald Reagan Presidential Library in Simi Valley, California, had originally borne a security classification, had been withdrawn from the archives for declassification review, and had never been reviewed. When I early on filed a Freedom of Information request for access to some of these documents, I was informed that the backlog before my request could be considered was over 12 years! At that point I seriously considered giving up on the study. I changed my mind only after my colleague Dr. Dwayne Day told me about a collection of declassified documents available from the Central Intelligence Agency called CREST, the CIA Records Search Tool (www.cia.gov/library/readingroom/collection/crest-25-year-program-archive). That collection contains more than 11 million pages of declassified documents. Because during the Reagan administration the Intelligence Community was an active participant in interagency space discussions, many of the critical papers I needed for my research could be found in the CREST database. Needless to say, I am very grateful to Dr. Day for introducing me to CREST and thereby allowing me to move forward with my research. Without his suggestion, this book likely would not have been possible.

Even with access to the CREST document collection, I am quite aware that in preparing this study I have not had access to all of the relevant primary source material in the Reagan archives. I have tried to compensate for this lack of access by seeking alternative sources of primary documents and a number of interviews with participants in Reagan administration policy activities. Still, as I have told many colleagues in the course of preparing this account, I feel that I have been trying to put together a 10,000-piece jigsaw puzzle with only 7000 pieces. I can only hope that the missing pieces are not critical to the validity of the study, but recognize that important information or perspectives may be missing. I of course accept responsibility if that should turn out to be the case.

I have a lot of people to thank. The many individuals who shared their recollections with me in formal interviews were essential to giving the narrative nuance and context. In addition, interviewees Hans Mark, Gil Rye, Courtney Stadd, Peggy Finarelli, and Darrell Branscome shared material from their files. So did David Thompson of Orbital Sciences Corporation (now Northrup Grumman Innovation Systems), David Lippy, founder in the 1980s of the Center for Space Policy, Professor Howard McCurdy from American University, Dr. Valerie Neal of the Smithsonian National Air and Space Museum, and Bud Evans, formerly at NASA. Rye, Stadd, Finarelli, and Evans read portions of the manuscript and provided valuable comments. My colleague Stephen Smith of the Kennedy Center Visitors Center read almost every chapter in draft and provided both helpful comments and continuing encouragement. At the Reagan Library, Jennifer Mandel helped me navigate the archives and apologized for the lack of access to unprocessed documents, and Michael Pinckney was very responsive in providing the illustrations in the book. Liana Sherman and Allyson Reneau provided essential research assistance. Elspeth Tupelo of Twin Oaks Indexing put together the comprehensive index.

As I finished the manuscript for my 2015 *After Apollo?* study, I wrote in its preface: "My wife Roslyn has provided the loving foundation of my life. Maybe now that this long-running opus is finished we can find more time to enjoy life together." That turned out to be a false hope; working on this sturdy has proven even more time- and attention consuming. All I can do is repeat the hope. In 1970, I dedicated my very first book *The Decision to Go to the Moon* to my wife; after almost a half-century, she deserves another recognition of my appreciation of her love and forbearance.

Washington, DC
May 2018

John M. Logsdon

Contents

Abbreviations

APP	American Presidency Project (http://www.presidency.ucsb.edu)
AWST	*Aviation Week & Space Technology*
CREST	CIA Records Search Tool (www.cia.gov/library/readingroom/collection/crest-25-year-program-archive)
NARA	National Archives and Record Administration, Greenbelt, Maryland
NHRC	NASA Historical Reference Collection, NASA Headquarters, Washington, DC
NYT	*The New York Times*
RRL	Ronald Reagan Presidential Library, Simi Valley, CA
WP	*The Washington Post*

List of Figures

CHAPTER 1

A Cowboy Comes to Washington

"Space truly is the last frontier."
Ronald Reagan, handwritten diary entry, June 11, 1985

On November 4, 1980, Ronald W. Reagan was elected to be the 40th president of the United States. Reagan won the presidency by only a slim majority of the popular vote, 50.7 percent, compared to President Jimmy Carter's 41 percent and third-party candidate John Anderson's 6.6 percent. But Reagan was the winner in 44 states, capturing a landslide total of 489 electoral votes compared to Carter's meager 49. Reacting to Jimmy Carter's dour four years in the White House, and before that Richard Nixon's failed presidency, voters "sought relief in the … mythic imagery of America's limitless potential and special mission. They summoned a cowboy hero out of the West, a nostalgic flashback out of the heart of America."[1]

As president, Ronald Reagan would face a challenging array of issues, among them double-digit inflation, a burgeoning government deficit, declining military power, and geopolitical and ideological challenges from the communist-led Soviet Union. But perhaps more significant, he would face what President Jimmy Carter in July 1979 had labeled a "crisis of confidence" among the American people. Addressing the nation, Carter had said: "It is a crisis that strikes at the very heart and soul and spirit of our national will. We can see this crisis in the growing doubt about the meaning of our own lives and in the loss of a unity of purpose for our Nation. The erosion of our confidence in the future is threatening to destroy the social and the political fabric of America." It would be up to Reagan, whose "idiosyncratic conservatism, which combined forward-looking optimism with his deep regard for America's heritage and the idea of American exceptionalism," to counter Carter's assessment of national "malaise."[2]

J. M. Logsdon, *Ronald Reagan and the Space Frontier*,
Palgrave Studies in the History of Science and Technology,
https://doi.org/10.1007/978-3-319-98962-4_1

One facet of that perceived lack of a positive national spirit was the depressed status of the U.S. space program. A decade after the triumphant first human footsteps on the surface of the Moon, president-elect Reagan was told that "the U.S. civil space program stands at a crossroads ... NASA [National Aeronautics and Space Administration] and the space program are without clear purpose or direction ... The year 1980 finds NASA in an untenable position."[3]

This study chronicles how President Ronald Reagan and his administration over their eight years in the White House reacted to this assessment. Aerospace historian Andrew Butrica has suggested that "Ronald Reagan's two terms as president saw the United States undertake more new and more large space initiatives than any previous administration since that of John Kennedy," and that the programs and policies initiated during the Reagan administration would "cause the 1980s to be remembered as a major turning point in space history." He concluded that "Reagan would pass on a lasting legacy to the nation's space program."

The following pages provide the basis for evaluating this judgment. It contains a detailed account of the Reagan administration's engagement with the U.S. civilian space program. The word *civilian* is important here. Probably the first space-related Reagan initiative that comes to many people's mind is his March 1983 proposal to create a defense against ballistic missile attack, the Strategic Defense Initiative (SDI). This initiative quickly became known as "Star Wars" after the 1977 movie with that title. While the SDI is an important part of Reagan's heritage, this study will not discuss it or, for that matter, other national security space efforts during the Reagan presidency, many of which remain classified. For one thing, SDI was not a space program per se; it was a national security initiative that incidentally involved stationing defensive systems in orbit. There are already a number of excellent accounts of the decision process that led to Reagan's defense against ballistic missiles initiative and his unwillingness to compromise it in order to achieve other agreements with the Soviet Union.[4] In contrast, there is no comprehensive account of the decisions made by President Reagan and his associates with respect to U.S. civilian and commercial space activities.

The Reagan administration made a number of decisions intended to reinvigorate the U.S. civilian space effort, and Ronald Reagan personally would turn out to be the most pro-space U.S. president, before or since. He valued almost everything associated with the space program, from its many accomplishments to the astronauts and others responsible for achieving them. While John F. Kennedy had sent Americans to the Moon in the 1960s, he pursued that path, not because of his own vision of an expansive future in space, but as a battleground in the Cold War competition with the Soviet Union. Reagan did have a space vision, one that saw space as the final frontier for American leadership. In one of his last space speeches, Reagan in September 1988 would say: "It is mankind's manifest destiny to bring our humanity into space; to colonize this galaxy; and as a nation, we have the power to determine whether America will lead or will follow. I say that America must lead."[5]

An Actor-Turned Politician

Ronald Reagan was 69 years old when elected, until then the oldest person ever chosen to occupy the White House. He was most widely perceived as an actor in Western and other low-prestige movies, although he had also starred in several well-regarded films. Reagan appeared in 52 Hollywood features between 1937 and 1964, and was a performer or narrator in the long-running television series *General Electric Theater* and for one year on *Death Valley Days*.

Reagan had several other identities; he had been a labor leader, serving eight years as president of the Screen Actors Guild. As a spokesman for General Electric (GE) between 1954 and 1962 he had traveled the country meeting employees in GE plants and giving as many as 14 speeches in a day's time. Once a liberal Democrat, he became increasingly conservative in his political views and switched his voting registration to the Republican Party in 1962. Reagan's October 1964 eloquent "A Time for Choosing" speech in support of presidential contender Barry Goldwater made him a leading spokesperson of the conservative wing of the Republican Party and led him to an increasingly active role in electoral politics. In 1966, Reagan was elected governor of California, the nation's most populous state, and served two four-year terms in that position. After leaving the governorship, he made an unsuccessful 1976 attempt at challenging incumbent President Gerald Ford for the Republican presidential nomination; in 1980, he battled with George H.W. Bush, Senators Howard Baker and Robert Dole, and several others before securing that nomination.[6]

Still, even after his many years as an active politician, Reagan's identity as an actor was what many thought of as they elected him to perform "the role of a lifetime." To one observer, "Reagan's career in Hollywood was almost as important in preparing him for Washington as his eight years in Sacramento" as California governor; "both made him more effective as a leader." Another biographer suggests with admiration that Reagan "made politics, and governing, too, into a branch of his old business, entertainment" and that "he knew how to be President."[7]

There are multiple portrayals of Ronald Reagan during his White House years. One of the most insightful of them comes from an oral history interview of Howard Baker conducted in 2004. Baker during the first Reagan term was a senator from Tennessee who as majority leader of the Senate dealt with the president on a frequent basis. Baker retired from the Senate in 1985, and in February 1987 was asked by Reagan to become his White House chief of staff, a position he held until July 1988. Baker described Reagan as a "unique" and "multidimensional" personality. He added that Reagan's personality "may be the most unappreciated part of the Reagan legacy, not just his achievements, which were extraordinary, or his political successes, which were obvious, or his foreign policy endeavors, which are historic." Baker continued

> He's quick and insightful. He was a quick study ... The least deserved thing about the Reagan legacy is that he wasn't very bright. He was very bright, very quick. The idea that he forgot stuff is sort of true, but is more in the nature of a delete key than it was actual forgetting. When things were done he just deleted them from his mind and went on to other things.
>
> He was the most unpassive person I ever saw in the Oval Office. He was quiet and well modulated ... He had strong views ... He coupled that with a willingness to ... have strong people around him and listen to them.
>
> He was also not afraid to delegate. As long as that delegation stayed within the parameters of his fundamental conviction on whatever the issue was, he gave people extraordinary latitude ... Only when you seemed to veer over the line of what he really wanted to do or really thought, did he pull you back.
>
> He knew who he was. He was comfortable in his own skin. He understood that people thought he wasn't intellectually agile and that he wasn't very smart, but he knew better.[8]

"A Man of Ideas"

That Ronald Reagan was a person of basic intelligence and a lively imagination is confirmed by many of those who have studied his life. Richard Reeves in his book *President Reagan*, subtitled *The Triumph of Imagination*, suggests that while "no one ever called Reagan an intellectual ... he did see the world in terms of ideas." Hugh Heclo adds "Reagan was a man of ideas born out of life experiences, even though he was not a thinker's idea of a thinker." Rather, he was "a public man seeking political power in the name of certain ideas." By the time he was elected president, Reagan had developed "a particular philosophy of history." At the center of that philosophy was a vision of the United States as the God-chosen nation which was "the last best hope of man on Earth" and "a shining city on a hill." Reagan's belief in divinely guided American exceptionalism was the fundamental conviction that propelled his quest for the political power needed to translate that belief into action. It also shaped his views on the U.S. space program.

In addition to his belief in the superiority of the American way of life, which set him far apart from Jimmy Carter's "crisis of confidence," Ronald Reagan came to the White House knowing, according to his wife Nancy, "exactly what he wanted to achieve ... His goals had been honed over a twenty-year period ... Economic recovery. Greater economic freedom. A stronger defense. Less government." Added to this agenda was a strong anti-communism. Reagan saw "communist states as simply the latest enemy on the offensive against the American idea of freedom, an enemy that needed to be defeated rather than accommodated."

Reagan was a conservative, but one with a strong interest in the future. Even though he had little feeling for or understanding of the underlying technology, he "shared an American enthusiasm for technological gadgets ... and an American proclivity for endorsing progress in the same breath in which he celebrated memories of things past. He was at once old-fashioned and forward-looking, and frequently sounded as if he wanted to go back to the future."[9]

In his eight years as president, Ronald Reagan persisted in advocating and advancing almost all aspects of his political philosophy. He was "quite capable of adjusting his message to different audiences and accommodating particular political realities," but there was a consistency in his position throughout. Reeves comments that Reagan "imagined a future, and he made some of it happen."[10]

Of course, Ronald Reagan had his critics, both as an individual and as an advocate of a set of ideas and the policies and programs that flowed from them. While one can properly question the wisdom of many of Reagan's decisions and their consequences, and while the benefits of his presidency were not equally distributed across all social strata, Ronald Reagan did not, as one critic suggests, "sleepwalk through history."[11]

Ronald Reagan and Space

Almost from the start of his time in the White House, Ronald Reagan indicated his strong support for the nation's space activities. His initial post-inauguration statement on space was the message he sent to the crew of the first space shuttle flight in April 1981: "Through you, today, we feel as giants once again. Once again we feel the surge of pride that comes from knowing we are the first and we are the best and we are so because we are free." Reagan's first high profile public appearance following his March 1981 attempted assassination was an April 28 speech before a joint session of the Congress. At the end of his speech, he again linked the first shuttle flight to American greatness, saying:

> The space shuttle did more than prove our technological abilities. It raised our expectations once more. It started us dreaming again.
>
> The poet Carl Sandburg wrote, "The republic is a dream. Nothing happens unless first a dream." And that's what makes us, as Americans, different. We've always reached for a new spirit and aimed at a higher goal. We've been courageous and determined, unafraid and bold.

Commenting on the president's remarks, *The Washington Star* editorialized that "the importance of the space program is, like space itself, immeasurable. Its rewards may not be immediate, but they will surely become as real as those that followed the voyages of Columbus. Mr. Reagan, more than any president since John Kennedy, seems to sense this." The editorial suggested that "Mr. Reagan seems to understand the adventure of it all, that what he called 'magic historical moments' happen only when a nation is determined to make them happen, and that a nation's greatness and survival spring from such things."[12]

Ronald Reagan and the Space Frontier

Ronald Reagan repeatedly linked space achievement to the mythical role of the Western frontier in U.S. history. Writing in *Popular Mechanics* in 1986, he declared: "Americans need frontiers." In a commencement address at the Air

Force Academy, he observed: "Our willingness to accept the challenge of space will reflect whether America's men and women today have the same bold vision, the same courage and indomitable spirit that made us a great nation. Where would we be if the brave men and women who built the West let the unknowns and dangers overwhelm them? ... Our freedom and well-being will be tied to new achievements and pushing back new frontiers." As he announced in January 1984 his decision to approve a space station, Reagan called outer space "the next frontier," adding "our progress in space—taking giant steps for all mankind—is a tribute to American teamwork and excellence." He continued: "America has always been greatest when we dared to be great. We can reach for greatness again. We can follow our dreams to distant stars, living and working in space."[13]

At the start of 1985, Reagan named the members of a congressionally mandated National Commission on Space; the purpose of the commission was to propose space goals for the coming decades. On June 11, 1985, Reagan hosted a White House lunch to hear about the Commission's work and the latest developments in space exploration. Reagan found that discussion "fascinating." Writing in his diary that evening, he declared "space truly is the last frontier" and "some of the developments there in astronomy etc. are like science fiction except they are real."[14]

One senior White House official during the Reagan administration was Robert "Bud" McFarlane, who between January 1982 and October 1983 served as Reagan's deputy national security adviser and then from October 1983 to December 1985 his national security adviser. In the latter capacity, he saw Reagan almost every working day. In a 2016 interview, McFarlane commented that space activities "triggered all the romantic instincts in Ronald Reagan, who was a real romantic. He was 'Kennedyesque' in that sense. Striving for the next frontier and all of the really impressive visionary statements of President Kennedy, Reagan shared." McFarlane noted that Ronald Reagan thought that "our country is at its best when it seeks to go beyond the current realm of knowledge, and space exploration fit that perfectly. Just about anything that would have been suggested that dealt with space would have appealed to Reagan. He didn't pretend to know a darn thing about it, but he liked it—a lot."[15]

It was not surprising that Ronald Reagan "liked" space activities. Reagan was fundamentally an optimist, and space exploration is an optimistic undertaking. Throughout his administration, Reagan in his speeches and statements linked space achievement to his core belief in American exceptionalism. Reagan repeatedly identified the space program as a central element of America's rightful place as the global leader and beacon of progress. He enjoyed meeting National Aeronautics and Space Administration's (NASA's) astronauts, and made more calls to them as they orbited the Earth than any president before or since. He attended many space-related events, made more space-related speeches than any other president, and frequently inserted references to the space program in his addresses on other topics.

Reagan and Extraterrestrial Life

There was also a second, little-known aspect of Ronald Reagan's personality that suggested he would look favorably on the nation's space efforts. As noted earlier, Reagan had a lively imagination. That imagination included a fascination with the possibility of discovering or even encountering extraterrestrial life. Reagan was a lifetime reader of science fiction and a fan of science fiction films. As a teenager, he had been engaged by the series of stories by Edgar Rice Burroughs which described the adventures of a Civil War veteran, John Carter, who is transported to Mars and becomes a warlord dealing with aliens resembling the beasts of antiquity as well as other more humanoid beings. Reagan's daughter Patti told one journalist that her father was intrigued by stories about unidentified flying objects (UFOs); he even claimed that he and his wife had at least one sighting of a UFO. Long-time Reagan confidant and advisor Edwin Meese in a 2016 interview confirmed Reagan's interest in extraterrestrials; Meese recalled that both he and Reagan kept up with the Buck Rogers comic series. At times, it seemed to some of his less-imaginative associates that Reagan even had difficulty distinguishing between fiction and fact with respect to space affairs.

Reagan's interest in alien life even spilled over into his approach to international diplomacy. During his first meeting with Soviet leader Mikhail Gorbachev in 1985, he startled his advisers, and probably Gorbachev, by suggesting that the United States and the Soviet Union would certainly cooperate if the Earth were threatened by an invasion by aliens being transported to this planet by Halley's Comet during its 1986 transit through the inner solar system. Reagan repeated this theme in various venues, telling the United Nations General Assembly in 1987 "in our obsession with antagonisms of the moment, we often forget how much unites all the members of humanity. Perhaps we need some outside, universal threat to make us recognize this common bond. I occasionally think how quickly our differences worldwide would vanish if we were facing an alien threat from outside this world." Reagan's final national security advisor, Colin Powell, became convinced that Reagan's notion of an alien invasion as a unifying force came from a 1951 science fiction film, *The Day the Earth Stood Still*; Powell "struggled diligently to keep interplanetary references out of Reagan's speeches." When the subject came up, "Powell would roll his eyes and say to his staff, 'Here come the little green men again.'"[16]

Space During the Transition

Soon after his election, Reagan appointed a transition team to advise him on the issues he would face with respect to the civilian space program. The transition team was led by George Low, a well-respected engineer and manager who had worked at NASA from its inception in 1958 through 1976, rising in 1969 to the position of deputy administrator and for a few months in 1970–1971, acting administrator. Low in 1980 was the president of Rensselaer Polytechnic

Institute. The transition team included 11 other senior individuals, many with long experience in the space sector. The team's detailed report set out the challenge for the new president in shaping a productive future for the U.S. space program.

> Budgetary problems and Space Shuttle cost increases, absence of long-term goals, lost opportunities in space sciences, and unresolved questions about the future of space applications have hurt NASA's vitality ... This condition is exacerbated by the growing space challenge from Europe, Japan, and the Soviet Union, and by the fact that no major research and development project is planned to follow the Space Shuttle. While U.S. space activity has peaked and faces an uncertain future, its competition is on the upswing.
>
> This unhealthy state of affairs can only be rectified by a conscious decision.

What kind of "conscious decision" with respect to the space program's future the new administration might make was not clear as Ronald Reagan and his associates prepared to assume power. In his letter transmitting the team's report, Low suggested "NASA and its civil space program represent an opportunity for positive accomplishment by the Reagan administration"; it would be some time before it would become clear whether the administration would take advantage of that opportunity.[17]

Kennedy or Nixon?

Space advocates hoped that as president Ronald Reagan would follow in the footsteps of John F. Kennedy by setting a challenging goal as the focus for the U.S. space effort. Reagan's transition team, for example, suggested as one option for Reagan's consideration "a high-challenge manned initiative, as a catalyst to rekindle the nation's vitality, involving a long term commitment to higher budget levels." One such initiative, the team suggested, would be "the establishment of a space operations center in low earth orbit ... Such a program, or others like it, would open future space vistas, in the 1990s and beyond, that can only be imagined today: a permanently manned outpost on the moon or human exploration of Mars." As he took over as Reagan's NASA administrator in 1981, James Beggs hoped that Reagan might approve a space station, a form of space operations center, as a "bold stroke." To Beggs, "Reagan, like President Kennedy before him, seemed genuinely entranced by the space program." He and others at NASA "were pleased to have a true believer in the oval office again, one who might make an Apollo-type decision as President Kennedy had done two decades earlier."[18]

There was an alternative possibility—that Ronald Reagan, with his desire to reduce government spending, would more closely resemble President Richard Nixon in his approach to the future in space. Nixon had gloried in the success of Apollo 11, the first lunar landing in July 1969, but had rejected the ambitious post-Apollo program, beginning with a space station and leading to human missions to Mars, that had been proposed to him just two months after Apollo 11. Nixon's judgment was that the nation was not willing to support a continued Apollo-like space effort. He continued the rapid decrease in NASA's budget that had begun under President Lyndon Johnson, and as a means of lowering space spending accepted NASA's proposals to cancel two planned Apollo missions to the Moon and to terminate production of the Saturn V Moon rocket, thereby giving up the capability to send humans beyond low Earth orbit. Instead, Nixon had in early 1972 approved development of the space shuttle as NASA's major post-Apollo effort, promising that the shuttle would "revolutionize transportation into near space by routinizing it. It will take the astronomical costs out of astronautics."

Richard Nixon set out his approach to the post-Apollo space program in a March 7, 1970, statement that this author has labeled the "Nixon space doctrine":

> We must realize that space activities will be part of our lives for the rest of time. We must think of them as part of a continuing process—one which will go on day in and day out—and not as a series of separate leaps, each requiring a massive concentration of energy and will and accomplished on a crash timetable. Our space program should not be planned in a rigid manner, decade by decade, but on a continuing flexible basis, one which takes into account our changing needs and our expanding knowledge.
>
> We must also recognize that our space expenditures must take their proper place within a rigorous system of national priorities. What we do in space from here on in must become a normal and regular part of our national life and must therefore be planned in conjunction with all of the other undertakings which are also important to us.[19]

In a sense, then, the "crossroads" faced by the space program that Reagan's transition team pointed out in its report was a choice between two future paths—one emulating the approach to space followed by John F. Kennedy, which had treated the space program as one of the nation's highest priorities and led to the remarkable achievement of sending 12 Americans to walk on the Moon, or one reflecting Richard Nixon's view that the space program should become "a normal and regular part of our national life."

As Ronald Reagan took the presidential oath of office on January 20, 1981, it was not at all clear which of those paths he and his administration would follow.

Notes

1. Richard Darman, *Who's in Control? Polar Politics and the Sensible Center* (New York: Simon & Schuster, 1996), 28.
2. Jimmy Carter: "Address to the Nation on Energy and National Goals: 'The Malaise Speech,'" July 15, 1979. Online by Gerhard Peters and John T. Woolley, *APP*, http://www.presidency.ucsb.edu/ws/?pid=32596. Although

this address became noted as "The Malaise Speech," Carter did not use that word in his talk. The description of Ronald Reagan is from Steven Hayward, "Ronald Reagan: Conservative Statesman," June 4, 2013, 4, The Heritage Foundation, Washington, DC, http://www.heritage.org/political-process/report/Ronald-reagan-conservative-statesman.

3. These sentences were contained in the NASA transition team report prepared for president-elect Ronald Reagan by a team led by George M. Low. Letter from George Low to Richard Fairbanks, Director, Transition Resources and Development Group, December 19, 1980, with attached transition team report, papers of George M. Low, Folsom Library, Rensselaer Polytechnic Institute, Troy, NY. The quoted passages are from 3–5 of the report.
4. Among these studies are Donald Baucom, *The Origins of SDI: 1944–1983* (Lawrence, KS: University of Kansas Press, 1992); Frances Fitzgerald, *Way Out There in the Blue: Star Wars and the End of the Cold War* (New York: Simon & Schuster, 2001); Nigel Hey, *The Star Wars Enigma: Behind the Scenes of the Cold War Race for Missile Defense* (Washington: Potomac Books, 2006); Edward Reiss, *The Strategic Defense Initiative* (Cambridge: Cambridge University Press, 1992); and Sanford Lakoff and Herbert York, *A Shield in Space? Technology, Politics, and the Strategic Defense Initiative* (Berkeley, CA: University of California Press, 1989).
5. Andrew J. Butrica, *Single Stage to Orbit: Politics, Technology, and the Quest for Reusable Rocketry* (Baltimore, MD; The Johns Hopkins University Press, 2003), 13. For a discussion of Kennedy's motivations in setting the lunar landing goal, see John M. Logsdon, *John F. Kennedy and the Race to the Moon* (New York: Palgrave Macmillan, 2010). Ronald Reagan: "Remarks at the Johnson Space Center in Houston, Texas," September 22, 1988. Online by Gerhard Peters and John T. Woolley, *APP*, http://www.presidency.ucsb.edu/ws/?pid=34875.
6. This study is, of course, not a comprehensive account of Ronald Reagan's life or of the Reagan administration. There are a large number of books that provide such an account and assessment. Among these, I have particularly depended on: Richard Reeves, *President Reagan: The Triumph of Imagination* (New York: Simon & Schuster, 2005); Steven F. Hayward, *The Age of Reagan: The Conservative Counterrevolution, 1980–1989* (New York: Three Rivers Press, 2009); H.W. Brands, *Reagan: The Life* (New York,: Doubleday, 2015); Lou Cannon, *President Reagan: The Role of a Lifetime* (New York: Public Affairs, 2000);W. Elliott Brownlee and Hugh Davis Graham, eds. *The Reagan Presidency: Pragmatic Conservatism & Its Legacies* (Lawrence, KS: University Press of Kansas, 2003); and Jacob Weisberg, *Ronald Reagan* (New York: Henry Holt and Company, 2016). The quote is from the last of these books, 34.
7. The phrase "role of a lifetime" is the subtitle of Cannon, *President Reagan*; David Gergen, *Eyewitness to Power: The Essence of Leadership, Nixon to Clinton* (New York: Simon & Schuster, 2000), 244; Reeves, *President Reagan*, xiii, xvi.
8. Oral history Interview with Howard Baker, August 24, 2004, Miller Center of Public Affairs, University of Virginia, Charlottesville, VA, https://millercenter.org/the-presidency/presidential-oral-histories/howard-h-baker-jr-oral-history-senate-majority-leader, 6–7.
9. Cannon, *President Reagan*, 8.
10. Reeves, *President Reagan*, xii-xiv; Hugh Heclo, "Ronald Reagan and the American Public Philosophy," in Brownlee and Graham, eds., *The Reagan*

Presidency, 17–37; Nancy Reagan with William Novak, *My Turn: The Memoirs of Nancy Reagan* (New York: Random House, 1989), 113.

11. For a critical assessment of the Reagan presidency, see Haynes Johnson, *Sleepwalking Through History: America in the Reagan Years* (New York: Anchor Books, Doubleday, 1991).
12. Howell Raines, "Reagan Sees Launching on TV at White House," *NYT*, April 13, 1981, A11. Ronald Reagan: "Address Before a Joint Session of the Congress on the Program for Economic Recovery," April 28, 1981. Online by Gerhard Peters and John T. Woolley, *APP*, http://www.presidency.ucsb.edu/ws/?pid=43756. "The President's Vision," *The Washington Star*, May 15, 1981, A13.
13. Ronald Reagan, "The Building of America," *Popular Mechanics*, July 1986, 107. Ronald Reagan: "Address at Commencement Exercises at the United States Air Force Academy in Colorado Springs, Colorado," May 30, 1984. Online by Gerhard Peters and John T. Woolley, *APP*, http://www.presidency.ucsb.edu/ws/?pid=39979. Ronald Reagan: "Address Before a Joint Session of the Congress on the State of the Union," January 25, 1984. Online by Gerhard Peters and John T. Woolley, *APP*, http://www.presidency.ucsb.edu/ws/?pid=40205.
14. Ronald Reagan, *The Reagan Diaries*, edited by Douglas Brinkley (New York: Harper Collins, 2007), 334.
15. Interview with Robert McFarlane, April 22, 2016.
16. Cannon, *President Reagan*, 179, 40–42; interview with Edwin Meese, October 12, 2016; Ronald Reagan: "Address to the 42d Session of the United Nations General Assembly in New York, New York," September 21, 1987. Online by Gerhard Peters and John T. Woolley, *APP*, http://www.presidency.ucsb.edu/ws/?pid=34823.
17. Letter from George Low to Richard Fairbanks, Director, Transition Resources and Development Group, December 19, 1980, with attached transition team report, papers of George M. Low, Folsom Library, Rensselaer Polytechnic Institute, Troy, NY. The quoted passages are from 3 to 5 of the report and Low's transmittal letter. Low was president of Rensselaer Polytechnic Institute in 1980. Other members of the transition team were Burton Edelson, COMSAT General; Frederick Haise, Grumman Aerospace; James Head, Brown University; Noel Hinners, National Air and Space Museum; Gerald Jenks, staff, U.S. House of Representatives; Gerald Kovach, staff, U.S. Senate; Robert Monks, The Boston Company; Richard Muller, Chrysler Corporation; Susan Perry, National Academy of Sciences; General Samuel Phillips, TRW; and John Young, American University.
18. Ibid, 6; Howard McCurdy, *The Space Station Decision: Incremental Politics and Technological Choice* (Baltimore, MD: The Johns Hopkins University Press, 1990), 41. For a discussion of John F. Kennedy's approach to space, see John M. Logsdon, *John F. Kennedy and the Race to the Moon* (New York: Palgrave Macmillan, 2010).
19. For a discussion of the "Nixon space doctrine" and other aspects to post-Apollo decisions, see John M. Logsdon, *After Apollo? Richard Nixon and the American Space Program* (New York: Palgrave Macmillan, 2015). The quoted passage is on p. 115.

CHAPTER 2

Getting Started

Moments after he was sworn in as the 40th president of the United States, Ronald Reagan declared, "We are going to begin to act, beginning today." This declaration was not just rhetorical; more than most incoming administrations, "the Reagan administration not only had clear goals but also had a clear strategic concept" of how it intended to achieve those goals. It also "knew who it wanted to help it get what it wanted … As in any administration, a number of people were loyal to the President as an individual, and some were loyal to the party of the President … More than any other administration, however, many individuals who composed this one were loyal to the ideas for which Ronald Reagan stood." As he became president, Ronald Reagan surrounded himself with individuals whom he thought would help him advance the agenda on which he had campaigned. How those individuals would interact among themselves, and with Ronald Reagan as president, was crucial to the way decisions would be made in the Reagan White House. While Reagan had well-developed ideas of what he wanted to achieve as president, he did not have nearly as clear a sense of the policies and program needed to achieve his goals. Indeed, he had limited interest in the details of policy formulation. He "rarely initiated a meeting, a phone call, a proposal, or an idea. He thought his staff would tell him anything he ought to know." Reagan's focus was "on the big picture, setting basic directions and letting aides handle the details."[1]

The Reagan Approach to Policy Choice

Ronald Reagan came to the presidency with a defined approach to policy choice, one from which he would seldom deviate during his years in the White House. One long-time associate described his approach, saying that Reagan "made decisions like an ancient king or a Turkish pasha, passively letting his subjects serve him … Rarely did he ask searching questions … He just sat back

J. M. Logsdon, *Ronald Reagan and the Space Frontier*,
Palgrave Studies in the History of Science and Technology,
https://doi.org/10.1007/978-3-319-98962-4_2

in a supremely calm, relaxed manner and waited until important things were brought to him. And then he would act, quickly, decisively, and usually, very wisely."[2]

In his eight years as governor of California, Reagan had given an important role in shaping his decisions to the members of his cabinet, particularly those in charge of the various executive departments and agencies comprising the California executive branch. He intended to follow the same approach as president. One of his biographers commented "as governor, Reagan's practice was to delegate authority to his agency heads while using them as a collegial body to provide input and advice … Six cabinet members met regularly with the governor; following their joint deliberations, Reagan would make his decisions." Reagan set out in his public speeches and private comments the general goals and objectives he wanted to achieve and trusted his associates to devise the best way to accomplish them. If there was disagreement among those associates, he wanted to hear their differing points of view; if there was a consensus, he most often approved it without modification. In meeting with his advisers, Reagan was more interested in hearing what others had to say than in asking questions, and he usually did not probe into the reasoning behind his advisers' views. He seldom announced his decision at the close of a meeting, preferring to spend some time thinking about what he had heard before making his choice. He would often end a meeting with one of his endless supply of stories or by saying as he left the room, "you fellas work it out and get back to me."[3]

Reagan himself had a clear perspective on his approach to policy choice:

> You surround yourself with the best people you can find, delegate authority, and don't interfere as long as the overall policy that you've decided upon is being carried out. In the Cabinet meetings—and some members of the Cabinet who have been members of other Cabinets have told me there have never been such meetings—I use a system in which I want to hear what everybody wants to say honestly. I want the decision made on what is right or wrong, what is good or bad for the people of this country…
>
> And when I've heard all that I need to make a decision, I don't take a vote. I make the decision.[4]

Reagan's hands-off approach to decision-making meant that options for his approval would emerge from the various agencies of government and his White House staff, much of the time only after extensive discussion. In that process, there was often as much conflict of views and interests as there was consensus. The Reagan administration over its eight years was characterized by sometimes bitter interagency conflicts and disagreements among the president's advisers. This was as much the case for the space sector as any other policy area. Reagan was personally a strong supporter of the U.S. space program, and was likely to approve almost every space-related proposal that reached his desk. The challenge for his most senior advisers was thus managing the conflictual policy development process so that those proposals that came before the president both reflected Reagan's broad objectives and constituted high-quality policy choices.

Given the Reagan approach to policy choice, it is not totally accurate to say that "Reagan decided." While certainly presidential approval is a necessary condition for a White House policy decision, often a president will merely ratify a decision made by his associates. This is the case in general, but particularly for Ronald Reagan, given his lack of attention to the details of policy choice and his willingness to approve without questioning a policy option with consensus support from his most trusted advisers. For many decisions, it is more accurate to say that "the Reagan administration decided." Even so, it is the president who bears the ultimate responsibility for his administration's choices.

Staffing the White House

Ronald Reagan depended on his White House staff to manage his cabinet-agency-focused decision process. In theory, they were to be the neutral link between the president and his agency heads, without themselves having independent influence on policy alternatives; they would communicate Reagan's objectives to the agency heads and ensure that the policy options developed for presidential choice reflected those objectives. In practice, those close to Reagan were often able to inject their own views into the policy development process. Given this crucial role, the backgrounds and views of the individuals appointed to key White House positions assumed high importance.

Senior White House Staff

At the center of the incoming administration were three men who themselves reflected the mixture of individuals who comprised the Reagan coalition. They were Edwin Meese III, James Baker III, and Michael Deaver; together they would be called the "troika." The three met with the president at the start of almost every day and often at its close, and most other White House and Executive Office of the president's staff reported to Reagan through one of the three (Fig. 2.1).

This structure had been worked out during the postelection transition. On November 14, just ten days after the election, Reagan had announced that Meese would serve as "counselor to the president"; in this role he would have overall policy development responsibility, both coordinating the interactions between the president and his cabinet secretaries and the heads of other executive agencies and overseeing the White House domestic and national security policy staffs. Baker would be Reagan's chief of staff, with responsibility for managing the White House, including press and congressional relations, speechwriting, personnel, and relations to various political constituencies. Deaver's role was not part of the November announcement, but it soon became clear that he would look after the president himself—his public image, his schedule, and links to his family, especially his strong-willed wife Nancy Reagan. Of the three, Deaver would be the least involved in policy development and decision-making, although he had frequent opportunities to make his views known to the president.

Fig. 2.1 President Ronald Reagan at his desk in the Oval Office meeting with his three top advisers, the so-called troika. From left to right, chief of Staff James Baker III, Reagan, Deputy Chief of Staff Michael Deaver, and presidential counselor Edwin Meese III. (Photograph courtesy of Reagan Presidential Library)

Meese and Deaver were long-time Reagan associates who had been on his staff in his eight years as California governor (1967–1975). Meese had served as Reagan's chief of staff from 1969 on; Deaver had been his director of administration. Both, in particular Meese, were committed to Reagan's conservative political philosophy. Baker, by comparison, was a mainstream Republican who had served in the Ford administration and had helped Gerald Ford in 1976 and then George H.W. Bush in 1980 oppose Reagan in the quest for the Republican presidential nomination. After winning that nomination and selecting Bush as his vice presidential running mate, Reagan invited Baker to take a lead position in his presidential campaign, together with Meese and Wall Street lawyer William Casey. Baker was a pragmatic politician who was much more familiar with "how Washington works" than Meese, Deaver, and indeed president-elect Reagan. Meese had expected to be Reagan's White House chief of staff, but came to accept the wisdom of having someone with Baker's Washington experience and political judgment as a counterpoint to the California true believers.

One student of presidential transitions has noted that "the ability to pull policy and politics together in a coherent manner would turn on the working relationship among Reagan's three top aides"; that relationship would be reflected in the day-to-day interactions of the large White House staff that the three, particularly Baker and Meese, soon created. During the first Reagan term, the relationship was described as resembling a coalition government, one characterized by a mixture of collaboration and often divisive conflict. That conflict was a result of different political perspectives, bureaucratic turf wars, and personal antagonisms; the tension between the administration's "ideologues," led by Meese, and its "pragmatists," led by Baker, was an everyday reality. Even so, Reagan's first-term team was effective in advancing much of the agenda that the president had brought to Washington. As will be discussed later in this study, this was not the case for most of Reagan's second term.

Other White House Staff

In order to support him in his role as coordinator of administration policy development, Meese appointed three key deputies. Richard Allen, a long-time foreign policy and national security adviser to Reagan, would serve as national security adviser and would head the staff of the National Security Council. Having Allen report to the president through Meese was a downgrading of the national security adviser position, which since the Kennedy administration had been filled with a powerful individual reporting directly to the president. Martin Anderson, a conservative intellectual and another long-time Reagan associate, would serve as domestic policy adviser. He would be supported by a newly created Office of Policy Development, staffed primarily with "true believers" in the Reagan agenda and intended to serve as an idea shop for how to translate elements of that agenda into innovative approaches to governance. Craig Fuller, a 30-year-old Californian who had been a public relations executive before joining the administration, was appointed head of the Office of Cabinet Affairs. Because President Reagan intended to draw heavily on his cabinet officers for advice, Fuller as coordinator of cabinet activities thus held a strategic position as the link between the Oval Office and the cabinet officers and their departments on which Reagan intended to depend.

Baker chose as the leaders of his staff individuals who were Washington veterans familiar with government operations and who shared his pragmatic outlook on politics and policy. In particular, Baker's deputy, Richard Darman, held the position of presidential staff secretary and in that role controlled all papers going into and out of the Oval Office. Darman at age 37 was already a veteran of the Nixon and Ford administrations, and was a keen student of policy and political action.

Two of the individuals who occupied key White House staff positions, Craig Fuller and Dick Darman, were personally predisposed to be favorable toward space program initiatives. Fuller in his pre-White House public relations position had worked with the Space Division of Rockwell International, the prime contractor for the space shuttle. He was an aviation enthusiast who had been a private pilot since he was 16 and who found the shuttle an exciting development that could open new possibilities in space. Fuller would play an important role in space policy development in the first Reagan term; he helped move forward positive Reagan administration decisions on issues such as increased commercial involvement in space and approval of space station development. Darman was also a space enthusiast; he even sent his young sons to space camp. With these two young men controlling much of the paper flow and policy interactions in the Reagan White House, space issues would be sure to get a full and sympathetic hearing.[5]

Stockman as Budget Director

Reducing government spending was a central promise during Ronald Reagan's presidential campaign. Reagan chose as his director of the Office of Management and Budget (OMB) a 34-year-old congressman from Michigan, David Stockman. In this position, Stockman would be point man in cutting the federal budget, a role he eagerly embraced. Stockman was "brilliant, confident and overconfident, lean and hungry," a "zealous libertarian conservative," a "wunderkid," and the "smartest member of Congress." Stockman had "hungered for OMB," turning down an initial offer during the transition to become Secretary of Energy. He was "a detail oriented" individual "who could master the clever schemes of the spendthrift bureaucracy and dissect the arcane codes of the federal budget."[6]

During the postelection transition period, Stockman was a major author of an "alarmist" essay about the economic situation, which he described as "a treatise on the risks of a fiscal shipwreck." In that essay, Stockman had lumped the National Aeronautics and Space Administration (NASA) together with a number of domestic programs such as Urban Development Action Grants, the Economic Development Administration, Department of Energy commercialization and information programs, and arts and humanities, suggesting that "most of these programs are ineffective or of low priority and could be cut by at least one-third." If Stockman followed through with cutting the NASA budget by such an amount, the space program's future would be in real jeopardy. Quizzed about his intent vis-à-vis NASA by Senator John Glenn (D-OH) during his January 8 confirmation hearing, Stockman replied, "I could probably find something in NASA to cut. We will take a careful look at that."[7]

Keyworth as Science Adviser

As late as early March, *Science* magazine reported that "the Reagan administration is having doubts about the need for a science adviser in the White House, and it is considering transferring to another agency" the functions performed by Office of Science and Technology Policy (OSTP), the staff office headed by the science adviser.[8] The Reagan administration ultimately decided to retain OSTP and the science adviser position; it took several more months to find a suitable candidate.

On May 19, the White House announced that the president had nominated for the position George A. Keyworth II, known to most as "Jay." The 41-year-old nuclear physicist was head of the physics division at Los Alamos National

Laboratory, the laboratory where the first atomic bombs had been designed and which was still a major research center for nuclear weapons and other advanced weaponry. Keyworth was a protégé of the controversial scientist Edward Teller, whom Ronald Reagan had come to know and respect while Reagan was California governor. Keyworth acknowledged that he owed his appointment to Teller, saying "bluntly the reason that I was in that office was that Edward first proposed me, and the president very much admired Edward." Keyworth's nomination encountered "unease, verging on hostility, from parts of the scientific community. He was an outsider whose candidacy had not been sponsored by scientists active on the science advisory circuits in previous administrations. Indeed, few had even heard of him."[9] Keyworth was to be an active, although only occasionally influential, participant in Reagan administration space policy debates and decisions.

New Leadership for NASA

The most avid suitor for one of NASA's top positions during the Reagan administration, either administrator or deputy administrator, was Hans Mark, who had been President Carter's undersecretary and then secretary of the Air Force and simultaneously director of the National Reconnaissance Office (NRO). Mark was born in Germany and lived for a time in Austria; his family had immigrated to Canada in 1939 and to the United States the following year. He became a U.S. citizen in 1945. After getting a doctorate in nuclear physics from the Massachusetts Institute of Technology, Mark pursued a combined academic and research career, dealing primarily with nuclear weapons. He became deeply embedded in the national security community, which recognized him as an individual with high intelligence, strong opinions, and great energy. He had worked for NASA beginning in 1969 before joining the Carter administration.

Mark, as he joined the Carter administration, was already a committed advocate of the space shuttle. Interacting with Mark during his time at NRO and as secretary of the Air Force was the director of NASA's Jet Propulsion Laboratory, Bruce Murray, who characterized Mark as a "Shuttle ... messiah." Visiting Mark in his Pentagon office, Murray took particular notice of "the detailed model of the Shuttle that dominated his desk. The only difference between this model and those in the NASA offices was that Mark's was painted Air Force regulation blue. It was no secret that Mark advocated an Air Force-controlled Shuttle system."[10] When, as he had anticipated, the shuttle in 1978 and 1979 ran into technical, schedule, and budget problems, Mark's advocacy and the steps he had taken at NRO to make shuttle use essential to intelligence efforts were important to the program's survival.

Mark goes into great detail in his book *The Space Station: A Personal Journey* about how he went about seeking a NASA leadership position in the Reagan administration. He recounts that by summer 1979 he began to have serious doubts about whether Jimmy Carter would be reelected, and was concerned about the fate of the space shuttle under Carter's successor. By mid-1980, after

Ronald Reagan had secured the Republican presidential nomination, he agreed to provide an in-depth briefing on the space shuttle program "to some of Governor Reagan's people." In particular, Mark wanted to meet with Richard Allen, whom he thought would be "the odds-on choice" to become President Reagan's national security adviser.

That meeting was arranged; over a September 11, 1980, dinner with Allen, Mark spent more than four hours describing "the status of the shuttle program." Mark did not inform his boss, Secretary of Defense Harold Brown, that he was going to meet with representatives of the challenger to the president Mark was currently serving. He justified his action because what he "wanted to do very badly was to stay on in Washington" to see the shuttle program through its initial launches. He hoped that "there might be a chance of being selected to one of the senior positions at NASA … If that could be arranged, then I would be in a position to help push the shuttle program through to completion."[11]

After Ronald Reagan was elected, Mark met with George Low, head of the president-elect's NASA transition team. Low asked him if he was interested in the top NASA job; Mark said that he was. Mark also met with former NASA and Department of Transportation official James Elms. He discovered that Elms was supporting another individual, James Beggs, as a candidate for NASA administrator. Beggs was in 1980 executive vice president of General Dynamics, a major defense and aerospace firm. He was a graduate of the Naval Academy and had a graduate degree in business administration from Harvard. Despite his lack of deep technical background, Beggs had served in a number of managerial positions in technology-intensive organizations. He had been a senior NASA official in the late 1960s and in fact in 1969 had hired Mark as director of NASA's Ames Research Center in the San Francisco Bay area. In 1970, Beggs became Richard Nixon's undersecretary of Transportation. Leaving the government in 1973, Beggs had spent a year working for eccentric billionaire Howard Hughes before joining General Dynamics. Elms "thought that it would be important to have someone with the political and leadership skills of Beggs to head NASA"; Mark, with his technical background, "would be the ideal deputy for Beggs."

It would take several months for the White House to agree with this assessment; for Mark, those months were "very confusing and frustrating." He was contacted in December by the president-elect's personnel staff and asked whether he was interested in the top NASA job; the trade journal *Aviation Week & Space Technology* at about the same time identified Mark as the leading candidate to be NASA administrator. Others being mentioned for the position included former NASA managers Sam Phillips and Rocco Petrone and Apollo 8 commander Frank Borman. Beggs's name as a candidate had not yet surfaced. *The Washington Post* on January 29 reported that Mark "was moving to the lead position in a hurry" for the top NASA job.[12]

But there was no word on a potential appointment from what was now the Reagan White House. Mark recognized that being a lifelong Democrat and having been a senior official in a Democratic administration were major obstacles to getting a position that required a presidential nomination. Mark was pleased to be invited to attend a social gathering in early March 1981; other guests included two of the top people in the new administration, both long-time Reagan associates, Deaver from the White House and Deputy Secretary of State William Clark. Mark "nurtured the hope (a false one, probably) that attending events such as this one might be helpful in my effort to secure a post at NASA."[13]

Compared to Mark's active campaigning, the path James Beggs followed to his nomination as Reagan's NASA administrator was more straightforward. Soon after the election his name as a potential candidate for the position was suggested (likely by Elms, who was a good friend) to Pendleton James, who was in charge during the transition of identifying candidates for senior administration positions. James called Beggs and asked him if he was interested in being considered as NASA administrator; Beggs replied, "Yes." During the transition, Beggs did not hear anything more from the White House. Then, after the inauguration, "one of the several people who worked for Reagan called me one day in February and asked me if I'd be interested in the Administrator's job." Beggs's response was, "Well, that depends. What's the president's commitment? What does he want to do with NASA?" Beggs wanted to know whether NASA would be a target in the administration effort to reduce government spending; if that were to be the case, Beggs said, "I don't think I'm your man. I can't close up the agency for you."

An early-March meeting between Beggs and President Reagan was arranged. Reagan told Beggs that "the space program was one of those things that the federal government ought to do." He was "was very, very interested in what they [NASA] were doing," but "he didn't know anything about [space] technology." Satisfied with Reagan's response, Beggs decided that he would accept the NASA job if it were offered.[14]

That offer came a few days later. By the time the White House called Mark to offer him the position as NASA deputy administrator, Beggs had already been contacted and had accepted the top position. Both Beggs and Mark agreed that their nominations should not be announced until after the first flight of the space shuttle. Mark identified as his "first order of business" to meet with Beggs; on March 21, he flew to St. Louis, where Beggs lived. The two agreed on their top priorities for NASA: making the space shuttle "a fully operational system" and persuading the Reagan administration "to adopt the construction of a permanently manned orbiting space station as the next major goal in space." Recognizing that the existing national space policy, still in force, although it had originated in the Carter administration, prohibited large new engineering initiatives, Beggs and Mark "determined that we would take the first opportunity we had to tell people what we had in mind with respect to the space station ... so as to smoke out the potential opposition. We also wanted to make certain that there would be no blanket proscription against ambitious goals in the space policy of the new administration."[15] Over the next two and a half years, the new NASA leaders would discover that there was indeed widespread opposition to developing a permanently occupied space outpost, and that getting a pro-space but financially conservative president committed to such an ambitious and expensive undertaking would present a real challenge.

On April 23, a few days after the shuttle's successful first flight, the White House announced that President Reagan would nominate Beggs and Mark for the top two NASA positions. Soon after, the two moved into temporary offices at NASA; although not formally able to conduct space agency business, they quickly became engaged in NASA's affairs. It took almost two months to schedule a confirmation hearing before the Senate Committee on Commerce, Science, and Transportation; the hearing took place only on June 17. It was a cursory affair. Senator Barry Goldwater(R-AZ), a few moments after the hearing began and before either Beggs or Mark had testified, moved "that we approve both of these gentlemen and send their nominations to the floor"; Senator Harrison Schmitt (R-NM), who was in the chair, replied "without objection that will be done." Schmitt, a former astronaut who had walked on the Moon during the Apollo 17 mission, asked Beggs, "What do you see as the next major step that it is reasonable for our space program to undertake?" Beggs responded: "It seems to me that the next step is a space station. That is the thing that will make a lot of other things possible in the future." Mark added: "I think that is the natural next step, the establishment of a permanent presence in space."[16]

The Senate confirmed Beggs's nomination on June 25 and Mark's nomination on July 8. Beggs was sworn in by Vice President George H.W. Bush on July 10, and Beggs swore in Mark later that day. NASA now had the leaders who would guide the space agency through the complex issues it would face in coming months and years.

Beggs and Mark were very different individuals. Beggs, who was 55 as he assumed the NASA position, was tall, well dressed, sometimes sporting a Stetson cowboy hat. He and especially his wife Mary had become active in Republican politics during their time in Missouri; they were quite comfortable operating in Washington political circles. Beggs predictably inserted quotes from Shakespeare in many of his speeches. By contrast, Mark, 52 years old, was a high-energy workaholic who liked to call very early morning meetings. He sported a military-style crew cut and was blunt spoken. Mark was much more at ease discussing technical issues than exchanging political gossip. Beggs and Mark never became social friends during their three years together managing NASA, yet together they were able to achieve most of what they had set out as their goals as they took over the space agency.

It would take some time for Beggs and Mark at NASA and those involved in space matters at the White House to settle in to a regular working pattern. But there were decisions to be made from almost the day of Reagan's inauguration. As the new president had said in his first address, "We are going to begin to act, beginning today." Even though Ronald Reagan was expected to be a pro-space president, in practice his administration's first space act was to cut the NASA budget.

Notes

1. Ronald Reagan: "Inaugural Address," January 20, 1981. Online by Gerhard Peters and John T. Woolley, APP, http://www.presidency.ucsb.edu/ws/?pid=43130; Charles O. Jones, *The Reagan Legacy: Promise and Performance* (Chatham, NJ: Chatham House Publishers, 1988), 24; David Gergen, *Eyewitness to Power: The Essence of Leadership: Nixon to Clinton* (New York: Simon & Schuster, 2000), 170–171; Jacob Weisberg, *Ronald Reagan* (New York: Henry Holt and Company, 2016), 48.
2. Martin Anderson, *Revolution* (New York: Harcourt Brace Jovanovich, 1988), 290.
3. John P. Burke, *Presidential Transitions: From Politics to Practice* (Boulder, CO: Lynne Rienner Publishers, 2000), 105.
4. Reagan is quoted in Martin Anderson and Annelise Anderson, *Reagan's Secret War: The Untold Story of His Fight to Save the World from Nuclear Disaster* (New York: Crown Publishers, 2009), 12–13.
5. Interview with Craig Fuller, December 14, 2016.
6. Richard Reeves, *President Reagan: The Triumph of Imagination* (New York: Simon & Schuster, 2005), 12–13; Steven F. Hayward, *The Age of Reagan: The Conservative Counterrevolution, 1980–1989* (New York: Three Rivers Press, 2009), 82–83; David Stockman, *The Triumph of Politics: Why the Reagan Revolution Failed* (New York: Harper & Row, 1986), 60, 54.
7. Stockman, *The Triumph of Politics*, 71, 74. This part of Stockman's treatise is quoted in *Defense Daily*, January 6, 1981, 6. The interchange with Glenn is quoted in a memorandum from Terry Finn to Robert Allnutt, January 14, 1981, Folder 2222, NHRC.
8. Colin Norman, "Science Adviser Post in Doubt," *Science*, March 6, 1981, 1026.
9. Gregg Herken and Richard Leone, *Cardinal Choices: Presidential Science Advising from the Atomic Bomb to SDI* (Palo Alto, CA: Stanford University Press, 2000), 199, 201. Colin Norman, "The Making of a Science Adviser," *Science*, November 12, 1982, 9.
10. Bruce Murray, *Journey into Space: The First Thirty Years of Space Exploration* (New York: W.W. Norton, 1989), 199. It is likely that Mark advocated only that the Air Force control the Shuttle fleet operating out of the west coast Vandenberg Air Force Base, not also the fleet operating out of NASA's Kennedy Space Center.
11. Hans Mark, *The Space Station: A Personal Journey* (Durham, NC: Duke University Press, 1987), 108–113.
12. Ibid, 116; *AWST*, December 15, 1980, 17; *WP*, January 29, 1981, A-5.
13. Mark, *The Space Station*, 120; interview of Hans Mark, May 24, 2016.
14. Oral history interview with James Beggs conducted by Kevin Rusnak, March 7, 2002. A copy of the interview is in the NHRC. Also, oral history interview of James Beggs conducted by Howard McCurdy, December 3, 1985. I am grateful to Professor McCurdy for sharing this interview and other material.
15. Mark, *The Space Station*, 121–122.
16. Committee on Commerce, Science, and Transportation, U.S. Senate, Hearing on James Montgomery Beggs to be the Administrator and Hans Michael Mark to be Deputy Administrator, National Aeronautics and Space Administration, June 17, 1981, 11, 22. A copy of the hearing report is in Folder 4235, NHRC.

CHAPTER 3

First Decisions

In order to achieve his highest priority goals of a strengthened nation defense, a reduction in the tax burden on individuals and corporations, and a balanced budget, President Reagan would have to make major cuts in the Federal government's nondefense spending, even as he also reduced individual and corporate taxes. Identifying those budget cuts was the new administration's first order of business.

The National Aeronautics and Space Administration (NASA) would not escape budget reductions. There was, however, an important constraint on what in the NASA budget could be cut. NASA was in charge of bringing the space shuttle to operational status, and the Reagan White House early on recognized that shuttle was needed for launching several key intelligence satellites and other Department of Defense (DOD) missions. One of the earliest Reagan administration space policy decisions was thus, for national security reasons, to wall off the shuttle program as a candidate for funding cuts. This left NASA's aeronautics (not discussed in this study) and robotic space science budgets as targets for budget reductions. The Office of Management and Budget (OMB) during 1981 proposed to cut some parts of the space science budget and put tight funding constraints on the space science enterprise overall. This pressure had a number of impacts on both short- and longer-term space science efforts.

The space shuttle program was also the focus of other early policy attention. First of all, the White House had to decide how it wanted to deal with the excitement of the first shuttle launch in the early spring. Once that launch was successful, a pressing question was whether the Reagan administration would continue the policy, first set during the Nixon administration and reinforced by President Jimmy Carter, that the space shuttle, with its promise of routine and lower-cost operation, would become at some point the sole means of access to space for all U.S. government missions. By late 1981, the decision to continue that policy was made; it would turn out to be a major policy mistake.

J. M. Logsdon, *Ronald Reagan and the Space Frontier*,
Palgrave Studies in the History of Science and Technology,
https://doi.org/10.1007/978-3-319-98962-4_3

Cutting the NASA Budget

President-elect Ronald Reagan in early December 1980 selected David Stockman as his OMB director. In this position, Stockman would be point man in cutting the federal budget. Although Stockman had during the postelection period identified the space agency as a candidate for significant budget reductions, by the time the Reagan administration took office he had softened his position. While still a member of the House of Representatives, one of Stockman's staff had called NASA, saying that Stockman wanted to attend the first launch of the space shuttle. NASA wrote back, telling Stockman, "You can be assured that you will be invited to join us for this historic event." On the day after Ronald Reagan was sworn in as president, Stockman told a National Press Club audience that he did not foresee "any major changes or major reductions in the space budget," that he was "a strong supporter of the Space Shuttle program," and that the shuttle was "a very constructive and very important investment for the country to make, not only because of its technological spinoff, but simply because of the boost that that [the Shuttle] gives to our economy and our aspirations."[1]

Budget reductions were made in the absence of any overall space policy framework to set out priorities and future goals; that framework would not appear until mid-1982. According to Stockman, the initial cuts being proposed, in space as well as in other areas of the budget, did not reflect "policymaking, at least in the normal sense. No basic policy options were appraised, discussed, and debated." Rather, the budget was spelled out at "breakneck speed" on a line-by-line basis in a series of "black books." Decisions about what to cut were made without any in-depth analysis.[2]

Initial Reductions

The proposed reductions in NASA's budget were also made in the absence of a presidential science adviser or anyone else on the senior White House staff with a background in space issues. The acting administrator of NASA, Alan Lovelace, was almost totally focused on getting the space shuttle cleared for its first launch. Stockman largely depended on the career OMB staff for advice on what might be cut without major damage to the overall space effort.

First reports coming out of Stockman's budget-cutting process were that the Galileo mission to explore Jupiter would be cancelled, and that the new start on the Venus Orbiting Imaging Radar mission (VOIR) that President Carter had approved as he left office would be rescinded. If sustained, these cuts would leave NASA with no new planetary missions started over a nine-year period, "decimating the Jet Propulsion Laboratory," NASA's center for planetary exploration. The OMB "black book" listing these proposed budget cuts commented that "strongest reaction would come from the space science community, since the planetary program would be particularly impacted, including the elimination of major projects for the Jet Propulsion Laboratory and a generally severe impact on U.S. planetary mission capability."[3]

Protests from senior members of Congress, coupled with the anticipated outcry from the space science community, convinced OMB to keep the Galileo program alive, at least temporarily. Other ways to cut the NASA budget had to be identified. OMB suggested that NASA itself decide on its priorities and identify programs that could be canceled or deferred. In response, NASA offered, among other budget cuts, to cancel its spacecraft contribution to the two-spacecraft International Solar Polar Mission (ISPM), a cooperative program with the European Space Agency (ESA). The mission was a joint effort between NASA and ESA in which each agency would contribute a spacecraft, with both launching aboard the space shuttle. One spacecraft would fly over the north pole of the Sun; the other, simultaneously, the south pole. Data from the mission would help understand the Sun's influence on the Earth's environment. The NASA offer to cancel the U.S. ISPM spacecraft was quickly accepted by OMB.[4]

Addressing a joint session of Congress on February 18, President Reagan proposed to cut $41.4 billion from the $739 billion Fiscal Year (FY) 1982 budget request that had been sent to Congress by outgoing President Carter. With respect to NASA, Reagan said: "The space program has been and is important to America, and we plan to continue it. We believe, however, that a reordering of priorities to focus on the most important and cost-effective NASA programs can result in a savings of a quarter of a billion dollars."[5]

More short-term NASA reductions were to come in the following weeks; there had not been time before Reagan's speech to tie up the loose ends and lower priority issues in the budget cuts the president was proposing. In addition, the administration's economic forecasts were in a state of high flux, so it was difficult to estimate the magnitude of reductions needed to balance the budget by 1984, Reagan's stated goal. By the time the formal FY 1982 budget request was sent to Congress on March 10, an additional $116 million had been cut from the NASA budget; the Reagan FY 1982 request was $6.1 billion, compared to the $6.7 billion requested by outgoing President Carter. Stockman noted that "NASA was hardly suffering. Even with the cut, its 1982 budget would be 11 percent higher than 1981." Most of that budget increase, however, would go to the space shuttle program. Both immediately and in the months to come, spending for space science would be constrained, with a number of consequences.[6]

Canceling ISPM Spacecraft Causes a Diplomatic Dust Up

Canceling the U.S. spacecraft contribution to the ISPM on budget grounds got the Reagan administration off on the wrong foot with respect to its international space relations.[7] ESA insisted that the cancelation was "totally unacceptable." The reaction of those involved in the mission in Europe was "outrage and incredulity. Outrage at the way the cancelation had been carried out, and incredulity that an international agreement would be canceled at all." ESA asked its individual member states to protest the U.S. decision in hopes of reversing it. On behalf of all 11 ESA member states, Italy, Switzerland, and Sweden on March 3 sent an *aide-memoire* to the Department of State, saying that "the governments of the member states strongly urge that the US government reconsider its position with a view to restoring full US participation in the mission."

This diplomatic intervention was not successful, and over the succeeding months other attempts to find a way to restore ISPM to a two-spacecraft mission also did not succeed. On September 4, 1981, NASA informed ESA that NASA would not be requesting funds for the second ISPM spacecraft in its FY 1983 budget submission. The reaction of Eric Quistgaard, ESA's director general, was emotional; he suggested that by its decision NASA had "destroyed" the basis for cooperation and that he was "profoundly dismayed that this decision was made." Despite Quistgaard's threat, NASA's cooperation with Europe continued, albeit with both ESA and individual European countries as more wary partners.[8]

Missing Halley's Comet

From the start of the Reagan administration, there was controversy over whether to undertake a U.S. mission to Halley's Comet in 1986 during its once-every-76-years passage through the inner solar system.[9] The primary advocate of a U.S. comet mission was Jet Propulsion Laboratory (JPL) Director Bruce Murray. Murray for several years had tried, without success, to get the Carter administration to approve a Halley mission. After Ronald Reagan was elected, Murray tried to convince Reagan's California friends, known as his "kitchen cabinet," and senior members of Reagan's staff to urge the president to direct OMB and NASA to insert a Halley mission into NASA's plans. Unsuccessful in that effort, he then tried both to wage a campaign through the press and to seek allies in Congress. John Noble Wilford of *The New York Times*, in an article headlined "Rousing U.S. Science to Meet Halley's Red Dawn," described the situation in which the Soviet Union, but not the United States, was mounting a Halley mission.

Murray and highly visible scientist Carl Sagan mobilized the rapidly growing membership of their just-created organization, The Planetary Society, in support of the Halley mission. On August 1, Sagan wrote to each of the society's 70,000 members, urging them to send an immediate letter or telegram to the president. Almost 10,000 did so, but most of the communications were routed to NASA unopened.[10]

Even so, the Reagan administration was seemingly responsive to these external pressures. On August 5, new science adviser Jay Keyworth wrote to NASA's James Beggs, asking him to "develop several options for carrying out a U.S. mission to Halley's Comet." He added, however, an important qualification. He hoped that "most, if not all, of these options would fall within existing budgetary guidelines." Beggs replied that "the idea of taking a close look at Halley's Comet has caught the public imagination," suggesting that "the Administration should make a positive response." Even so, Beggs gave higher priority to restoring the U.S. spacecraft to the ISPM than to the Halley mission, telling *The Washington Post* that "if there is any extra money in the upcoming budget for pure exploration he would rather spend it on a second spacecraft to swing around the north or south pole of the sun than spend it on a photographic flyby of Halley."[11]

Beggs wrote to Keyworth on September 16, enclosing the information on comet mission options. He told Keyworth that "following completion of our budget review at NASA, we have not included a Halley Comet mission in the FY1983 recommended budget. This is primarily because in the prevailing climate of fiscal restraints, the cost is not justified considering the probable science return ranked against competing science priorities." Beggs then asked the White House, in what was tantamount to an ultimatum, to indicate whether it really wanted a Halley mission. He said that "if I do not receive a positive Administration commitment to a Halley's Comet mission by October 1, I will have to terminate all efforts in support of such a mission."[12]

That White House commitment was not forthcoming. On September 30, NASA science chief Andrew Stofan wrote to Murray, saying that with "deepest regrets" he was directing JPL "to discontinue all activities" associated with a Halley mission. The United States in 1986 would not send a spacecraft to greet Halley's Comet, while Japan, Europe, and the Soviet Union each mounted such a mission.[13]

Survival Crisis of Planetary Exploration

The controversy over a mission to Halley's Comet was a microcosm of a broader disagreement between NASA and the White House budget and science offices with respect to NASA's planetary exploration efforts.[14] On September 15, 1981, NASA submitted to OMB an FY 1983 budget request of $7.6 billion; this was well above the $6.5 billion ceiling OMB had given NASA several months earlier. Beggs identified cuts that could bring the budget down to $7.1 billion, but argued that a budget below that level would require major reductions in the space shuttle program, an action which he knew was not acceptable to the White House, or "dropping out of one or more major program areas."[15]

Beggs, on September 29, told David Stockman that meeting the OMB guidelines while maintaining "viable programs in some areas" would mean closing down "other major programs that NASA has operated since its inception." He declared that the planetary exploration program was at the top of the list of efforts that NASA would give up, if forced to accept the major budget cuts OMB was proposing. Beggs offered the following rationale for his position:

> The planetary exploration program is one of the most successful and viable NASA programs. However, it is our judgment that in terms of scientific priority it ranks below space astronomy and astrophysics. Planetary exploration is much more

> highly dependent on launch vehicles, and it is our opinion that the most important missions that can reasonably be done within the current launch vehicle capability have, more or less, been done. The next step in planetary exploration is to do such things as landing missions and sample return missions, and these require full development of the Shuttle and the ability to assemble elements in earth orbit before sending the assembled spacecraft on its way. In our judgment, it is ultimately better for future planetary exploration to concentrate on developing the Shuttle capabilities rather than to attempt to run a "subcritical" planetary program given the current financial restrictions we face. Of course, elimination of the planetary exploration program will make the Jet Propulsion Laboratory in California surplus to our needs.[16]

Beggs was playing budgetary hardball. The shuttle program was sacrosanct due both to its association with the public appeal of humans in space and to its links to national security. The planetary program was NASA's other widely known activity. In addition, it had its roots at JPL, part of the California Institute of Technology (CalTech) in Southern California, the home base of the president and many of his top advisers. Beggs's calculation was that shutting down the planetary program would not be an acceptable option to the White House, and thus that NASA would get a budget allocation adequate to keep going both the planetary program and the other activities to which NASA had assigned higher priority.[17]

The budget review process moved forward on its usual pace over the next two months. NASA received its tentative budget allowance from OMB late in November. Looming over the budget decisions was the specter of the financial demands of the space shuttle, which was turning out to be far from the inexpensive system promised when the program was approved a decade earlier. OMB had reduced the overall NASA budget by $1.3 billion from the NASA request, to $6.3 billion. The planetary budget had been reduced to $118 million, and included no funds for either Galileo or VOIR.[18]

Beggs appealed the OMB allocations to Stockman on November 30. He told the budget director that "as someone who has devoted his entire professional career to working for American pre-eminence in space and aeronautics, I cannot accept the proposition that national economic imperatives compel the draconian funding reductions you have proposed on programs which have had such an extraordinary history of success." Beggs's appeal set the stage for the final decisions on the fate of the planetary exploration program. The focal point for those decisions would be the White House Budget Review Board, comprised of Ed Meese, James Baker, and David Stockman. This group was the last stop before taking a budget dispute to President Reagan. The Board scheduled a meeting on the NASA appeal on December 9.[19]

The advocates of canceling or deferring planetary exploration efforts had gained an ally during the fall in the person of presidential science adviser Keyworth. *The Washington Post* reported on December 2 that Keyworth "has recommended halting all new planetary space missions for at least the next decade—an idea he said the White House seems to be buying." In a submission to the Budget Review Board, Keyworth indicated that "I totally concur" with OMB's decision to cancel Galileo and VOIR, because those missions would "revisit the planets at much higher cost without commensurate additional scientific payoffs." He suggested that "the shuttle offers us a new capability to expand our horizons through ... new astrophysical initiatives," and that "NASA is not in principle opposed to this philosophy. Their basic concern is over continued stability at the Jet Propulsion Laboratory." Keyworth argued that "*the cut in planetary exploration represents an example of good management.* If 'business as usual' were to continue in planetary exploration, an unjustifiable increase in the overall space program would result."[20]

As the Budget Review Board meeting approached, NASA had few allies in the inner circles of the White House who could block the proposed budget cuts, and with them the end of a significant U.S. program of planetary exploration. If help was to arrive, it would have to come from outside.

The Division of Planetary Sciences of the American Astronomical Society was meeting during the week of October 12, as initial reports of the NASA-OMB controversy over the future of the planetary program surfaced. This meant that the scientists who would be most affected by the termination of the planetary program were gathered in one place. Not surprisingly, their response was outrage. Eugene Levy, chairman of the top scientific advisory body for solar system exploration, was particularly vocal. "At this moment," he commented, "not one of us knows whether, a year from now, the U.S. will have a program of solar system exploration."[21]

At the meeting there was significant disagreement over how to respond to the threat of program termination. While some thought it appropriate for working scientists to be active advocates in favor of their area of science, others believed that the integrity of the scientific community would be compromised by such open advocacy. All agreed that a letter reflecting the community's concerns should be sent to the most senior White House official identified as having policy responsibility for space, Ed Meese. Accordingly, on October 14, David Morrison, outgoing chairman of the Division of Planetary Science, and Carl Sagan, in his role as president of The Planetary Society, wrote to Meese "to ask your support to ensure the survival of planetary exploration in the United States." They argued that "a thousand years from now our age will be remembered because this is the moment we first set sail for the planets," and told Meese that "we and millions of Americans will appreciate any help you give to the enterprise of the planets."[22]

There was no organized campaign of public protest over the potential termination of the planetary program. The vehicle for mobilizing public protest would have been The Planetary Society. The dramatic images from the Voyager flybys of Jupiter and Saturn, the high public profile of Sagan and his public television series "Cosmos," and an effective direct mail membership campaign had led to the society's membership mushrooming from 25,000 to 70,000 within a little more than a year. The Planetary Society membership had just been mobilized in August 1981 for a letter-writing campaign in support of a

U.S. mission to Comet Halley. The idea of another mobilization of the society's members was considered in early December, but the combination of the lack of payoff from the earlier campaign and the difficulty and costs of gathering enough support to influence executive branch decisions in the short run led to the abandoning of the idea for such a campaign.[23]

The planetary program had become identified with CalTech's JPL, and had brought worldwide attention and prestige to the university. In addition, the annual fee paid by NASA to CalTech for managing JPL had become an important component of the overall CalTech budget.

The trustees of CalTech in January 1981 had created a "Trustees Committee on JPL." That committee had a number of members of national reputation and influence; it was chaired by Mary Scranton, wife of former Pennsylvania governor and Republican presidential aspirant William Scranton, and herself an individual with high-level political connections. Scranton in early December 1981 reported to Murray that Keyworth's public statements on canceling the planetary program had provided "a rallying point around which to arouse interest and sympathy." She contacted Republican senators Charles Percy, Charles Mathias, and Mark Hatfield and Vice President George H.W. Bush. She also spoke with Fred Bernthal, top assistant to Senate Majority Leader Howard Baker. The vice president had already been briefed on the JPL situation by prominent California Republican Robert Finch. Scranton asked Bush to "look at the political problem that cancelation of such program might bring to the Republican party in the future."

Indeed, Finch had met with Bush on December 3; in a letter the following day, he told the vice president that "I was very agitated" at what Keyworth had said to the press, since it "could only be read as the administration's turning its back on the space exploration heritage given to us post-Sputnik by President Eisenhower. What we have built up in a few decades is the visible showcase of national vision and high technology in one of the few areas where the U.S. still has clear cut-leadership in the world community." Finch added a political note, saying, "If Dr. Keyworth is successful, we will abandon space exploration for practical purposes and hand this issue to the Democrats."[24]

Concern over JPL's future had earlier been brought to White House attention by Arnold Beckman, chairman of Beckman Instruments and another CalTech trustee. Beckman had written Meese on October 5, saying that the administration's FY 1982 and FY 1983 budget reduction "threatens to create total chaos and a rapid disintegration of a 5000 person, $400 million Southern California enterprise … There are obvious implications to the support of the President and to his Party should the Administration permit such a catastrophe to take place." As reports of OMB's budget recommendations surfaced in early December, Beckman once again wrote to Meese, urging him "not to allow the emasculation of the technical and scientific capabilities of the Jet Propulsion Laboratory."[25]

CalTech president Marvin Goldberger made an early December trip to Washington in support of JPL. He met, among others, with a group of senators interested in the planetary program and other CalTech activities. In particular,

Goldberger urged Howard Baker to express his support for a continued program of planetary exploration. Goldberger was a Democrat, and although he had good connections with centrist Republicans such as Baker, he had limited ability to influence the conservative Californians in the Reagan inner circle.[26]

The various approaches to Senator Baker bore fruit. On December 9, he wrote to Ronald Reagan, saying, "I urgently request that $270 million be restored to the NASA budget for FY 1983 to continue the Galileo mission as originally planned." Baker originally intended to hand his letter directly to the president, but did not do so, instead leaving the letter with the president's staff. Baker called the White House on both December 9 and December 10 to make sure that Reagan had indeed seen the letter and to "underscore his interest." Baker stressed that the letter was not "a pro forma request nor a matter of parochial Tennessee interest." The Baker letter was routed to David Stockman for action; it is not clear whether in fact it ever reached the president. However, the fact that the top Republican in the Senate was supporting the planetary program was not lost on Stockman. The budget director by that time had come to recognize "the resourceful intransigence of the congressional politicians" and the limits to cutting the budget that such intransigence presented.[27]

The Budget Review Committee met on December 11. Keyworth suggested a compromise in which $80–90 million would be added to NASA's planetary exploration budget in order to avoid canceling the Galileo mission. This addition, he noted, "would permit the stability and excellence of the Jet Propulsion Laboratory to be continued." It would also provide the planetary science community time to come up with a more-modest, less-expensive program for the 1980s and beyond, one that the Reagan administration could support. The Budget Review Board asked NASA "to consider this alternative and report back immediately." It also hoped that OMB and NASA could settle the issue and thus that "an appeal to the President … be avoided." NASA accepted Keyworth's suggestion, and thus the controversy did not have to be taken to the president for resolution.[28]

As a result of Baker's intervention and the activity of influential Californians, the immediate possibility of the demise of the U.S. program of solar system exploration had passed. But the program had hung on by its fingertips; no new planetary mission was approved, for no funds for VOIR were restored to the NASA budget. What was gained was a year's breathing space, and the opportunity for NASA and the planetary community to come forward with a downsized planetary exploration program that could gain the support of the Reagan administration. It was able to do so, and planetary exploration did not become a matter of top-level White House concern for the remaining seven years of Ronald Reagan's time as president.

President Reagan and the Shuttle's First Flight

In the first months of the Reagan administration, in addition to OMB activity in reducing NASA's budget there was also a top-level focus on the first flight of the space shuttle. The shuttle's initial flight was approaching, and the White

House was planning how to take maximum advantage of interest in that flight to advance Ronald Reagan's overall agenda. Cabinet secretary Craig Fuller, already familiar with the shuttle from his days as a public relations person working with North American Rockwell, attended a White House senior staff meeting a few weeks after the inauguration that discussed likely headline-grabbing events for the rest of 1981. Fuller suggested that among the top stories of 1981 would be the initial launch of the shuttle; most of his colleagues, less familiar with the shuttle program, were surprised at that suggestion. Office of Policy Development staffer Danny Boggs in a February 24 memo suggested that the first flight of the space shuttle would "represent America's return to progress, not lethargy, in space exploration." He noted that "NASA has suggested that the President may wish to participate in the pride and success Americans will feel in the flight." Boggs endorsed the NASA suggestion, since "a restoration of pride and prestige was one very strong part of the President's appeal in the recent campaign. The President's personal participation ... provides a unique opportunity to emphasize the limitless possibilities of American ingenuity and enterprise." National Security Adviser Allen weighed in from a less-political perspective. He suggested that "the extent of Presidential involvement should be carefully weighed, and he should also be aware of the bureaucratic tugging that is going on in connection with the space shuttle." Allen added: "The public relations aspect should also be carefully investigated, as the national security and civilian uses are both important and neither should dominate."[29]

Allen also pointed out the need for a "communications hook-up to keep the President informed" and the need for "prepared statements in case of an emergency." With respect to direct presidential involvement, he suggested a "prelaunch visit by President to inspect the [Kennedy Space Center] complex," a "phone call to astronauts prior to launch," and a "visit to the landing site" to greet the crew upon their return. NASA had recommended that "the President appear at the landing rather than the lift-off for safety reasons ... the landing is considered the 'safe' 10% of the voyage."[30]

Discussion of the president's direct involvement in the shuttle mission came to an abrupt halt on March 30, as Ronald Reagan was seriously wounded by a bullet fired by would-be assassin John Hinckley. Even so, discussions of how to relate the White House to the shuttle flight continued. Staffers from the National Security Council (NSC) and the Office of Science and Technology Policy (OSTP) recognized "the need to be sensitive to the President's situation," but also pointed out the opportunity "to focus on the magnificence of this event"; they were concerned that "the White House will not obtain maximum benefit from the first Shuttle Launch," which would be a media event "not unlike the Apollo landing on the moon." The staffers suggested that a call from the convalescing president to the shuttle crew, veteran astronaut John Young and first-time flyer Robert Crippen, "would be well received" and "would remind the nation that the President is involved and in charge." They also noted that "we need to proceed with planning for a White House reception for the two Shuttle astronauts and families. After a successful flight this would be good press and would have the effect of raising national spirits ... To involve the White House in this extravaganza would have many positive aspects with few down sides."

The launch was originally scheduled for April 10, but was scrubbed shortly before liftoff and rescheduled for the morning of April 12, which was Ronald Reagan's first full day back in the White House after being released from the hospital. Reagan awoke at 6:50 a.m. to watch the 7:00 launch; he was reported to have exclaimed: "It's a spectacular sight." Reagan did not call the crew during their two-day mission; instead, Vice President George Bush made the call from his White House office. Shortly after launch the White House issued a presidential message to the crew, saying, "Through you, today, we feel as giants once again. Once again we feel the surge of pride that comes from knowing we are the first and we are the best and we are so because we are free." As Young and Crippen landed the shuttle orbiter *Columbia* two days later, Reagan wrote in his diary: "Our astronauts landed and what a thrill that was. I'm more & more convinced that Americans are hungering to feel proud & patriotic again." The White House issued a formal welcoming presidential message to the shuttle crew, saying, "Your brave adventure has opened a new era in space travel" and noting that "today the world watched us in triumph. Today our friends and adversaries are reminded that we are a free people capable of great deeds."[31]

Young and Crippen visited the White House on May 19. In the Oval Office, they received medals recognizing their achievement, and then lunched with the president and a number of astronauts and other space-related people under a tent in the White House Rose Garden. In his remarks, President Reagan once again returned to the theme of the space program as an example of American exceptionalism, saying that through shuttle's crew "we've all been part of the greatness pushing wider the boundaries of our freedom" and "their deeds reminded us that we, as a free people, can accomplish whatever we set out to do. Nothing binds our abilities except our expectations, and given that, the farthest star is within our reach."[32]

As one of Reagan's biographers has noted, his speeches and other public statements were "the most important component of Reagan's leadership style" and the way "he brought others serving in his administration into alignment with his beliefs." Another suggested that in his public remarks, Reagan was "the voice of optimism and national destiny, saying, as he always had, that Americans were God's chosen, the world's last best hope."[33] By linking the success of the initial shuttle mission to broad concepts of American greatness, power, and pride, Ronald Reagan both evidenced a broad sense of the relevance of U.S. space achievement to the central themes of his presidency and sent a message to those serving under him that he would be positively disposed to space proposals that reinforced those themes. It would be up to his associates, as it often was in other policy areas, to work out the details of the Reagan approach to space. That process had begun in the early days of the Reagan presidency.

Shuttle Policy Review

Even as the space shuttle was preparing for its first mission, the White House was slowly moving to initiate a major review of the shuttle's future role. Less than a month after the inauguration, two holdovers from the Carter administration, acting director of OSTP Ben Huberman and NSC space staff member Colonel Michael Berta, pointed out to National Security Adviser Allen that there were "several space policy issues which merit your attention." One issue was "how best to organize within the US government in the future to deliver Shuttle launch services for national security, civil, scientific, and commercial applications." Another was the implication of sole reliance on the space shuttle for government launches, especially those placing national security satellites into orbit.

Huberman and Berta were particularly concerned about the potential vulnerability of the shuttle in times of military conflict, if it were to become the only U.S. government means of access to space. They asked: "How will the Shuttle be used in times of crisis or war to launch satellites? ... Will the nation accept the conscious decision to expose Shuttle flight crews to Soviet antisatellite attack?" They suggested that it was necessary to examine whether there needed to be an expendable launch vehicle (ELV) capability, using either existing or new boosters, as a backup to the shuttle. They noted that to date "the tenuous nature of the Shuttle program has prevented a frank appraisal of these considerations." Even as the shuttle began flying, the wisdom of the policy decision to make it the only means of access to space was being questioned.

In handwritten comments on the memo, Allen responded that "I think this is a high priority matter. We are very interested in the national security aspects of space." He added that "a memo—informational—should be prepared for the President ... outlining the issues—and what's at stake. Use the Shuttle flight as the 'hook' for the memo. We need to get the WH [White House] people interested in the Shuttle. Is there a briefing—a movie—or something similar we could use to generate interest by inviting 20–25 top colleagues—thus getting them involved?"[34]

On April 6, as the first shuttle launch was imminent, Allen told senior White House officials that it was "important to review some of the policy issues surrounding the space transportation capability of the Shuttle program," since "many of our very important national security systems depend greatly on the Shuttle's performance." Allen observed that "it is appropriate to review both the status of the Shuttle program and some of the space policy issues attendant to the Shuttle era." He indicated that he would "soon" call a meeting to initiate such a review.[35]

"Soon" turned out to be a relative term. The meeting to discuss shuttle policy issues was not held until June 10. Attending were National Security Adviser Allen, Secretary of the Air Force Verne Orr, Hans Mark from NASA, plus from the White House, Martin Anderson, Edwin Harper, deputy director of OMB, William Schneider, OMB associate director for national security and international affairs, Richard Darman, and Jay Keyworth, Reagan's designee as science adviser.

According to Mark's diary notes, three decisions were made at the June 10 meeting:

1. Persuade the President to reaffirm the national commitment to continue the Shuttle program. To do this we will put together a "show and tell" operation for the NSC with the President in the chair…
2. Set up a decision making mechanism in the NSC to resolve disputes between NASA and DOD…
3. Set up a launch vehicle study under the direction of Jay Keyworth. I suggested this partially to buttress Jay's position since he is the new boy on the block.[36]

The next step was to get presidential approval for a statement of support for the space shuttle. On August 3, President Reagan chaired a meeting of the National Security Planning Group, a subset of NSC members, to discuss space issues. Invited to the meeting were Vice President George Bush; the secretaries and deputy secretaries of defense and state, Caspar Weinberger and Paul Thayer for DOD and Alexander Haig and William Clark from State, respectively; CIA Director William Casey; and from the White House, presidential counselor Ed Meese, James Baker, Richard Darman, and Richard Allen. (There is some suggestion that new NASA Administrator James Beggs was also present.) As preparations for the high-level meeting were being made, Air Force Secretary Orr pointed out that combining a presidential statement of support for the shuttle with guidelines for a proposed "detailed study on policy issues" would "serve to detract from and weakens the needed presidential commitment to the STS [Space Transportation System was another designation for the space shuttle]." Orr suggested separating "a presidential statement which is intended to reaffirm the national commitment to the STS from a presidential directive that would establish national policy and direct major studies." This suggestion was accepted. Participants at the meeting agreed that a "policy statement [would] be issued stating the Administration's support for the Space Shuttle." Following Orr's suggestion, a separate, more general space policy review would be also be initiated. That review is discussed in the following chapter.[37]

Embracing a Flawed Policy

Even though the decision that President Reagan would issue a statement in support of the space shuttle program was made on August 3, it took more than three months to get the statement into final form and signed by the president. Opposition to a strong endorsement of the shuttle as the sole U.S. means of access to space came from the Air Force and especially the National Reconnaissance Office (NRO), which was already very leery of becoming solely dependent on a single launch system. The new director of the NRO, undersecretary of the Air Force Edward "Pete" Aldridge, suggested in a 2009 interview that even after only one shuttle launch it was clear that some of the optimistic statements about

Fig. 3.1 President Reagan in mission control at NASA's Johnson Space Center in Houston, Texas, on November 13, 1981, talking to space shuttle astronauts Joe Engle and Richard Truly as they orbited the Earth on the second space shuttle flight. Seated to the left of Reagan is astronaut Dan Brandenstein. Standing behind the president are (l. to r.) astronaut Terry Hart, NASA Deputy Administrator Hans Mark, NASA Administrator James Beggs, and Johnson Space Center Director Chris Kraft. In the background between Mark and Beggs is legendary flight director Gene Kranz. (NASA photograph)

shuttle operations were "going to be way lacking, that the turnaround time was not seven days, it was much, much more than that." The first two shuttle orbiters, *Columbia* and *Challenger*, "were so heavy that they couldn't meet DOD demands … The cost was not one third the cost of an expendable [launch vehicle]; it was more likely equal at best, and possibly much higher than that … So we started getting worried." That worry was to increase in following years.[38]

The shuttle support statement was finally issued on November 13, 1981, as National Security Decision Directive (NSDD)-8. On that day, the president was in Houston, talking to the astronauts aboard the second space shuttle mission (Fig. 3.1). The statement said that "the United States is committed to a vigorous [space] effort that will ensure leadership … The Space Transportation System (STS) is a vital element." The statement reaffirmed NASA's lead role in managing the shuttle, saying that "the United States will continue to develop the STS through the National Aeronautics and Space Administration in cooperation with the Department of Defense to serve all authorized space users." It also reaffirmed the shuttle's central role in providing U.S. access to space, saying that "the STS will be the primary launch system for both United States military and civil government missions. The transition to the

Shuttle should occur as expeditiously as possible." The word "primary" did leave a little leeway for those advocating an expendable backup capability to the shuttle. Addressing concerns of the national security community that it might have to compete with NASA and commercial missions for launch priority, the statement was explicit in saying that "launch priority will be provided to national security missions, and such missions may use the Shuttle orbiters as dedicated mission vehicles." If the shuttle was used to launch a highly classified payload such as an intelligence satellite, it would not have to share shuttle payload bay space with nonsecurity community users.[39]

The notion that the space shuttle would become the U.S. government's sole means of launching payloads into space had originated as part of the rationale that had led the Nixon administration in 1972 to approve development of a large and capable shuttle system. That approach had been embedded in national space policy during the Carter administration, with a May 1977 Presidential Directive (PD)-37 statement that the shuttle would "service all authorized space users—domestic and foreign, commercial and governmental." From that point on, both NASA and the DOD had planned to phase out their use of ELVs, once the shuttle was demonstrated to be fully operational. This phase out would lead to the shutdown of ELV production lines unless private operators decided to assume responsibility for marketing and operating ELVs on a commercial basis.

Reagan's November 1981 statement thus reaffirmed the existing policy of shuttle dependence. Over the next five years, the high-cost and technical problems of shuttle operations, culminating in the January 1986 *Challenger* accident in which the shuttle's seven-person crew died, would demonstrate that, however impressive the achievements of the space shuttle may have been, making it the sole means of U.S. access to space was a flawed decision, leading to a major space policy mistake.

Notes

1. Letter from Terrance Finn to Rep. David Stockman, October 15, 1980, Folder 2222, NHRC; *Aerospace Daily*, January 22, 1981, 100; "Paying for the Future," *The Washington Star*, February 2, 1981.
2. David Stockman, *The Triumph of Politics: Why the Reagan Revolution Failed* (New York: Harper & Row, 1986), 81ff, 105.
3. *AWST*, February 9, 1981, 9; Craig Covault, "NASA Assesses Planning with Preliminary Budget," *AWST*, February 16, 1981, 19–21; Bruce Smith, "NASA Funding Cuts Could Lead to Science Data Cuts," *AWST*, February 16, 1981, 22.
4. "Planet Exploration Dwindles in 'Hit List' on NASA Budget," WP, February 5, 1981, A7; "Spacelab, Solar-Polar Curtailed," *AWST*, February 23, 1981, 18–19.
5. Ronald Reagan: "Address Before a Joint Session of the Congress on the Program for Economic Recovery," February 18, 1981. Online by Gerhard Peters and John T. Woolley, APP, http://www.presidency.ucsb.edu/ws/?pid=43425.
6. "NASA Budget Reduced to $6.1 Billion," *AWST*, March 16, 1981, 24; Stockman, *The Triumph of Politics*, 151.

7. What follows is drawn from Joan Johnson-Freese, "Canceling the US Solar-Polar Spacecraft: Implications for International Cooperation in Space," *Space Policy*, February 1987, 24–37.
8. John M. Logsdon, "U.S.-European Cooperation in Space Science: A 25-Year Perspective," *Science*, January 6, 1984.
9. Most of this discussion is extracted from John M. Logsdon, "Missing Halley's Comet: The Politics of Big Science," *Isis*, Volume 80, No. 2, June 1989, 254–280.
10. Carl Sagan to members of The Planetary Society, August 1, 1981, NHRC.
11. Letters from George A. Keyworth to James Beggs, August 5, 1981 and James Beggs to George Keyworth, August 17, 1981, NHRC. Thomas O'Toole, "U.S. May Send Robot Spacecraft for Sample of Halley's Comet," *WP*, August 15, 1981, A3.
12. Letter from James Beggs to George Keyworth, September 16, 1981, NHRC.
13. Letter from Andrew Stofan to Bruce Murray, September 30, 1981, NHRC.
14. For a more detailed account of this disagreement, see the author's essay with the same title in Roger D. Launius, ed., *Exploring the Solar System: The History and Science of Planetary Exploration* (New York: Palgrave Macmillan, 2013), 45–76.
15. Letter from James Beggs to David Stockman transmitting NASA's FY 1983 budget recommendations, September 15, 1981, NHRC.
16. Letter from James Beggs to David Stockman, September 29, 1981, NHRC.
17. *AWST*, June 24, 1981, 56; interview with James Beggs, February 2, 1989.
18. These figures are drawn from the material prepared by NASA to appeal the OMB allocations and transmitted to the White House by a letter from NASA Comptroller C. Thomas Newman to Craig Fuller, December 5, 1981, NHRC.
19. Letter from James Beggs to David Stockman, November 30, 1981, NHRC.
20. Philip J. Hilts, "Science Board to Advise Presidential Proposal," *WP*, December 2, 1981; George Keyworth, paper prepared for Budget Review Board, "Selected White House Views. Department: NASA. Issue: Planetary Exploration," December 8, 1981, NHRC; interviews with Jay Keyworth, April 4, 1989, and January 25, 2017.
21. *Science News*, October 24, 1981, p. 260.
22. Letter from David Morrison and Carl Sagan to Edwin Meese, October 14, 1981, NHRC.
23. Interview with Lou Friedman, May 17, 1988.
24. Letter from Mrs. William W. Scranton to Marvin Goldberger, December 6, 1981, NHRC; letter from Robert Finch to George Bush, December 4, 1981 and memorandum from Craig Fuller to Ed Meese, December 8, 1981, both in Box 6, Outer Space Files, RRL.
25. Letter from A.O. Beckman to Edwin Meese, October 5, 1981; draft of letter from A.O. Beckman to Edwin Meese, December 10, 1981, prepared by JPL and provided to Beckman by Bruce Murray, NHRC.
26. The results of Goldberger's trip are summarized in a letter from Bruce Murray to Arnold Beckman, December 10, 1981, NHRC.
27. Letter from Howard Baker to the President, December 9, 1981, NHRC; Stockman, *The Triumph of Politics*, 13.
28. Budget Review Board Decisions, National Aeronautics and Space Administration, December 11, 1981, NHRC. The date on this document places the date of the Budget Review Board meeting in question. Originally scheduled for

December 9, most evidence suggests it was postponed until December 15. This means either that the date on this document is incorrect or that, after Baker's intervention with the president, the Budget Review Board met on the NASA appeal on December 11.

29. Interview with Craig Fuller, December 14, 2016; memorandum from Danny Boggs to Ed Gray, "Presidential Participation in First Flight of *Columbia*," February 24, 1981, Box 82, Papers of Danny Boggs; Memorandum from Richard Allen to James Baker, "Space Shuttle Launch," March 2, 1981, Box 7, Papers of Michael Baroody, both in RRL.
30. Memorandum from Richard Allen to James Baker, "Planning for the First Shuttle Launch," March 16, 1981; memorandum from Patricia Rogers to Gregory Newell, "Meeting with NASA Officials," March 10, 1981, both in Box 1, Outer Space Files, RRL.
31. Howell Raines, "Reagan Sees Launching on TV at White House," *NYT*, April 13, 1981, A11; Douglas Brinkley, ed., *The Reagan Diaries* (New York: Harper Collins, 2007), 13; "Text of Message from President to Astronauts," *NYT*, April 15, 1981, A21.
32. Ronald Reagan: "Remarks at a White House Luncheon Honoring the Astronauts of the Space Shuttle *Columbia*," May 19, 1981. Online by Peters and Woolley, APP, http://www.presidency.ucsb.edu/ws/?pid=43837.
33. Jacob Weisberg, *Ronald Reagan* (New York: Henry Holt & Company, 2016), 116; Richard Reeves, *President Reagan: The Triumph of Imagination* (New York: Simon & Schuster, 2005), xvi.
34. Memorandum from Ben Huberman and Michael Berta to Richard Allen, "Space Policy Issues," February 10, 1981, Box 30, Subject Files, Executive Secretariat, National Security Council, RRL.
35. Memorandum from Richard Allen to Various Addressees, "The Shuttle and Space Policy," April 6, 1981, Box 37, Papers of Martin Anderson, RRL.
36. Hans Mark, "Daily Diary," June 10, 1981. I am grateful to Dr. Mark for sharing this and several other excerpts from his diary.
37. Memorandum From Verne Orr to Allen Lenz, "Space Shuttle Policy," July 28, 1981, RAC 14, Papers of George Keyworth, RRL; memorandum from Richard Allen to Addressees, "National Security Planning Group (NSPG) Meeting on Monday, August 3, 2:00–3:00 p.m."; memorandum from Richard Allen to the President, "National Security Council Meeting of August 3, 1981, on the Space Shuttle," Meeting Files 11–20, Executive Secretariat, National Security Council, RRL. The suggestion that James Beggs was at the meeting is in Hans Mark, *The Space Station: A Personal Journey* (Durham, NS: Duke University Press, 1987), 131.
38. Interview of Edward C. "Pete" Aldridge by Rebecca Wright, May 29, 2009, NASA Headquarters Oral History Project, https://www.jsc.nasa.gov/history/oral_histories/NASA_HQ/Administrators/AldridgeEC/AldridgeEC_5-29-09.htm.
39. National Security Decision Directive 8, "Space Transportation System," November 13, 1981, https://www.reaganlibrary.gov/sites/default/files/archives/reference/scanned-nsdds/nsdd8.pdf.

CHAPTER 4

An Initial Reagan Space Policy

As senior Reagan administration officials concerned with the space program met on June 10, 1981, they had decided that once he took office, Reagan's science and technology adviser George A. "Jay" Keyworth II would lead an administration-wide review of space policy, with a focus on space transportation issues. Keyworth had an ambitious concept for the review he would head. Even though he was yet to be confirmed by the Senate, Keyworth as science advisor-designee was not shy about publicly expressing his policy views. In late June he declared that the review would lead to "the ideas and plans that will set the course for our country's activities in space for years to come," and that his Office of Science and Technology Policy (OSTP) would play "an important role" in the review.

Keyworth's confirmation hearing was held on July 20. He testified that there was "considerable concern" in the Reagan administration regarding the future course of the U.S. space program, and that because the issues to be reviewed were "sufficiently complex, including "the 'turf' of various agencies," the space policy review would take until the end of 1981 to complete. He underestimated the time required by six months; the results of the review, in the form of a new National Space Policy, were announced on July 4, 1982, as Ronald and Nancy Reagan watched the space shuttle *Columbia* land at California's Edwards Air Force Base. By that time, much to Keyworth's chagrin, there had been an organizational "coup" within the White House, and Keyworth and OSTP had lost the lead role in shaping U.S. space policy.[1]

Review Begins

Ronald Reagan approved the space policy review at the August 3 National Security Planning Group meeting described in Chap. 2. The formal directive initiating the review was issued on August 10; it confirmed that the review would be broad in scope. The directive said: "The President has directed

J. M. Logsdon, *Ronald Reagan and the Space Frontier*,
Palgrave Studies in the History of Science and Technology,
https://doi.org/10.1007/978-3-319-98962-4_4

Dr. George Keyworth to examine, in coordination with appropriate agencies, whether new directions in space are warranted. The study should address what capabilities may be needed in order to ensure U.S. leadership, to ensure that the Space Transportation System is managed in the most effective manner to meet the future needs of space users, and to ensure continued satisfaction of national security needs." The review would be carried out under the direction of Keyworth and his staff, but would take place under the auspices of the National Security Council (NSC), rather than OSTP, since it would deal not only with the civilian space program but also with the highly classified national security uses of the shuttle and with other national security space programs. This would turn out to be a crucial development in terms of determining the lead role in developing space policy options for the president.[2]

Before the space policy review could begin, "terms of reference" to guide the effort had to be developed. After the June 10 meeting that had decided a review was needed, work had begun on drafting such terms of reference; the draft was circulated for comment on July 17. The study would "evaluate the national security, arms control, political and economic implications of the Shuttle as the only [U.S.] launch capability," and "assess the current defense and civil launch vehicle backup strategy." In addition, the study would "assess the feasibility of using the Shuttle as an integral part of a weapon system," including "to what ends would we be willing to use the Shuttle in an ASAT [anti-satellite] mode?" It would "review the present policy that NASA should continue to be responsible for overall management and operations of the Shuttle," and examine alternative management approaches. More broadly, the review would also examine Carter administration space policy directives in the context of Reagan administration priorities, keeping in mind "the long term commercial and national security interests of the nation." The study would also "assess the process by which decisions that affect these interests are coordinated within the US Government."[3]

The final terms of reference were issued on September 8, marking the formal start of the policy review. Within OSTP, Keyworth assigned staff responsibility for the effort to Victor Reis, his new assistant director for national security. Reis had come to OSTP on leave from the Lincoln Laboratory of the Massachusetts Institute of Technology. He had no earlier experience with respect to space issues. There was no one else on Keyworth's senior staff conversant with those issues, and the space policy portfolio was assigned to Reis because of its national security aspects. Reis got a rapid education on civilian space issues from the staff of the Office of Management and Budget (OMB), among others; most OMB space staff had been in their positions for a number of years and represented policy, program, and budget continuity within the Executive Office of the President.[4]

Reis soon took active charge of the review. He, like Keyworth, was willing to discuss with external audiences his views on policy issues even as they were under review inside the government. For example, he told a November 1981 meeting that "maturing of the Space Transportation System (STS) is the nation's top priority in space, even if it means that other programs must take a 'step backward' … Over the next decade, if we have to take a step backward in other programs by making the Shuttle into a STS, we will be a lot better off." Reis told his audience: "I don't understand why you need a permanent man in space" when "you will have very close to a manned permanent presence in space when the STS starts working."[5]

These comments by Reis signaled that National Aeronautics and Space Administration (NASA) was not likely to achieve all of its objectives in the space policy statement that the review would produce. Those objectives included preventing "the inclusion of a negative statement" such as the one that had appeared in the Carter administration 1978 space policy saying that "it is neither feasible nor necessary at this time to commit the US to a high-challenge, highly-visible space engineering initiative comparable to Apollo." According to NASA's Hans Mark, "this was a damage-limiting objective that we felt was absolutely essential if any progress was to be made at all." With respect to this goal, NASA found that there was "no real controversy," since there was no doubt that "most members of the new administration were generally more in favor of the space program than the members of the previous one."

A second NASA objective was having the policy include "a strong statement in support of the shuttle program." Here, "the situation was more complicated," since even after the issuance of National Security Decision Directive (NSDD)-8, there was "substantial opposition to the policy that the shuttle would be the primary U.S. launch vehicle. Both the representatives of the Defense Department and the Commerce Department (representing commercial interests) argued that it might be better to have a mixed fleet than to depend exclusively on the shuttle." These objection presaged a continuing controversy within the Reagan administration, one that would not be fully resolved until the aftermath of the January 1986 *Challenger* accident.[6]

A third NASA objective, "if possible," was to include in the policy statement "an endorsement of the proposed space station program." This proved to be unachievable. What NASA discovered in the course of the policy review was that there was widespread opposition to moving forward in the near term with an orbital outpost. Reis's negative statement regarding the station reflected the views of science adviser Keyworth, who in public statements at the time of the review called a space station a "mistake" and a "step backwards." The Defense Department (DOD) was also opposed, feeling that NASA's initiating a major new program would divert its attention and best engineering talent away from Department of Defense's high-priority objective of making the space shuttle fully operational. Concern about the increase in the NASA budget required to develop a space station also motivated OMB opposition. In support of the station proposal, NASA had begun what the White House perceived as "a carefully organized campaign," with "hundreds of letters and telegrams" coming to OSTP in support of the initiative and with a variety of press articles predicting presidential approval of the station. Keyworth thought such a campaign "completely improper. It does not exactly endear people in this Administration to the initiative."[7]

Missing from NASA's objectives for the new space policy was a presidential commitment to a major new space exploration goal such as a lunar base or even human missions to Mars. The new NASA leadership favored a step-by-step approach in developing capabilities to achieve such a goal, and to them, a space station was what James Beggs frequently called "the next logical step" in that approach. Anything beyond a station, Beggs and Mark judged, was too great a jump to be politically feasible. The benefits and risks of setting a long-range goal for NASA's efforts would remain a continuing policy issue throughout Ronald Reagan's time in the White House.

Keyworth in his comments during his July confirmation hearing had indicated that the review would be completed by the end of December 1981. But by the turn of the year, there was no sign that the review was near completion; instead, it had become characterized by conflicts among the participating agencies on a number of the issues under review. In January 1982, Reis reported that the review was "moving along—however ponderously—toward a late spring draft" of a policy statement.[8]

Policy Review Completed: SIG (Space) Created

In January 1982, there was a major change in the organization of the White House senior staff. Richard Allen had turned out to be a rather ineffectual national security adviser, not at all in the tradition of Nixon's Henry Kissinger or Carter's Zbigniew Brzezinski, powerful individuals who had a major role in shaping foreign and national security policies. Allen viewed his NSC role more as a coordinator of policy proposals originating from the various executive departments and agencies rather than as an initiator of plans and policies. The fact that he reported to President Reagan through Ed Meese, rather than directly, also diminished his influence. Allen also had continuing conflicts with imperious Secretary of State Alexander Haig. In late 1981, Meese had ordered a management review of the NSC, sensing that the body under Allen was not serving the president's needs. Then Allen got caught up in a minor scandal. All of these factors led to Allen's forced resignation on January 4, 1982.

Allen was replaced by William Clark, often called "Judge" because Ronald Reagan had appointed him to the California Supreme Court. Before that appointment Clark had preceded Ed Meese as Governor Reagan's chief of staff. He was also a personal friend of Ronald Reagan; the two shared a love of horseback riding. Clark during 1981 had been deputy secretary of state, even though he had practically no experience in foreign or national security policy. As part of the shakeup of the NSC, Clark insisted on reporting directly to the president, not through Meese; he thus became the fourth member, along with Meese, Baker, and Deaver, of Reagan's inner circle of advisers, meeting with the president almost every morning to deliver an intelligence briefing. Clark was as much, if not more, committed to Reagan's overall philosophy as

Ed Meese, and was strongly anti-communist. Clark and James Baker would end up in frequent conflict once Clark moved to the White House.

Moving to the NSC along with Clark was Robert "Bud" McFarlane, who had been one of Secretary of State Haig's advisers. In contrast to Clark, McFarlane was well steeped in international affairs, having served on the NSC staff during the Nixon and Ford administrations. Clark and McFarlane created a much more activist NSC than had been the case under Richard Allen, one that would manage policy reviews and identify potential presidential policy initiatives, including those dealing with space issues.[9]

Space matters within the NSC staff were the traditional purview of a senior Air Force officer detailed from the DOD. This role had been filled by Michael Berta during the final years of the Carter administration; Berta had stayed on during the first year of the Reagan presidency. His replacement, arriving in April 1982, was Lieutenant Colonel Gil Rye; Rye would soon assume a central role in shaping Reagan administration space policy. James Beggs recalled in a 2002 interview that Rye "gave us very significant support … He was able to talk … the senior [White House] staff into doing things that were very helpful." Hans Mark, who had known Rye when he was secretary of the Air Force, characterized him as "persistent and astute" and capable of "bureaucratic legerdemain."[10] Before coming to the NSC, Rye had been chief of the Air Force space plans division in the Pentagon. Even prior to taking on his new responsibilities, Rye had been interested in NASA and its programs. He was the only person from the DOD who had attended a November 1981 NASA symposium discussing the agency's very preliminary space station plans.

As will be discussed below, President Reagan was scheduled to attend the July 4 landing of the space shuttle *Columbia* at Edwards Air Force Base in California. One of Rye's early actions at the NSC was to suggest to Clark that Reagan's appearance be used as a forcing function to bring the space policy review to a conclusion, so that its findings could be released in conjunction with the president's visit. Clark agreed to Rye's suggestion that the policy review should be ready for presidential approval before the July 4 shuttle landing.

As Rye arrived at the NSC, the space policy review team was "mired in dissent" and "unable to reach a consensus on several policy statements." For example, some in the group were attempting to "walk back" the strong presidential endorsement of the central role of the space shuttle that had been contained in the November 1981 NSDD-8, and there was no agreement on a continuing space policy coordinating mechanism. Rye asked OSTP's Reis to provide him with a draft of the policy statement. Rye took the draft and "polished it off, identified the remaining unresolved issues along with options and recommendations."[11]

Most significantly, Rye's revisions led to the NSC seizing away from OSTP the lead role on space policy within the Reagan White House. The draft Reis provided to Rye maintained the approach to coordinating space issues that had been followed during the Carter administration and up to that point in the Reagan administration—an ad hoc coordinating body, the Policy Review

Committee for Space, chaired by the director of OSTP and convened only when needed. Rye and Clark's deputy McFarlane reasoned that since over half of the U.S. space budget was spent on national security-related efforts and since even NASA's programs had important foreign policy aspects, it should be the NSC, rather than the science-oriented OSTP, which should have the lead in preparing major space policy issues for presidential decision. In addition, they were interested in enhancing the influence of the NSC, influence that had diminished during Richard Allen's brief tenure.

McFarlane and Rye seized upon the idea of creating a standing Senior Interagency Group (SIG) for Space, chaired by the national security adviser and with a NSC staff person as its executive secretary, as a means for carrying out the space policy development and coordinating role. Clark agreed to support that idea. Since the presidency of Lyndon Johnson, there had been SIGs operating under the auspices of the NSC. They focused on specific national security issues and were comprised of senior officials concerned with those issues. On February 26, 1981, three such Reagan administration SIGs had been created: one on foreign policy, chaired by the secretary of state; one dealing with defense policy, chaired by the secretary of defense; and one on intelligence, chaired by the director of central intelligence. Underneath most SIGs were Interagency Groups (IGs) composed of less senior officials concerned with a specific policy issue or geographical area. The process of initiating an SIG to deal with a particular issue area proliferated during the Reagan administration; by its close, 22 SIGs had been created. After presidential approval, a recommendation emerging from SIG deliberations would be transformed into a NSDD issued over the president's signature.[12]

Rye discussed the idea of creating such a body for space policy with NASA's Mark, who "urged Rye to work as hard as he could to secure the chairmanship of SIG (Space) for Judge Clark," on the grounds that Clark was an "exceedingly astute politician" who would "see the political value that the initiation of the space station program might have for the president." Mark reports that NASA administrator James Beggs was skeptical of the wisdom of such a move because it might threaten the civilian character of NASA, but was persuaded to support it. Rye's proposal would reduce the influence of the two leading skeptics with respect to NASA's space station initiative, OMB and OSTP, an attractive prospect to NASA. The DOD also supported Rye's initiative, thinking it would give the Defense Department more influence over national space policy than if the science adviser were in charge of formulating that policy.[13]

Rye thus revised the draft policy statement to say: "To provide a forum to all Federal agencies for their policy views, to review and advise on proposed changes to national space policy, and to provide for the orderly and rapid referral of space policy issues to the President for decisions as necessary, a Senior Interagency Group (SIG) on Space shall be established." The SIG (Space) would be chaired by the president's national security adviser; its members would be the secretaries of defense, state, and commerce, the director of the Central Intelligence Agency, the chairman of the Joint Chiefs of Staff, the director of the Arms Control and Disarmament Agency, and the NASA administrator. Thus, five of the eight members would come from the national security community. Representatives of OMB and OSTP would have only "observer" status, a clear downgrading from their prior space policy roles. Clark on June 21 forwarded the draft statement for interagency and White House review, revised to include the SIG (Space) proposal.[14]

Comments were quick in coming. Keyworth returned a copy of the draft statement with Rye's language replaced by the provisions that had been in the draft that Reis had provided, with the science adviser chairing any future policy review that might be established. Keyworth was seconded by OMB director David Stockman, who thought that "it would be unnecessary and undesirable to establish a special purpose" SIG. Stockman was "frankly concerned that program advocates could cause problems for the President by seeking policy-level commitments to major new program initiatives outside of the fiscal discipline imposed by the budget review." Stockman argued that the coordinating mechanism suggested by Keyworth should be reinserted in the policy statement, saying that "what Jay and I have in mind is reconvening, when necessary, the same type of mechanism" that had been used to prepare the draft policy.

White House Director of Cabinet Affairs Craig Fuller, who was becoming increasingly involved in space issues, proposed an elaborate approach to space policy coordination. It was Fuller's view that it was "extremely important that space policy be given high-level and visible attention by the administration." He proposed that "a coordinating group of senior officials be formed that would guide the analysis, policy review and decision making process with respect to space," providing to Ronald Reagan "alternatives that are consistent with the overall policy determinations" that Reagan had established. If this approach were adopted, suggested Fuller, SIG (Space) "would be the forum for national security policy considerations." An already-existing Cabinet Council on Commerce and Trade "would be the policy making forum for review of the many civil space issues related to commercialization and other matters." Overseeing all of this activity would be a "Senior White House Space Policy Coordinating Group" composed of Ed Meese, James Baker and his deputy, Richard Darman, Judge Clark, Stockman and his deputy Ed Harper, and Fuller.[15]

Fuller's mention of the Cabinet Council on Commerce and Trade reflected the emerging pattern of Reagan administration decision-making. As a means of limiting the number of people involved in developing policy options for presidential choice to a manageable size, while still maintaining President Reagan's preference for a cabinet-style government, presidential counselor Ed Meese on February 13, 1981, had created five cabinet councils; the Cabinet Council on Commerce and Trade had been one of the five. Each council would consist of those cabinet members with a particular stake in the issues covered by the council. Each council could create interagency working groups composed of representatives of member departments and agencies to develop policy positions for the council's deliberations and eventual transmittal to the president for decision. The result of this interagency approach was often a slow-moving process, as members of cabinet council working groups argued their agency's position, as papers reflecting those positions and compromises among them were prepared, and as the Cabinet Council principals debated options before they were presented to the president. This would turn out to be the case in 1983 and after as the Cabinet Council on Commerce and Trade became increasingly involved in space-related issues.[16]

The suggestions of Keyworth, Stockman, and Fuller were not accepted. Rather, the version of the policy statement that was approved, reportedly during a June 28 meeting of the NSC at which President Reagan was briefed on the policy review's conclusions, retained the proposal to establish a SIG (Space). There was agreement that, before the president was formally asked to sign the policy statement, he would be made aware of the disagreement over establishing SIG (Space).[17]

Rye prepared several papers for Judge Clark to take to President Reagan, who on July 1 had traveled to his ranch in the mountains behind Santa Barbara for the July 4 holiday. One paper was an explanatory memorandum on the disagreements regarding establishing the SIG (Space). The memo noted that all of the executive branch departments and agencies that had participated in the policy review had endorsed that action, but that OMB and OSTP had favored "a different approach … Dave Stockman and Jay Keyworth recommend formation of a lower-level, ad hoc, policy advisory group chaired by Jay Keyworth." Rye's memo noted that the agencies involved in space matters "felt strongly that the President must visibly demonstrate his leadership and his appreciation for the importance of the U.S. Space Program and the critical issues lying ahead through formation of one high level standing group … It would not be the intent of SIG (Space) to make decisions that would circumvent the budgetary process, but rather to coordinate the future course and direction of the U.S. Space Program."

A second paper was a decision memorandum that asked: "Should a NSDD on National Space Policy be issued?" The memorandum summarized the contents of the proposed policy statement, saying that it

> provides broad goals and principles for the conduct of U.S. space programs; reaffirms the Shuttle as the primary U.S. launch system; places first priority on making the Shuttle an operational and cost effective system; directs the expeditious deployment of a U.S. anti-satellite system; authorizes research and planning for space weapons as a hedge against potential Soviet breakout; and directs close coordination of the military, intelligence and civilian sectors of the U.S. space program. Most importantly, the NSDD directs the formation of a Senior Interagency Group (SIG) for Space … Within the White House staff, there is a difference of opinion on the implementation of the space policy … Your July 4th speech for the Shuttle landing is being written to announce your new space policy.

A copy of the policy statement was attached to the memorandum. President Reagan was given the choice of "OK" or "No" to the recommendation "that you sign the attached NSDD to be dated July 4." On July 3, Reagan wrote "RR" next to the "OK" option and signed the policy statement.[18]

With Ronald Reagan's approval of what became NSDD-42, "National Space Policy," the NSC, through its chairmanship of SIG (Space), took over the White House space policy–coordinating role that had belonged to OSTP. What had transpired in the final stages of preparing the directive was characterized by a disappointed Vic Reis as a NSC "coup."

The National Space Policy statement was classified "Top Secret" since it dealt with intelligence and military issues as well as civilian topics. The policy provided "the broad framework and the basis for the commitments necessary for the conduct of the United States space program." Its first declaration was "the Space Shuttle is to be a major factor in the future evolution of United States space programs." Goals of that program would include: "strengthen the security of the United States"; "maintain United States space leadership"; and "expand United States private-sector investment and involvement in civil space and space-related activities." This last objective was the first time a statement of national space policy had called out the importance of private sector space activities; the directive also said that "the United States will provide a climate conducive" to expanded private sector space activities. Those activities would be "authorized or regulated by the government." The policy's intent was to "preserve United States preeminence in critical major space activities"; this language echoed science adviser Keyworth's belief that it was too expensive and not necessary for the United States to have across-the-board space preeminence. The policy did not endorse space station development, but in a concession to NASA it allowed the space agency to "continue to explore the requirements, operational concepts, and technology associated with permanent space facilities." The final paragraph of the directive established SIG (Space).[19]

Ronald Reagan Meets the Space Shuttle

Ronald Reagan had visited shuttle mission control in Houston in November 1981, during the second shuttle flight, and had talked then to the astronauts in orbit. But he had never actually seen a space shuttle. The fourth flight of the shuttle was launched on June 28, 1982; the week-long mission would land on July 4. Since early May press reports had suggested that President Reagan would be present as the shuttle returned to Earth at Edwards Air Force Base in the California high desert. NASA had told the White House, "The three complete shuttle orbiters, the *Columbia*, the new *Challenger* and the old *Enterprise* [used for approach and landing tests but not intended for orbital missions] will all be within towing distance of each other, as a possible backdrop for a presidential space policy address." The image of the president joining up to a half million people in welcoming *Columbia* back to Earth on the holiday celebrating U.S. independence was irresistible to senior presidential adviser Michael Deaver and his associates concerned with Ronald Reagan's image, and plans were made for Ronald and Nancy Reagan to attend. On the early morning of July 4, the Reagans took a short helicopter ride from their ranch to the Air Force base.[20]

The Independence Day shuttle landing was described as "a patriotic spectacle." Standing on a flag-bedecked platform next to the test shuttle orbiter *Enterprise*, the Reagans searched the sky for a first glimpse of the returning orbiter. Not seeing it, the president asked astronaut Bob Crippen, pilot on the initial shuttle flight in April 1981, where the shuttle was at that point; Crippen replied "over Hawaii." According to Rye, this may have been the point at which Reagan "fully understood the technological achievements of the space program." Just after 9:00 a.m., twin sonic booms announced the arrival of *Columbia* over Edwards Air Force Base. A crowd estimated at 500,000 were also present; 10,000 of them, each with a small American flag, were near the presidential platform. The shuttle glided to a smooth landing, and a few minutes later the Reagans were driven to *Columbia* so that they could greet the two crewmembers, T.K. Mattingly and Hank Hartsfield, as they descended from the vehicle (Fig. 4.1). Then the president, his wife, plus Mattingly and Hartsfield, returned to the elevated platform for his remarks.

Fig. 4.1 President Reagan and his wife Nancy greet crewmembers T.K. Mattingly and Hank Hartsfield shortly after shuttle orbiter *Columbia* landed at Edwards Air Force Base in the California high desert on July 4, 1982. (Photograph courtesy of Reagan Presidential Library)

At the end of a runway, the second shuttle orbiter, *Challenger*, sat atop its 747 shuttle carrier aircraft, ready to begin its transfer to the Kennedy Space Center in Florida. Reagan addressed the audience:

> I think all of us, all of us who've just witnessed the magnificent sight of the *Columbia* touching down in the California desert, feel a real swelling of pride in our chests.
>
> In the early days of our Republic, Americans watched Yankee Clippers glide across the many oceans of the world, manned by proud and energetic individuals breaking records for time and distance, showing our flag, and opening up new vistas of commerce and communications. Well, today, I think you have helped recreate the anticipation and excitement felt in those home ports as those gallant ships were spotted on the horizon heading in after a long voyage.
>
> Today we celebrate the 206th anniversary of our independence. Through our history, we've never shrunk before a challenge. The conquest of new frontiers for the betterment of our homes and families is a crucial part of our national character, something which you so ably represent today. The space program in general and the shuttle program in particular have gone a long way to help our country recapture its spirit of vitality and confidence. The pioneer spirit still flourishes in America. In the future, as in the past, our freedom, independence, and national well-being will be tied to new achievements, new discoveries, and pushing back new frontiers.
>
> The fourth landing of the *Columbia* is the historical equivalent to the driving of the golden spike which completed the first transcontinental railroad. It marks our entrance into a new era. The test flights are over. The groundwork has been laid. And now we will move forward to capitalize on the tremendous potential offered by the ultimate frontier of space. Beginning with the next flight, the *Columbia* and her sister ships will be fully operational, ready to provide economical and routine access to space for scientific exploration, commercial ventures, and for tasks related to the national security.
>
> * * *
>
> Way out there on the end of the runway, the space shuttle *Challenger*, affixed atop a 747, is about to start on the first leg of a journey that will eventually put it into space in November. It's headed for Florida now, and I believe they're ready to take off. *Challenger*, you are free to take off now.

As the huge airplane carrying *Challenger* lifted off and flew by the president, dipping a wing in salute, a band played "God Bless America." Reagan remarked "This has got to beat firecrackers." He told NASA administrator Beggs: "That's the most fun I've had since I got this job."[21]

An unclassified "fact sheet" summarizing the National Space Policy was released to the press and public as the president spoke. It had taken 18 months, but the Reagan White House now had in place not only a statement of its space goals and policies, but also a new mechanism—the SIG (Space)—through which already-evident policy disputes would be addressed.

Notes

1. *Defense Daily*, June 29, 1981, 325, and July 21, 1981, 105.
2. Memorandum from Richard Allen to Addressees, "Space Policy Review," August 10, 1981, RAC 14, Papers of George Keyworth, RRL.
3. The draft terms of reference are attached to a memorandum from Allen Lenz, staff secretary, National Security Council, to various addressees, July 17, 1981. A copy of the memorandum is reproduced in John M. Logsdon et al., eds. *Exploring the Unknown: Selected Documents in the History of the U.S. Civil Space Program*, NASA SP-4407, Vol. IV, Accessing Space, NASA SP-4407 (Washington: Government Printing Office, 1999), 329–333.
4. Interview with Vic Reis, March 22, 2016.
5. Ibid; Dennis Stone, "Reis of OSTP Outlines Shuttle Role in Space Policy," *Astronautics & Aeronautics*, January 1982, 17; memorandum from Victor Reis to George Keyworth, "National Security and Space Plans," January 13, 1982, RAC 14, Papers of George Keyworth, RRL.
6. A copy of the October 1978 Presidential Directive-42 containing this statement can be found at https://fas.org/irp/offdocs/pd/pd42.pdf. Hans Mark, *The Space Station: A Personal Journey* (Durham, NC: Duke University Press, 1987), 144.
7. Ibid; R. Jeffrey Smith, ""Squabbling Over Space Policy," *Science*, July 23, 1982, 332.
8. Ibid; memorandum from Victor Reis to George Keyworth, "National Security and Space Plans," January 13, 1982, RAC 14, Papers of George Keyworth, RRL.
9. For a comprehensive discussion of the role of the national security adviser during the Reagan administration, see John P. Burke, *Honest Broker? The National Security Adviser and Presidential Decision Making* (College Station, TX: Texas A&M Press, 2009).
10. Interview of James Beggs by Kevin Rusnak, March 7, 2002, NHRC; Mark, *The Space Station*, 147.
11. Letter from Gil Rye to Hans Mark, October 25, 1984, commenting on an early draft of Mark's *The Space Station*. Quoted with the permission of Gil Rye. I am grateful to Howard McCurdy for providing a copy of this letter. Interview with Gil Rye, April 25, 2016.
12. Burke, *Honest Broker?*, 203.
13. Interview with Gil Rye, April 25, 2016; Mark, *The Space Station*, 146; Smith, "Squabbling Over Space Policy," 333.
14. Memorandum from William Clark to addressees, "National Space Policy" with attached draft policy, classified Top Secret, June 21, 1982, Box 5, Papers of Edwin Meese, RRL.
15. Memorandum from David Stockman to William Clark, "National Space Policy," June 25, 1982, RAC 14, Papers of George Keyworth, RRL; memorandum from Craig Fuller to William Clark, "National Space Policy," June 26, 1982, Files of Valerie Neal, Smithsonian National Air and Space Museum. I am grateful to Dr. Neal for making her files available for my research.
16. For a discussion of the initial operation of the Cabinet Councils, see John Burke, *Presidential Transitions: From Politics to Practice* (Boulder, CO: Lynne Rienner Publishers, 2000), 144–153.
17. *Defense Daily*, June 29, 1982, 322. A comprehensive list of National Security Council meetings during the Reagan administration does not show a June 28,

1982, meeting, so the final version of the space policy statement may have been approved in another forum. See http://www.thereaganfiles.com/document-collections/national-security-council.html.
18. Memorandum from William Clark for the President, "NSDD on National Space Policy," July 2, 1982, with attached memorandum on "Implementation Issue," Box 5, Papers of Edwin Meese, RRL. These two documents were attached to an August 13, 1982, memorandum to Meese from Robert McFarlane, indicating that the argument over creating SIG (Space) went on even after the National Space Policy statement was signed.
19. A copy of NSDD-42, portions of which remain classified, can be found at https://fas.org/irp/offdocs/nsdd/nsdd-42.pdf.
20. *AWST*, May 3, 1982, 11.
21. Howell Raines, "Reagan Affirms Support for U.S. Space Program," *NYT*, July 5, July 5, 1982, 8; Ronald Reagan: "Remarks at Edwards Air Force Base, California, on Completion of the Fourth Mission of the Space Mission of the Space Shuttle Columbia," July 4, 1982. Online by Gerhard Peters and John T. Woolley, APP, http://www.presidency.ucsb.edu/ws/index.php?pid=42704. Interview of James Beggs by Kevin Rusnak, March 7, 2002, NHRC.

CHAPTER 5

SIG (Space) Gets Started

Reaction within the space community to the July 4 statement of National Space Policy was less than enthusiastic. The National Aeronautics and Space Administration (NASA) leadership believed that "they received politically safe encouragement from a President who supports the space program," but not "the strong presidential leadership that will be necessary for exploitation of the space shuttle to its fullest potential." *Science* magazine suggested that the policy represented "a small but important shift in the direction of the U.S. space program, toward increased military control of activities in space and increased involvement of the private sector in space ventures." The trade magazine *Aviation Week & Space Technology* characterized the July 4 event as "better than nothing," observing that "nothing in the speech or policy was likely to provoke any quarrels from space partisans, or from opponents for that matter. Continued space activity, unspecified; expanding private sector involvement; international cooperation in space, strengthening U.S. security—all as wholesome as motherhood and the flag. Essentially, the new policy is a codification of the status quo." A more critical reaction came from liberal California congressman George Brown, who suggested that the statement was "at best a treading water policy and at worst a strictly military-inspired space policy." Brown's suggestion that having the Senior Interagency Group for Space [SIG (Space)], with the majority of its members coming from the national security community, be in charge of formulating space policy represented the "militarization" of the space program would persist through the remaining six plus years of the Reagan administration.[1]

Even though with the creation of SIG (Space), presidential science adviser Jay Keyworth would no longer be in charge of space policy for the White House, he had chaired the review that had produced the new policy statement, and thus it fell to him to be the primary public defender of the policy. On August 4, 1982, he testified before the House Subcommittee on Space Science and

J. M. Logsdon, *Ronald Reagan and the Space Frontier*,
Palgrave Studies in the History of Science and Technology,
https://doi.org/10.1007/978-3-319-98962-4_5

Applications, saying that the policy represented "a strong affirmation of our nation's historic commitment to space … The United States is the world leader in space, and our policy is—in simplest terms—to remain so." Keyworth noted that the space shuttle "occupied a central position in developing the new space policy" and that "the United States is fully committed to maintaining world leadership in space transportation," even in the face of foreign competition. He acknowledged that there had been "many criticisms pertaining to the lack of mention of any specific programs in the space policy," such as a space station or an additional shuttle orbiter. "This omission was deliberate," he said, since the role of the policy was to provide "guidance for overall direction and scope, for coordination between sectors, and for balance." With respect to accusations that the policy furthered the "militarization of space," Keyworth suggested that while "recognizing the important role that space plays" in assuring national security, the policy did not "imply that national security uses of space in any way inhibit or are developed at the expense of civil or commercial uses."[2]

Keyworth Is Not Happy

While publicly science adviser Keyworth defended the new national space policy, he was in fact less than pleased with its contents. Responding to a question during his congressional testimony, Keyworth defended the tone of the president's July 4 speech, saying that "I think we could have written the speech for the President on the 4th of July in a certain way better." But, he explained, "in the months preceding the 4th of July speech, the amount of pressure that was exerted through the media in trying to push the President to announce the fifth Orbiter or the space operations center [another name for a space station] caused much consternation among us in the White House; it certainly influenced the way in which that speech was written." Keyworth's admission of the internal controversy that had preceded Reagan's speech was more candid than normal for a senior White House official, and reflected the lingering tensions among those who had been involved in preparing for Reagan's appearance at the shuttle landing.[3]

Indeed, by the time of Keyworth's August 4 testimony, the fact that the policy had emerged only after bureaucratic conflicts inside the Reagan administration was widely known. *Science* magazine in its July 23 issue had reported that while "at first glance, the policy appears to be a bland recitation of existing ideas for the exploitation of space … to those who participated in its creation … the policy represents the outcome of an enormous struggle among nine agencies with frequently conflicting interests." NASA sought as a goal of the policy to preserve "United States preeminence in space activities," but the Office of Science and Technology Policy (OSTP) had insisted on the more limited statement that the goal was to preserve U.S. preeminence only in "critical space activities." Rather than have the policy say that the government "will promote and encourage expanded private sector investment," the Office of Management and Budget (OMB) insisted on the more limited commitment that it would only "provide a climate conducive" to such investment. Facing increasing budget deficits, "anything that implied a budget implication was fought by OMB almost to the point of lunacy." The designation of OSTP and OMB as merely "observers" within SIG (Space) "generated great protests by Keyworth and Stockman."[4]

Conflicts among members of the White House staff were a regular feature of the Reagan administration; the tensions related to the national space policy and Reagan's July 4 speech were thus not all that unusual. In the view of Gil Rye, the National Security Council (NSC) director for space, the account of the conflicts over the content of the space policy statement that appeared in *Science* was likely linked to the unhappiness of Keyworth and his associates in OSTP regarding their loss to the NSC of the leadership role in space policy development. In a July 9 memorandum to his boss, Reagan's national security adviser William "Judge" Clark, Rye reported that he was "distressed" to learn that Keyworth had held a background press conference on the new space policy with representatives of 10–12 newspapers and magazines and that "unfortunately, we were not informed of the conference." Rye had been interviewed by Jeffrey Smith, the author of the *Science* article, who was already aware of the conflicts between NSC, OSTP, and OMB in the final stages of preparing the policy statement. Rye hoped that Smith "did not receive this information during Jay Keyworth's press conference, but obviously it came from a White House source."

Rye's complaints were not limited to Keyworth's press conference. He noted that he had just learned that the OSTP director intended to prepare a "National Space Plan" for implementing the new space policy, thereby "becoming involved in NSC prerogatives relating to national security, not to mention the responsibilities of the new SIG (Space) for coordinating the overall U.S. Space Program." Rye also objected to negative public comments by Keyworth and his staff, particularly Vic Reis, regarding NASA's hope to gain presidential approval of a space station. To Rye, the "role of the White House Staff is to act as an honest broker on such issues and not prejudge a future Presidential decision"; this was a bit disingenuous on Rye's part, since he was becoming a committed advocate of the space station. He suggested to Clark that "we must reach an understanding with OSTP that the new SIG (Space) represents the focal point for coordinating [the] U.S. Space Program and that any action by his office or discussions with the press concerning space policy is divisive." Rye concluded his memo by noting, "I am uncomfortable with bureaucratic intrigues, but I suppose it comes with the job."[5]

Initial SIG (Space) Activity

If indeed the new National Space Policy was intended only to provide "guidance for overall direction and scope" of U.S. space efforts, then more specific policy, program, and budget decisions would be required for its implementation. In SIG (Space), the policy had created a new forum for developing those policy decisions, and that body soon began its work.

Rye was particularly eager to get the SIG (Space) process up and running. On July 14, 1982, only ten days after its creation had been announced, he proposed to Clark that the first SIG (Space) meeting be scheduled for early September. He noted that he "would have preferred to hold the first meeting in August, but unfortunately the UNISPACE 82 Conference that will be held in Vienna in August complicates schedules." He asked Clark to approve forming a working group to prepare for the September meeting. Forming an interagency working group would be the first step in almost every SIG (Space) activity.[6]

Initial SIG (Space) Meeting

The first meeting of SIG (Space) took place on September 23, 1982. It turned out that for many Cabinet-level agencies, it was their number two official, rather than the department secretary, who would usually be the agency representative in SIG (Space) deliberations; space policy issues often did not have a high enough priority to merit the attention and time of the agency's top official. There were 21 individuals in attendance. Most senior among them were Deputy Director of the Central Intelligence Agency (CIA) John McMahon; Undersecretary of State William Schneider; Deputy Secretary of Defense Frank Carlucci; Michael Bayer, associate deputy secretary of commerce; Lt. General Paul Gorman, assistant to the chairman of the Joint Chiefs of Staff; Norm Terrell, assistant director of the Arms Control and Disarmament Agency; and James Beggs and Hans Mark from NASA. From the White House came Deputy OMB Director Joseph Wright; Ron Frankum, deputy science adviser; Fuller from the Office of Cabinet Affairs; and from NSC, Clark, his deputy Bud McFarlane, and Rye. There were only two agenda items: approval of the SIG (Space) Terms of Reference and approval of three topics for initial SIG (Space) consideration that the working group had selected from among eight topics that Rye had originally proposed. Those areas were space launch policy, the space station, and remote sensing.

In his introduction to the meeting, Clark stressed that "the President is firmly committed to a vigorous U.S. Space Program which will demonstrate our resolve to exercise leadership in space" and that SIG (Space) was "a recognition of the true interagency nature of the U.S. Space Program and the importance that space will have for the future of our nation." He recognized some concern among SIG (Space) members that the group would become "bogged down in addressing detailed programs," and assured those in attendance that the NSC had "no intention for this to occur." Rather, he said, SIG (Space) would be a "high level mechanism for: resolving major policy issues in order to establish the appropriate framework for budgetary or detailed program decisions to be made; and addressing major areas in which we recognize that the President's personal involvement will be required." Clark emphasized "the President's desire to be an active participant in major decisions and be presented with budget decisions that are reflective of a disciplined policy resolution process." By indicating that policy choices should drive budget decisions, the NSC through SIG (Space) was posing a challenge to OMB's "first among equals" role in the White House decision-making process.

The scope of SIG (Space) activities was an area of particular concern for the representatives of the national security community, who were used to planning their space activities with minimal White House involvement. Rye had alerted Clark in advance of the meeting that the Department of Defense (DOD) and the elements of the Intelligence Community, represented by the CIA, were concerned "about the linkage of policy issues to the budgetary process which implies the SIG (Space) will make programmatic decisions." In particular, the national security space community was concerned that SIG (Space) might recommend that some of its budget be reallocated to finance major NASA initiatives such as the space station.

In preparing CIA's McMahon for the meeting, his staff had warned him of "major philosophical differences" between the NSC and the agencies involved in the preparatory working group with respect to how intrusive in agency decisions the NSC staff (in essence, Rye) would try to be. At the SIG (Space) meeting, Carlucci from Defense insisted that it was important to "stick to policy issues." Reacting later to the minutes of the meeting, prepared by Rye, McMahon commented that they reflected a "contravening view" from what he and others had actually said. He added that "it was agreed at the meeting that the SIG would devote itself to major policy issues and *not* get involved in programmatic issues." He noted that "that point was heatedly debated." Clark's introductory remarks at the meeting had been intended to ameliorate this concern, but it would soon reappear as SIG (Space) dealt with the space launch issue as its first order of substantive business.[7]

The participants in the meeting endorsed the SIG (Space) Terms of Reference and the three topics for initial study. The Terms of Reference spelled out four functions for SIG (Space): "a. Periodically review the policy implications involving the implementation of NSDD-42; b. Provide a forum to all Federal agencies for their policy views; c. Review and advise on proposed changes to national space policy; and d. Provide for orderly and rapid referral of space policy issues to the President as necessary."

Brief white papers outlining the scope of the three study topics were available to the meeting participants; they had been drafted by the preparatory working group. With respect to space launch policy, SIG (Space) was to "determine the future U.S. national space launch policy, especially with regard to (1) the increasing foreign space launch capabilities and competition, (2) U.S. commercial launch systems and operations, and (3) development and maintenance of a capability to satisfy U.S. Government current and projected requirements." With regard to the space station, the question was "what policy issues must be identified and resolved in order to establish the basis for an Administration decision on whether or not to proceed with development of a permanently-based, manned space station?" With regard to remote sensing, SIG (Space) was only to "assess the policy implications of current and projected U.S. and foreign civil remote sensing satellite activities," not get involved in an ongoing debate over Landsat privatization. (That debate is discussed in Chap. 12 of this study.)[8]

NSC's McFarlane chaired a follow-on meeting of the less senior-level IG (Space), held on October 1. The status of the working groups being formed to examine the three issues was discussed. Recognizing that a decision to develop the space station was a major policy initiative, and that there was already evident significant controversy regarding such a decision, the SIG (Space) plan was to spend the whole next year laying the basis for a fall 1983 presidential decision on whether to proceed. A more pressing issue, one that required a decision in time for its being reflected in the Fiscal Year (FY) 1984 budget then being formulated, was "the composition of the future US launch fleet," with particular attention to NASA's request to OMB for White House approval of a fifth space shuttle orbiter. At the IG meeting, Hans Mark characterized an additional orbiter as "providing for possible attrition in advance."

This was a considerable narrowing of the agreed-upon launch policy topic. Mark told the meeting that a working group chaired jointly by NASA and the DOD would prepare an issue paper regarding the fifth orbiter for SIG (Space) consideration. (The debate over whether to build a fifth orbiter is discussed in the following chapter.)[9]

A National Space Strategy?

In commenting on the NSC's plans for SIG (Space), OMB director Stockman had suggested that "what we really need is a careful reassessment of our military and civil space programs to ensure that the allocation within each agency of currently planned funding is consistent with the highest priorities of the President." He suggested that "the SIG (Space) could make a significant contribution by assuring that the space programs of each agency represent, in the aggregate, the most coherent and effective overall space program to meet the Administration's objectives within current funding constraints."[10]

The call for a comprehensive assessment of the national space effort was very much consistent with the NSC's ambition for SIG (Space), but the constraint that it would have to respect "current funding constraints" was not. It is not clear whether President Ronald Reagan in the months following the issuance of the new National Space Policy made a specific call to his advisers for a recommendation on how to ensure that the U.S. space program retained its leadership position. That kind of targeted request was not characteristic of Reagan's approach to the presidency; Reagan's long-time policy adviser Martin Anderson comments that Reagan "made no demands, and gave no instructions." However, Rye recalls that "in multiple meetings that I attended, President Reagan repeatedly asked for more initiatives to capitalize on the potential of space and demonstrate U.S. leadership."[11]

Rye and his bosses at the NSC thus decided in late 1982 to create an overall space leadership strategy; such an initiative could involve significant budget increases. In arguing to Clark that the NSC should develop such a comprehensive strategy, Rye observed that "the American people overwhelmingly support the U.S. Space Program and feel we should be doing more in spite of current economic difficulties." Such an attitude, suggested Rye, provided a "unique opportunity" for Ronald Reagan "to not only *substantively* reinvigorate and focus the Space Program, but also to rekindle the spirit of the American people and the pride in our technology and innovative capabilities."

McFarlane raised a cautionary note to Rye's enthusiasm, telling Clark that "the concept of launching a study designed to develop a strategy for a major new effort in space is great, but if we cram it down everyone's throat ... we will have a bunch of enemies on our hands before we get started ... On something this big we are better off to try and build at least a measure of consensus going in." McFarlane suggested that a prudent course to pursue was getting agreement in advance that such a study was worthwhile from key White House staff, in particular Ed Meese, and then from agency leaders such as Cap Weinberger at DOD, George Shultz at State, William Casey at the CIA, and Jim Beggs at NASA. Clark agreed with this cautious approach. He met with Meese and gained his agreement that creating a national space strategy was a positive idea; meetings with Weinberger, Shultz, and Casey, however, revealed a notable lack of enthusiasm for such a comprehensive effort, presaging the agency opposition that would be present throughout the space strategy effort. As he prepared for a November 10 meeting at which Clark would formally inform him of the study effort, the CIA's McMahon was advised that the study proposal "is solely a NSC initiative" and that the initial reaction within the Intelligence Community was "one of great concern."[12]

The result of the NSC initiative was a National Security Study Directive (NSSD) 13-82, "National Space Strategy." The directive was approved by President Reagan on December 15, 1982. Clark sent the NSSD to Vice President George Bush, who, while not formally a member of SIG (Space), wanted to stay informed regarding major space issues, and to the members of the SIG (Space). He noted that "the President requests that agencies provide their highest priority to this effort"; this admonition may have been intended to head off anticipated agency resistance to helping prepare a strategy that could modify the current distribution of programs and related budgets and lead to organizational changes. The ambitious effort would be coordinated by the IG (Space), chaired by McFarlane. The strategy was due to be completed by May 31, 1983.[13]

Thus was launched what Gil Rye would describe in a 2016 interview as a "drama," as the NSC attempted to forge a single, coherent strategy for an expanded national space effort and as the various national security agencies carrying out space activities resisted such a centralized approach, insisting that there could be no such thing as a single strategy for those efforts, since their space activities were in service of broader and diverse policy goals. The agencies, of course, also viewed the NSC study directive as unwanted interference into their individual prerogatives and relationships; this was classic case of tensions between a central executive body and bureaucratic organizations.

NSSD 13-82

The NSSD 13-82 study directive setting out the terms of reference for developing the national space strategy was classified "top secret" because it dealt with both intelligence activities and space weapons.[14] The ambitious goal of the study was "a broad action-oriented plan for a more vigorous and focused U.S. Space Program." That plan would include "three alternative space programs: current, enhanced and significantly enhanced." The two premises underlying the study effort were that "the Soviet Union has initiated a major campaign to capture the 'high ground' of space," and that "regardless of Soviet activities, the space medium offers significant potential for the enhancement of civil, commercial and national security capabilities." Topics for study would include the organization, roles, and responsibilities of various government space agencies; "the adequacy of the current funding mix and priorities among the civil, defense, and intelligence sectors of the U.S. Space Program"; "new areas of private sector investment in space which the Administration should stimulate"; "the need for revitalizing our efforts to explore the planets and perform scientific experiments useful to mankind"; "the need for significant increases in basic technology"; and "the need for a more vigorous manned space program." (Additional study topics related to defense and intelligence activities remain classified.)

Reaction to NSSD 13-82 was swift and skeptical. On January 6, 1983, CIA Director Casey wrote to Clark, questioning the viability of the mandated study. Rye reported to Clark on January 7 that he was "receiving less than enthusiastic support for organizing the study and getting it underway," and that DOD had "reservations." Clark responded that "the president signed a directive ordering a study. The study is non-negotiable except in details. I'll say so in a follow-up memo."[15]

As the study was getting underway, Rye continued to argue for its multiple benefits. He suggested to Clark that a "major Administration commitment to the U.S. Space Program":

- Rekindles the spirit and pride of the American people in a period of high unemployment and economic difficulty
- Substantively enhances our civil, commercial, and national security capabilities
- Is not burdened with major arms control and unsavory social connotations.

Rye also noted a more political benefit from such a commitment, suggesting that "if John Glenn is the Democratic Presidential nominee [in 1984], such an initiative could be used to neutralize his astronaut image," and adding that, "traditionally, the Democrats (Kennedy and Johnson) have been associated with a strong space program. We have the opportunity to change that perception."[16]

With senior staff-level opposition to the space strategy effort from both the DOD and the Intelligence Community, in the first months of 1983 the NSC initiative to create a national space strategy was thus off to a rough start. On March 6, as they accompanied President Reagan on a flight returning to Washington aboard Air Force One, Clark discussed with Weinberger the need to end DOD resistance to the study; Weinberger agreed that DOD would "proceed ahead with the study and instruct his staff to fully support the effort." In a letter recording this conversation, Clark thanked Weinberger for his agreement and told him, "we plan to convene a SIG (Space) meeting on approximately March 25" to discuss the status of the strategy study.

Clark did call such a meeting, telling all SIG (Space) members what he had earlier said to Weinberger, that "the President is anxious to complete the NSSD 13-82 study as rapidly as possible," since "he wishes to consider options for a more vigorous and focused U.S. Space Program before this summer." Clark noted that as part of a revised study approach, NASA, the DOD, the Department of Commerce, and the CIA would each develop "three alternative strategies based on timing." The first alternative would identify "capabilities to fully implement the National Space Policy, NSDD 42, by the year 1998." Doing this would lead to a "broad 15-year plan that begins with the *current* (FY 84-89) program and extends it an additional 10 years." Clark added: "Agencies should then compress the schedules and develop two additional strategies whereby NSDD 42 is fully implemented (1) in 10 years (i.e., an enhanced space program covering the period FY 84-93) and (2) in 5 years (i.e., a significantly enhanced program covering the FY 84-89 period.)."[17]

The SIG (Space) meeting actually took place on March 28. In his introductory remarks, Clark told the group that it was "essential that interagency squabbling over details for executing the study be terminated and that we move out positively and aggressively." All present agreed to proceed with their assigned studies and to submit them to the NSC by May 31 for integration.[18] Rye indicated to McFarlane that once the agency inputs were received, it was his intent to chair a small drafting group to put together "a coherent product." He had asked the SIG (Space) Space Station Working Group, which was working in parallel to the space strategy effort, to accelerate its schedule so that its report could be integrated into the overall strategy document. (The activities of this working group are discussed in Chap. 7.) Rye, still optimistic that it might be possible to increase the overall budget for space activities, suggested to McFarlane that he "ask OMB to begin thinking about contingency plans for accommodating a Presidential Decision to increase space funding in both the civil and national security areas."[19]

The NASA report and a combined DOD/Intelligence Community report were submitted by the May 31, 1983, deadline. The commercialization report was missing; even in mid-July, as he was finalizing a first draft of the space strategy report, Rye told Clark that "in spite of my constant, yet gentle, urging Craig Fuller has not delivered on the commercialization section." Fuller was proceeding at his own pace in developing an administration position on space commercialization, and was not particularly concerned with meeting the NSC schedule (Fig. 5.1).

Fig. 5.1 Cabinet Secretary Craig Fuller and National Security Council Director for Space Gil Rye were key actors in a variety of space decisions during the first Reagan term. In this May 14, 1984, Cabinet Room photo, Rye (center) and Fuller (to Rye's left) interact with the president, Secretary of State George Shultz (on left), and Chief of Staff James Baker (on right). NASA Administrator James Beggs looks on in the background. Baker is showing the group a white glove left behind by singer Michael Jackson during a White House visit earlier in the day. (Photograph courtesy of Reagan Presidential Library)

As the agency reports arrived at the NSC, there was a decision to broaden the scope of the space strategy report. On June 6, Clark decided to incorporate into the report the results of two other ongoing studies in addition to Hodge's space station effort: the space-related portions of NSSD 6-83, the "Study on Eliminating the Threat Posed by Ballistic Missiles" that had been initiated after President Reagan's March 23, 1983, speech on the Strategic Defense Initiative; and National Intelligence Estimate 11-1-83 on Soviet Space Programs. The NSC plan was to have a draft report combining all inputs available for agency comments during July, with the aim of its review by SIG (Space) in early August and by early September presidential review and approval of its recommendations. This would be followed by issuing a National Security Decision Directive setting forth the National Space Strategy.[20]

Rye had a draft of the strategy ready by July 12. President Reagan was asked to make two major decisions (plus a third on commercialization initiatives if the Fuller input was received in time for review). One decision was whether to approve the space station. The other was a "funding strategy" for accelerating the space program; four options were offered. One was a "baseline" no-growth approach; the other three called for a 4 percent, 8 percent, or 12 percent real growth in the space budget. The report suggested that the president consider the 8 percent growth option; that percentage was "somewhat arbitrary, although many substantive space initiatives for NASA, DOD, and DCI [Director of Central Intelligence] can be accommodated" with that level of growth. Rye noted that "the actual percentage growth may be dictated by the outcome of the White House BRB [Budget Review Board] process, unless the President would want to fence off this money at the outset." This comment and the bias in the report toward 8 percent growth suggest that Rye was a bit out of touch with budget realities. By mid-1983, OMB director Stockman was persistently warning the president that the United States was facing bankruptcy if government spending were not gotten under control. Even Ronald Reagan's top priority of a defense budget buildup was facing an uncertain future. There was little chance that a significant increase in the space budget was in the cards.[21]

The draft space strategy also contained a section on possible organizational changes related to space. It did not propose presidential decisions related to these changes, since they were viewed as "internal prerogatives" of the involved organizations. One change discussed was NASA creating a new internal structure to assume responsibility for managing space shuttle operations, separate from the research and development activities that were central to the agency. Rye noted that "Jim Beggs has volunteered a commitment" to this organizational change. The other shift, much more controversial (and outside the scope of this study), was the possible creation in the DOD of a Unified Space Command which would assume the responsibilities of the separate services for space activities. Involved in this shift might also be a reorganization of the secretive National Reconnaissance Office in order "to continue the acknowledged US space leadership in intelligence collection and operational support to the DOD." Rye observed that he was "disappointed … in the inability of DOD to critically examine itself to determine if there are realistic options to bring greater focus and direction to the military space program" and that this was a "very sensitive subject." Indeed, it was this topic that had been and would continue to be a major barrier to agreement on a single national space strategy.[22]

The draft space strategy was distributed to members of SIG (Space) in mid-July. Agency comments on the draft were generally negative, and the NSC decided to convene a meeting of IG (Space) to discuss how to proceed. That meeting took place on July 28, 1983; those present decided that "it would be best to proceed to the President with two reports," one on the space station and one on a national space strategy. The space station report would be presented to the president first, with the space strategy report deferred until "later this fall." When SIG (Space) met on August 10, it approved this separation of the space station report and the space strategy report. The hope to have President Reagan soon announce a more ambitious and more comprehensive approach to the U.S. space program would at a minimum be deferred.[23]

Conclusion

While President Ronald Reagan in principle may have favored a more vigorous U.S. space effort, in practice he was faced with a budget deficit that was growing at an alarming rate despite his oft-repeated pledge to end deficit spending. The 1981 tax cut and reductions in government spending had not produced the hoped-for economic stimulus. Rather, the federal deficit was rapidly increasing. Thus, the possibility of an enhanced national space program with a significantly increased budget anticipated in the December 1982 NSSD 13-82 and reflected in the July 1983 draft national space strategy was in essence stillborn. A national space strategy, but one far less sweeping in scope than the NSC had hoped to produce, would not be issued until August 1984. Meanwhile, more specific policy discussions on the space station and on privatization and commercialization initiatives moved forward. In addition, issues related to the space shuttle, which had been declared operational after its fourth mission in July 1982, were at the forefront of Reagan administration space policy concerns.

Rye's July 1983 suggestion in the draft space strategy that the space program receive an 8 percent budget increase would turn out to be the high point of Reagan administration optimism regarding future space efforts. The influence of the "Nixon doctrine" that space had to compete for priority "with all of the other undertakings which are important to us" served as a counterweight to the strategy's budget proposal, and over the remaining years of Ronald Reagan's time in the White House would prove decisive. Whatever new initiatives might be undertaken in space would have to be carried out without requiring a significant increase in the space budget.

Notes

1. "Reagan Policy Expected to Aid Space Station Definition Work," *AWST*, July 12, 1982, 25–26; William H. Gregory, "Better than Nothing," *AWST*, July 12, 1982, 13: Brown is quoted in the same publication, July 19, 1982, 17; R. Jeffrey Smith, "Squabbling over the Space Policy," *Science*, July 23, 1982, 331.
2. U.S. House of Representatives, "National Space Policy," Hearing before the Subcommittee of Space Science and Applications, August 4, 1982, 15–17.
3. Ibid., 24.
4. R. Jeffrey Smith, "Squabbling Over the Space Policy," *Science*, July 23, 1982, 331.
5. Memorandum from Gilbert Rye to William Clark, "Relationship with OSTP re Space Policy," marked "Sensitive," July 9, 1982, Box 30, Subject Files, Executive Secretariat, National Security Council, RRL.
6. Memorandum from Gilbert Rye to William Clark, "Senior Interagency Group for Space," July 14, 1982, Box 30, Subject Files, Executive Secretariat, National Security Council, RRL.
7. "Minutes, Senior Interagency Group for Space – SIG (Space)," September 23, 1982; memorandum from Gilbert Rye to William Clark, "Background for First SIG (Space) Meeting," September 17, 1982; memorandum from

John McMahon to Deputy Assistant to the President for National Security Affairs, "First SIG (Space) Meeting," October 1, 1982, all in Box 30, Subject Files, Executive Secretariat, National Security Council, RRL; memorandum from Director, Intelligence Community Staff to Deputy Director of Central Intelligence, "First SIG (Space) Meeting," September 17, 1982, Document CIA-RDP84B00049R000501280003-8, CREST.

8. The Terms of Reference and the three white papers are attached to Memorandum from Robert McFarlane to SIG (Space) Members, "First IG (Space Meeting)," September 28, 1982, CIA-RDP84M00127R000200030028-4, CREST.
9. Memorandum from Director, Intelligence Community Staff to Deputy Director of Central Intelligence, "IG (Space) Meeting, 1 October 1982," October 16, 1982, Document CIA-RDP84 M00127R000200030028-4, CREST.
10. Memorandum from David Stockman to William Clark, "SIG (Space) Meeting," September 20, 1982, Box 30, Subject Files, Executive Secretariat, National Security Council, RRL.
11. Martin Anderson, *Revolution* (New York: Harcourt Brace Jovanovich, 1988), 289. Personal communication from Gil Rye to author, March 7, 2017.
12. Memorandum from Gil Rye to William Clark, "American Attitudes Toward Space," November 2, 1982 and memorandum from Bud McFarlane to Judge Clark, October 23, 1982, both in NSSD 13-82 File, Executive Secretariat, National Security Council, RRL; memorandum from (author redacted) to Deputy Director of Central Intelligence, "Your Meeting with Judge Clark on 10 November," November 9, 1982, CIA-RDP83M00914R000800160002-9, CREST.
13. Memorandum from William Clark to the Vice President and members of SIG (Space), "National Space Strategy," December 15, 1982, CIA-RDP85M00363R001202700026-6, CREST.
14. A copy of NSSD 13-82 can be found at https://www.reaganlibrary.gov/sites/default/files/archives/reference/scanned-nssds/nssd13-82.pdf. Portions of the directive remain classified.
15. Memorandum from Gil Rye to William Clark, "Weekly Report," January 7, 1983, NSSD 13-82 Files, Executive Secretariat, National Security Council, RRL.
16. Memorandum from Gil Rye to William Clark, "Politico-Economic Dimensions of the U.S. Space Program," January 13, 1983, NSSD 13-82 File, Executive Secretariat, National Security Council, RRL.
17. Memorandum from William Clark to Members, Senior Interagency Group for Space, "SIG (Space) Meeting," March 22, 1983, CIA-RDP85M00364R000400550065-0, CREST.
18. Talking Points for SIG (Space) Meeting, March 25, 1983, NSSD 13-82 File, Executive Secretariat, National Security Council, RRL; Minutes, Senior Interagency Group for Space, March 28, 1983, White House Situation Room, CIA-RDP85M00364R001101630035-5, CREST.
19. Memorandum from Gil Rye to Robert McFarlane, "IG (Space) Meeting – May 3, 1983," NSSD 13-82 File, Executive Secretariat, National Security Council, RRL.
20. Memorandum from Gil Rye to Robert McFarlane, "First Draft of the National Space Strategy Report (NSSD 13-82)," July 12, 1983, NSSD 13-82 File, Executive Secretariat, National Security Council; memorandum from William Clark to Members, Senior Interagency Group for Space, "Schedule for National

Space Strategy," Box 30, Subject Files, Executive Secretariat, National Security Council, June 6, 1983, both in RRL.

21. Memorandum from Gil Rye to Robert McFarlane, "First Draft of the National Space Strategy Report (NSSD 13-82)," July 12, 1983, NSSD 13-82 File, Executive Secretariat, National Security Council, RRL.
22. Ibid and Intelligence Community Staff, "Talking Points for DCI Meeting with Judge Clark on 6 July 1983," CIA-RDP85M00363R001102510003-3, CREST.
23. Memorandum from Robert Kimmitt, Executive Secretary, National Security Council to SIG (Space) Member Agencies, "Third SIG (Space) Meeting," August 10, 1983, CIA-RDP85M00364R000400550053-3, CREST.

CHAPTER 6

Space Shuttle Issues: Round One

During the second half of 1982, the Reagan administration focused on implementing the new National Space Policy embodied in National Security Decision Directive (NSDD)-42. An immediate focus was taking steps toward achieving the highest priority objective spelled out in the new policy: making the space shuttle "a major factor in the future evolution of United States space programs." Other shuttle-related provisions of the policy included making the shuttle "fully operational and cost-effective in providing routine access to space" and providing "capacity sufficient to meet appropriate national needs." With the completion of the fourth shuttle flight on July 4, the National Aeronautics and Space Administration (NASA) had declared the shuttle "operational"; this designation meant that the vehicle could begin to carry out a mixture of NASA research flights, revenue-producing missions for commercial and foreign customers, and missions carrying national security payloads. The head of the NASA office managing the shuttle program, James Abrahamson, later commented that this declaration was "arbitrary," having "nothing to do" with the actual status of the program, which still was very much a development effort with constant learning and change.[1]

It is not surprising, then, that the first issue addressed using the Senior Interagency Group for Space [SIG (Space)] mechanism established by NSDD-42 was the future of the space shuttle; more specifically, the question for immediate decision was whether to approve production of another shuttle orbiter, increasing the fleet to five orbiters, rather than the four-orbiter fleet that President Jimmy Carter had approved in 1977. The Nixon administration had approved the space shuttle with the anticipation that five orbiters would be built, but President Carter and his advisers had limited the shuttle fleet to four orbiters. Office of Management and Budget (OMB) Director James McIntyre on December 23, 1977, told NASA Administrator Robert Frosch that "the decision was clearly to support a four orbiter option … The president stated his explicit concern that no action be taken that might be interpreted as a

J. M. Logsdon, *Ronald Reagan and the Space Frontier*,
Palgrave Studies in the History of Science and Technology,
https://doi.org/10.1007/978-3-319-98962-4_6

commitment now by the Government to build a fifth orbiter." The word "now" did leave the door open for a future decision to add a fifth orbiter, and in 1982 NASA attempted to get approval for such a decision.[2]

As he wrote to Ed Meese in May 1982, setting out his priorities for NASA, James Beggs had given almost as much emphasis to producing a fifth orbiter as he did to gaining approval for space station. He told Meese that "a fifth Orbiter is necessary to assure dependable launch services in the event we lose or extensively damage an Orbiter, something that cannot be ruled out over the course of several hundred flights. Four Orbiters are essential to conducting the presently planned missions." Beggs added: "I happen to think we will need a fifth one to meet increased demand as well." As they put together NASA's budget request for Fiscal Year (FY) 1984, which would go to the White House OMB in September 1982, Beggs and his deputy Hans Mark had decided to defer seeking presidential approval for a space station until the next budget round; they judged that the time was not politically or fiscally ripe for seeking such approval. Instead, they requested $200 million as a down payment on a fifth shuttle orbiter. By making the fifth orbiter a budget as well as a policy issue, the NASA request put the OMB and the SIG (Space) on a collision course, with the question being whether fifth orbiter approval was primarily a budget issue with policy aspects or a policy issue with budget implications. Conflict between the two decision-making paths soon appeared.[3]

Even as the White House addressed the fifth orbiter issue between October 1982 and early 1983, the Reagan administration was also dealing with several other policy issues associated with the space shuttle, ranging from what price to charge commercial customers for a shuttle launch to whether the shuttle might actually carry fare-paying passengers and to whether NASA would accept private funding to produce an additional orbiter in exchange for commercial marketing rights for the shuttle fleet overall.

Even after only a few shuttle flights, public criticisms of the system's performance were also beginning to appear. For example, a November 1981 article in *The New York Times* headlined "Shuttle's Critics Call for a Broad Reassessment" characterized the shuttle as "a complex system at an early stage of development" and "more complicated, sensitive, and idiosyncratic" than anticipated, and thus "not a good buy." Leading space scientist James van Allen suggested that the shuttle might turn out to be "a technical success but a financial monstrosity … ahead of its time, by perhaps 50 years."[4] While these early criticisms did not undermine the general enthusiasm surrounding early shuttle flights, they suggested that not all was well with the shuttle program, a reality that in the following years would become increasingly evident, at least to those close to the program.

Pricing the Space Shuttle

A first step during the 1970s in creating the framework for shuttle use by other U.S. government agencies and commercial and foreign entities had been setting the price for such use. There were two categories of shuttle users for which

NASA had to set a price: one was the Department of Defense (DOD) and other government agencies and the other was commercial and foreign users. NASA in 1977 had set an initial price for a dedicated shuttle mission for commercial and foreign users at $22.6 million in FY 1975 dollars; that cost was to be adjusted for inflation and would be pro-rated depending on how much of the shuttle's capacity a particular mission would actually use. The most likely commercial mission for a shuttle, launching communications satellites, would not require all of the shuttle's payload bay space or weight-lifting capability, and thus the launch price would be significantly less than $22.6 million. NASA guaranteed the shuttle price for the first three years of operational flights, which turned out to be from late 1982 through 1985. It was NASA's original intent to revise the price to reflect the first three years of experience, with the goal of setting the revised price high enough to recoup the total costs of commercial shuttle operations over a 12-year period.

Among other considerations, in order to attract existing commercial space operators to quickly transition to using the shuttle, the 1977 price was purposely set to be substantially lower than the cost of launching an equivalent payload on an expendable launch vehicle such as a Delta or Atlas. Eager to maintain the commitment of DOD (and National Reconnaissance Office, which could not be publicly mentioned since its existence was classified) to use the shuttle to launch national security payloads, NASA set an even lower price of $12.2 million per flight for defense missions, an amount that covered only the estimated costs of the items consumed during the launch. NASA guaranteed that very low price for the first six years of shuttle operations, through September 30, 1988. These prices were set on the basis of what turned out to be very unrealistic expectations of the actual costs of shuttle operations.[5]

The 1977 pricing policy was very successful in enticing commercial users to book launches of their communication satellites on the space shuttle, since they were being offered a price significantly lower than they had been paying for a launch on a Delta or Atlas. However, that price turned out to be concessionary; actual costs of initial shuttle launches turned out to be much higher than the price NASA was charging. This was not surprising, since NASA in these early launches was still gaining experience with shuttle operations, and the vehicle was turning out to be much more expensive to operate than NASA had hoped.

The high cost per launch soon became the focus of criticism. In a February 1982 report, the General Accounting Office (GAO) found that the estimated average cost of an early shuttle launch had increased by 73 percent over prior forecasts. The GAO suggested that flights during the first three years of shuttle operations were likely to cost more than $60 million in FY 1982 dollars, rather than the $38.3 million that NASA was by that time charging. *Science* magazine described the situation as "budgetary hemorrhage," since "the contracts and agreements are signed, and for 3 years NASA is locked into the older prices." NASA was forced to absorb in its budget, already under pressure from OMB's attempts to cut overall government spending, the cost above reimbursement from commercial customers of each shuttle flight carrying a commercial payload. From the GAO's perspective, the cost of shuttle operations was squeezing out new NASA science and application missions, and NASA was in danger of becoming primarily a trucking company rather than a research and development agency. The gap between mission costs and user reimbursements would be even greater for future flights carrying DOD payloads. This would, observed the GAO, in effect be a subsidy to the shuttle's non-NASA users at the expense of NASA's own activities.[6]

Revising the Shuttle Price

The GAO suggested that NASA immediately revise its original pricing policy, but NASA did not do so, since that would have meant renegotiating existing contracts. Rather, NASA in spring 1982 proposed a new approach to shuttle pricing for the three-year period after 1985. NASA's Beggs wrote to OMB director David Stockman on April 12, 1982, telling him that "development problems and delays in buildup of the flight rate have extended the transition period," and thus "the timing of the announced pricing objectives of recovering the average full cost over the original mission model must be deferred." Beggs said that "for the three year period following the expiration of the current fixed-price charge in 1985, I would plan to base the reimbursement charges to non-U.S. Government users on the principle of recovering all costs incurred in support of their launches above the basic costs of operations to meet NASA and national security requirements." Rather than share in the overall costs of shuttle operations, commercial users would pay only for the marginal costs of their missions, still allowing a comparatively low price.[7]

Stockman replied to Beggs on June 14, indicating that OMB was "generally supportive" of the approach to shuttle pricing he had proposed, but only under some specific conditions. OMB told NASA to seek commercial customers for shuttle flights only to the degree that there was "excess capacity" beyond U.S. government requirements, with the price at which this capacity would be made available set "to minimize the overall cost to the Federal Government of meeting its own needs in the long run." What was of particular concern to OMB was ensuring that the price to commercial users "does not in itself lead to the demand for funding of additional capacity by the USG [U.S. Government]," such as "additional orbiters." In addition, the price should "not discourage or compete unfairly with possible private sector initiatives that would invest wholly or partially in Space Shuttle technology in order to service growth in the commercial and foreign user market." There was at this time at least one proposal, from a company called SpaceTran, to develop an additional space shuttle orbiter through private funding. In essence, OMB was directing NASA to avoid marketing shuttle services to commercial users in a way to create additional demand for those services. NASA was intending to undertake just such a marketing effort.[8]

Based on these NASA-OMB exchanges and the experience of the first three shuttle test missions, NASA in June 1982 announced significant changes in both shuttle expectations and future pricing. The number of anticipated shuttle missions during the first 12 years of shuttle operations was cut to 312 flights; it had been 572 flights in 1976, 560 in 1977, and 487 in 1979. Beggs later remembered that when he arrived at NASA in 1981, the plan was to launch the shuttle 40 times a year, but because of the problems encountered in early shuttle flights, "everything from tiles to the main engines," he and his deputy Hans Mark cut the target flight rate to 18 launches per year, even though NASA still said publicly that it was 24 flights annually. Beggs added: "My feeling at the time ... was that probably we would never launch more than twelve." NASA announced that for the three years after 1985 the base price to a commercial user for a shuttle flight would increase by 85 percent, from $38.3 million in FY 1982 dollars to $70.7 million.[9]

Competition Emerges

One reason for not attempting at this point to set an even higher price for the shuttle was the emergence of foreign competition to the shuttle as a launcher for commercial communication satellites. Under the auspices of the European Space Agency, a new expendable launch vehicle named *Ariane*, optimized for launches to the geostationary orbit where most communication satellites were located, had been developed; its first launch had been in December 1979. In 1980, a consortium of European aerospace firms, banks, and the French space agency CNES (*Ariane* was primarily a French-motivated project) formed a quasi-private company called Arianespace to oversee *Ariane* production and launching and to market the launcher on a worldwide basis. Arianespace set a goal of launching 30 percent of the world's commercial payloads; that objective set it in direct competition with the space shuttle for launch contracts. The first commercial customer for an *Ariane* launch was a U.S. firm, GTE.

The U.S. government response to *Ariane*'s emergence was chauvinistic; the notion of a European competitor to what had been a U.S. monopoly in providing launch services was troubling to the White House and Congress as well as to NASA. Threatened by Arianespace competition, NASA thus set its new shuttle price to be competitive with what Arianespace was offering. Success in the competition with *Ariane* could also be used as a justification for adding additional launch capacity by building another orbiter.

The issue of shuttle pricing for commercial payloads would continue to be controversial in subsequent years, especially as the U.S. private sector attempted to commercialize existing expendable launch vehicles such as Delta, Atlas, and Titan as alternatives to both the shuttle and *Ariane*. The controversy would be ended only in 1986 as the shuttle was prohibited from launching commercial payloads in the aftermath of the *Challenger* accident.

Who Should Fly on the Shuttle?

The space shuttle was designed so that it could carry into orbit not only highly trained astronauts but also individuals in normal health who would not require extensive training before a flight. As he had indicated his approval of the space shuttle on January 5, 1972, President Richard Nixon told NASA leaders James

Fletcher and George Low that one of the features he liked about the vehicle was "that ordinary people would be able to fly in the shuttle."[10] This possibility would come under active review early in the Reagan administration.

The Shuttle as a Spaceliner?

There were within the Reagan White House individuals with unconventional ideas related to space activities. One of those ideas was an early version of space tourism—that the space shuttle might carry enough fare-paying passengers to offset a meaningful portion of its operating costs. Danny Boggs of the Office of Policy Development raised the idea with Beggs in spring 1981, even before Beggs formally took over as NASA administrator; Beggs did not think it feasible. The idea resurfaced in April 1982 in interactions between Boggs and OMB deputy director Ed Harper. Boggs commented to Harper: "I hope that we would not … lightly ignore the revenue potential of passengers on the space shuttle." NASA had told Boggs that the shuttle could in principle carry up to four passengers on each flight once it became operational. Boggs had concluded that "on a straight auction basis, I find it extremely difficult to believe that we could not get over one million dollars a seat for many years, and possibly $5 to $10 million for one or more seats on the first flight."[11]

Harper contacted NASA administrator Beggs a few weeks later, suggesting that "in light of the increasing revenue needed to meet costs," NASA should evaluate "the possibility of paying passengers being carried on the Shuttle." He asked Beggs to provide information on "the possibilities, from a strictly operational point of view, for carrying non-mission-related passengers on flights." Beggs replied on June 9, 1982, noting that "our capability to carry sufficient paying passengers to make the revenue meaningful is limited with perhaps no more than a dozen opportunities in the next three to four years." He suggested that "the interests of the United States might be better served by using these opportunities in support of foreign policy, as is the USSR practice in recent years using their Space Station to host foreign nationals."[12]

Expanding Shuttle Flight Opportunities

While the idea of flying paying passengers on the shuttle never gained traction, NASA in 1982 did modify its policy with respect to who might fly on a shuttle mission. NASA's 1977 pricing policy had allowed a customer purchasing more than 50 percent of a shuttle flight to nominate a "payload specialist" for that flight. A payload specialist was an individual chosen by the customer, rather than being selected as an astronaut through the rigorous NASA recruitment process. Such an individual was expected to fly only once and would not have to go through the several-year training period before being cleared for flight. As of 1982, no shuttle customer had taken advantage of this opportunity. Several U.S. and foreign scientists whose experiments had been selected for placement in the European-provided Spacelab, a pressurized laboratory which would fly in the shuttle's payload bay on a few missions, were selected as payload specialists, but their flights would be not be on a commercial basis.

Beggs in 1982 reviewed the 1977 policy on payload specialists and found it "overly restrictive." In October 1982, in advance of the shuttle's first operational flight, NASA announced that "the minimum required payload factor" would be eliminated and that "flight opportunities for Payload Specialists will be made available on a reimbursable basis to all classes of Space Shuttle major payload customers, including foreign and domestic commercial customers." The new policy would go into effect for flights beginning in 1984.[13]

Similar to the rationale behind the revised pricing policy, an influential reason for this policy shift was the competition from Arianespace for commercial launch contracts. Arianespace enjoyed several advantages in this competition. While NASA required commercial customers to pay in advance the fee for launching their payloads, Arianespace offered flexible pricing and payment arrangements. Moreover, in marketing an *Ariane* launch Arianespace could fly potential customer airliner to view a launch in French Guiana on the northern coast of South America aboard the supersonic *Concorde*. Allowing a space shuttle customer to select someone to actually go into orbit was a very attractive counter to the Arianespace marketing approach. As the policy shift was announced, the newsletter *Aerospace Daily* noted that "the opportunity to fly a specialist with the payloads provides a marketing attraction that *Ariane* will not be able to match." In forwarding NASA's proposed policy change to Clark at the National Security Council (NSC) for White House approval, Rye had noted that "the new policy will allow virtually any user to provide a specialist. The change in policy is an attempt by NASA to obtain more customers for the Shuttle ... Increased demand for the Shuttle will drive the operating costs down which will benefit all users." OMB's hope to limit shuttle flights to available capacity was overridden.[14]

It was this change in payload specialist policy that allowed McDonnell Douglas engineer Charles Walker to fly into space three times in 1984 and 1985 with a company-funded experiment, also allowing NASA to tout its role in facilitating commercial space activities. Before the use of the shuttle to launch commercial satellites was ended after the *Challenger* accident, there were two occasions on which a non-U.S. payload specialist flew into space with his country's satellite. In June 1985, Sultan bin Salman bin Abdulaziz Al-Saud, a member of the Saudi royal family, accompanied *Arabsat-1B* into orbit. In November of that year, Mexican engineer Rudolfo Neri-Vela flew into space with the *Morelos-B* satellite. Walker, Al-Saud, and Neri-Vela were thus the only commercial payload specialists to complete a space shuttle flight; a fourth such specialist, Hughes Aircraft engineer Greg Jarvis, was one of the seven crew members to perish during the January 1986 *Challenger* accident.

The issue of flying others than experimenters and representatives of shuttle customers remained under discussion. Senator Jake Garn (R-UT), who chaired the Senate subcommittee that controlled NASA's budget, from 1981 on had been asking when he could take a shuttle flight. Beggs was also getting requests for a seat on a future shuttle flight from parties as diverse as *National Geographic* magazine and singer and space enthusiast John Denver. In his June 1982 response to Harper at OMB, Beggs had noted that he had "charged the NASA Advisory Council with the task of assessing flying non-crew passengers and … under what circumstances."[15] The Council response to that task will be discussed in Chap. 14.

A Fifth Orbiter?

Addressing the question of whether President Reagan should approve the production of a fifth space shuttle orbiter, pursue a less-costly approach to keeping orbiter production capability in being, or abandon that capability altogether would provide the first test of the new policy development process set up by NSDD-42. A meeting of SIG (Space) to address the fifth orbiter issue was set for December 3, 1982. There were two issue papers prepared for discussion at that meeting. One, prepared by the SIG Space Launch Policy Working Group, identified three options:

1. Close out orbiter production capability;
2. Maintain orbiter production capability by producing major orbiter parts, but not an assembled orbiter; and
3. Continue full orbiter production.

A fourth option, moving to a next generation, advanced technology, orbiter, was also briefly discussed, but was deemed not a credible choice. The paper recommended "continuing the orbiter production base" and suggested that "FY84 funding support for the production of the fifth orbiter" be approved.

The second paper was prepared by the IG (Space) Working Group on Space Launch Policy. It was more bureaucratic in its recommendations, which fell a bit short of suggesting a fifth orbiter, saying that "a responsive and viable orbiter production and repair capability should be maintained" and that "NASA … should continue to be charged with identifying the appropriate programmatic options to efficiently meet this objective." The paper did note that "the option currently proposed by NASA, i.e., the production start of a fifth orbiter in FY-84, satisfies all concerns and issues posed by the Working Group."[16]

After only five shuttle flights in 1981–1982, the high optimism of prior years among shuttle advocates that there would be increasing demand for shuttle launches once the vehicle entered service, demand that could lead to the need for even more than five orbiters, had been replaced with a more realistic assessment. Rather than 40 or more shuttle flights per year, as had been forecast in the late 1970s, NASA by late 1982 was publicly saying that later in the 1980s there would be a maximum of 24 flights per year, while planning internally and ordering consumable items like the shuttle's external fuel tanks to support only 12–18 flights. Even at the 24-flight rate, both papers concluded that the likely future demand for space shuttle launches could be satisfied by a four-orbiter shuttle fleet, and that the primary reason for either building a fifth orbiter or keeping the shuttle production line in being was as a backup to make sure that a fleet of four orbiters would continue to be available, even if one of the existing vehicles were lost in an accident or taken out of service for an extended period of time.

The December 3 meeting of SIG (Space) was contentious; there were "widely differing views on the central issue." The State Department, OMB, and Office of Science and Technology Policy (OSTP) voted for Option 1, to discontinue orbiter production. The DOD, the Joint Chiefs of Staff, and the Central Intelligence Agency (CIA) supported option 2, producing structural spares. NASA, the Department of Commerce, and the Arms Control and Disarmament Agency (ACDA) supported option 3, producing a fifth orbiter, and Deputy National Security Adviser McFarlane, who chaired the meeting, indicated that the NSC also favored that option. Faced with a lack of consensus, McFarlane proposed that a paper which accurately reflected the varying agency views be prepared as the basis for President Reagan to make a choice among the options. The deadline set for the option paper was December 9.[17]

As the SIG (Space) paper was being prepared, NASA-OMB negotiations over the FY 1984 budget continued. *Aviation Week & Space Technology* reported that the initial budget approved by OMB had "rejected a fifth shuttle orbiter," while the NSC had "endorsed procurement of a fifth shuttle orbiter, creating a conflict with OMB and forcing President Reagan to make the final decision on procurement of the $1.5 billion spacecraft." This report was not accurate; in fact, by December 8 NASA and OMB had reached a compromise, both indicating their willingness to accept option 2, producing shuttle spares. This compromise seemed to avoid the issue of a fifth orbiter being brought before Ronald Reagan for resolution as he finalized his FY 1984 budget proposal.[18]

Even with this compromise in place, the SIG (Space) process continued, producing the issue paper that McFarlane had requested. That paper noted that a "decision for or against the fifth Orbiter is in reality a decision whether or not to truncate the production program" and raised the question of whether "without a production base, can an operationally viable and responsive system, capable of absorbing problems and contingencies, be assured to meet U.S. launch requirements?" The paper provided the budget implications of each option identified by SIG (Space) and the pros and cons associated with each choice. It reflected the OMB-NASA compromise on the fifth orbiter, indicating that DOD, the Joint Chiefs, and NASA now supported either Option 2 or 3. The State Department, CIA, OMB, and OSTP had changed their votes to support option 2; Commerce and ACDA still preferred option 3. The paper ended by recommending option 2, producing spare shuttle assemblies, as being "in the best interests of the United States."[19]

Reflecting on the outcome of the fifth orbiter debate, Rye commented that "this was the first test case for the SIG (Space) and I believe it demonstrated its utility. Had the SIG (Space) not focused policy-level attention on this important issue, OMB refusal of funding would have terminated the total space launch capacity of the Nation." By pursuing option 2, "the President can demonstrate his determination to maintain space leadership with a 'robust' four Orbiter fleet."[20]

The Debate Continues

Even though the basic issue of how to proceed with respect to the fifth orbiter question had been resolved, there was continuing White House conflict on how to bring the process to closure. At the most senior level, Judge Clark disagreed with White House Chief of Staff James Baker and Presidential Counselor Ed Meese over whether the issue had been dominantly a policy question or a budget question, and thus what should have been the relative influence of SIG (Space) and OMB over the final decision. At the staff level, the issue revolved around the wording of a directive reflecting the fifth orbiter decision.

As agreement on option 2, producing shuttle structural spares, emerged both in the NASA-OMB compromise and the SIG (Space) options paper, Rye prepared a decision memo that asked President Reagan to "approve the SIG (Space) recommendation to maintain Orbiter production capability" and to "sign the attached National Security Decision Directive (NSDD) to document your decision." Rye had sent a preliminary version to this material to OMB and OSTP for comment; rather than comments, he received back from the two offices a totally new text for the NSDD. That text had been developed by OMB's associate director Fred Khedouri in consultation with NASA's Beggs, who was working with OMB to negotiate a final NASA FY 1984 budget and was willing to take Khedouri's side on the wording of the directive. Rye found the new text "unacceptable," because "it doesn't recognize that one of the major advantages of Option 2 is that it preserves Option 3 should demand increase" and that "it doesn't emphasize the major concern of SIG, i.e., maintaining Shuttle production capability." To Rye, the OMB's text was "written almost as Option 1 and could be interpreted as closing out any subsequent consideration of a 5th Orbiter." Rye commented that "unless OMB is going to be constructive in their comments, we should reconsider circulating these documents for consideration."[21] It was of course just such coordination that SIG (Space) had been created to achieve.

Apparently, NSC ignored the comments from OMB and OSTP, and Clark forwarded for presidential action the original decision memo and draft NSDD that Rye had prepared. This provoked a memo from OMB's Stockman to Meese suggesting that "the draft NSDD attached to the Clark memorandum does not reflect the consensus view of the SIG policy officials," that "it includes several troublesome features, including matter that was not even the subject of discussion by the SIG," and that "the most severe problem is that it characterizes the decision as a decision to keep open the shuttle production line." Such a decision was "completely at variance with the NASA/OMB budget agreement." Stockman suggested that the Rye version of the NSDD "appears to direct NASA to expand the demand for shuttle launch services in order to justify a fifth orbiter" and that "NASA will interpret it as a mandate to subsidize the Shuttle even more than at present, thereby making all private ventures uncompetitive with NASA." Stockman attached the alternate NSDD text that had been developed by Khedouri and Beggs to his memo, saying that it "had the concurrence of NASA, OSTP, OMB, State, and Commerce. We are virtually certain that DOD would have no objection to it." That version of the NSDD, he added, "confines itself to the direct policy question at hand: does the four orbiter fleet meet the objectives of the space policy?"[22]

On the morning of December 17, 1982, Meese, Baker, and Clark met with the president to review the situation with respect to the fifth orbiter. The result was indecisive; Reagan judged that the decision package was not yet ready for his approval. This outcome provoked a memorandum from Clark to Meese and Baker suggesting that "the question of continuing Shuttle Orbiter production capability is a policy issue with budgetary implications" Clark noted an apparent "misconception concerning NASA's budget" which, on the basis of the NASA-OMB agreement, would include $100 million to implement option 2 in the SIG (Space) issue paper. Clark suggested that "the point here is that the SIG (Space) has not attempted to specify the total amount of funds necessary to implement the option or the pace at which the option will be implemented. These are budgetary considerations and are obviously under the purview of OMB." Clark noted that there were "major disagreements between OMB and the other SIG (Space) members concerns the wording of the NSDD." The version of the NSDD preferred by SIG (Space) "preserves the option for subsequent procurement of a Fifth Orbiter should optimistic estimates of demand materialize. OMB desires to foreclose that option, i.e., they desire for the President to imply in the NSDD that there will be *no* future consideration for production of a Fifth Orbiter." To Clark, "this is a policy decision and *not* a budget decision ... It would clearly be unwise for the President to foreclose any option concerning the Shuttle program. Indeed, it would be a disservice to the Nation if he did so." Clark closed his memo by saying, "we should have the President sign the directive at the earliest opportunity."[23]

Baker and Meese responded to Clark on December 22, saying that "we conclude that the question raised regarding orbiter production capability is, in fact, more of a budget determination than a policy matter." They suggested that the OMB-drafted version of the NSDD was "more appropriate for NSC to forward to the President." That version, they argued, "reflects the decision and conclusions reached by OMB and SIG (Space). It does not rule out the fifth orbiter option. Nor does it in any way discourage an aggressive effort to enhance the commercial and military use of space—which we agree should be a priority." Baker and Meese agreed that "prompt action should be taken," but that "if the NSC is not satisfied with the draft NSDD ... which has been cleared, then we would suggest NSC should redraft an NSDD ... and have it circulated for comment and reviewed with the President early next week." In a slap at the SIG (Space) approach to decision-making, they commented that following this approach would "help establish a clear decision making track that avoids placing NSC in the position of resolving budget issues raised by various agencies."[24]

Bureaucratic conflict continued; Clark did not accept the version of the NSDD attached to the Meese-Baker memorandum. Instead, it fell to OMB's Khedouri and NSC's Rye to find a compromise between the two versions of the NSDD that would satisfy all parties to the dispute. The two met on January 4, 1983, and agreed on a text that, from Khedouri's perspective, eliminated "all of the seriously objectionable elements" of Rye's earlier drafts. In letting Craig Fuller, who would carry the revised NSDD to Meese for transmittal to President Reagan, know of his agreement with Rye, Khedouri asked Fuller to "verify that the document presented for signature is in fact identical" to the one on which he and Rye had agreed. He added that "I do not wish to undergo another episode of sending in comments that are totally ignored." The lack of trust between OMB and SIG (Space) engendered by the disputes of the past month was evident. Fuller did forward a copy of the revised NSDD to Baker and Meese on January 5, noting that it "contains the modifications we were seeking"; they in turn sent the directive to the president for his approval.[25]

The directive memorandum gave Ronald Reagan three options: (1) "terminate production"; (2) "maintain Orbiter production capability for major components"; and (3) "full production of a fifth Orbiter." The cover memo recommended that the president approve option 2. It was signed by Clark with the handwritten notation that it had been "cleared with Baker/Meese." When Reagan reviewed the memorandum on January 10, he did not accept that recommendation. He sent the memo back to Clark with the notation "Bill—I want to discuss option 3." With this comment, Reagan "signaled a desire to support a production policy that goes beyond the current capability/availability requirements using current assumptions," thereby communicating to his inner circle a "desire to support a fifth shuttle orbiter."[26]

It was Craig Fuller who finally put forth the language which would resolve the issue. Fuller in a January 27, 1983, memorandum to Meese suggested that the conclusion reached during the SIG (Space) review that a four-orbiter fleet would meet projected needs "can be challenged if one is to predict greater commercial use and military use of space will occur beyond what is now projected." With respect to the broader issue of space leadership, a central Reagan theme, Fuller suggested that building only enough orbiters to meet current projections of demand may "not show the kind of vision associated with the notion of leadership. There is, as always, the question of how high a price one wants to pay for vision and leadership." In addition, "there is some question about whether or not the spares program will really keep the production line available." Fuller suggested amending the version of the NSDD that had resulted from the Rye/Khedouri compromise in early January to have the president say, "I will continue to evaluate the needs for expanding Shuttle orbiter availability."[27]

Meese apparently agreed and told OMB and NSC to revise the NSDD to reflect Fuller's line of reasoning. This was done at a level above Rye and Khedouri; on February 2, the number three person on the NSC staff, Rear

Admiral John Poindexter, sent a revised NSDD to Meese, saying that he had worked with OMB Deputy Director Joe Wright (who had recently replaced Ed Harper) on the revision. On February 3, Ronald Reagan signed the document, which was issued as NSDD 80. The directive was titled "Shuttle Orbiter Production Capability." It said:

> The Senior Interagency Group (SIG) for Space, established to implement my National Space Policy, has examined the issue of whether Shuttle Orbiter production capability should continue beyond the four Orbiters currently scheduled for delivery.
>
> After reviewing the various options presented by the SIG (Space), I have decided that maintaining Orbiter production capability is in the best interests of the Nation. This objective will be achieved through the production of structural and component spares necessary to insure that the Nation can operate the four Orbiter fleet in a robust manner. These spares will provide the necessary assurance that Shuttle operations will continue in the face of minor problems, modifications or other periods of extended Orbiter outages.
>
> While this decision does not constitute approval for procurement of a fifth Orbiter in FY 1984, it does partially preserve this option for a future decision should optimistic estimates of demand materialize and other conditions dictate. It is my intent that the full potential of the Shuttle concept as originally envisioned is achieved and commercialization of space becomes a reality.
>
> The National Aeronautics and Space Administration is directed to submit to the Office of Management and Budget for inclusion in the Fiscal Year 1984 Budget a request for procurement of structural and component spares that will carry out this decision.[28]

Conclusion

The third paragraph of this directive incorporated Fuller's January 27 suggestion that it should reflect Ronald Reagan's optimistic view of the future potential of space activity. With this addition, the almost two-month-long bureaucratic battle between NSC and OMB over what the directive should say was brought to a conclusion that was acceptable to the senior leadership of the two groups and to Ronald Reagan's inner circle. It gave NASA a path forward that in 1986 would prove key to recovering from precisely what it was intended to protect against—the loss of a shuttle orbiter due to a catastrophic accident. The outcome of the conflict over the fifth orbiter issue was thus a prudent compromise.

With the shuttle declared operational and a policy toward a fifth orbiter decided, NASA leaders Beggs and Mark could now turn their attention to resuming the push for their top priority hope, getting Reagan administration approval for space station development as the agency's major engineering project for the coming years. In the debate over the NASA budget for FY 1984, they had focused their advocacy on getting presidential approval for a fifth shuttle orbiter, but also fought successfully to keep in the budget modest funds to continue the study of space station missions and concepts. They had earlier decided to defer their push for station approval for another year, hoping for a presidential commitment as the FY 1985 budget proposal was formulated in the final months of 1983. The space station had been in NASA's plans since the space agency's inception; waiting another year for a White House go-ahead was a risk Beggs and Mark were willing to take.

Notes

1. A copy of NSDD-42, portions of which remain classified, can be found at https://fas.org/irp/offdocs/nsdd/nsdd-42.pdf. Oral history interview of James Abrahamson by Jennifer Ross-Nazall, July 23, 2012, https://www.jsc.nasa.gov/history/oral_histories/NASA_HQ/Administrators/AbrahamsonJA/abrahamsonja.htm.
2. Hans Mark, *The Space Station: A Personal Journey* (Durham, NC: Duke University Press, 1987), 71–73. T. A. Heppenheimer, *Development of the Shuttle, 1972–1981* (Washington, DC: Smithsonian Institution Press, 2002), 349.
3. Letter from James Beggs to Edwin Meese III, May 21, 1982, NHRC. Mark, *The Space Station*, 152.
4. Philip Boffey, "Shuttle's Critics Call for a Broad Reassessment," *NYT*, November 17, 1981.
5. John Noble Wilford, "'Bargain' Prices Set for the Space Shuttle," *NYT*, January 13, 1977, 21. Information regarding Shuttle pricing policy in this and subsequent paragraphs is drawn from House of Representatives, Committee on Science and Technology, "Hearings – Space Transportation System," May 17–18, 1977 and NASA, *Space Transportation Reimbursement Guide*, JSC-11802, May 1980.
6. General Accounting Office, *NASA Must Reconsider Operations Pricing to Compensate for Cost Growth on the Space Transportation System*, Report MASAD-82-15, February 23, 1982. Also, Mitchell Waldrop, "NASA Struggles with Space Shuttle Pricing," *Science*, April 16, 1982, 278–279.
7. Letter from James Beggs to David Stockman, April 12, 1982, Folder 12463, NHRC.
8. Letter from David Stockman to James Beggs, June 14, 1982, Box 30, Subject Files, Executive Secretariat, National Security Council, RRL.
9. Interview of James Beggs by Kevin Rusnak, March 7, 2002, NHRC. Mitchell Waldrop, "NASA Cuts Flights, Sets New Shuttle Price," *Science*, July 2, 1982, 35.
10. George M. Low, Memorandum for the Record, "Meeting with the President on January 5, 1972," in John M. Logsdon et al., eds., *Exploring the Unknown: Selected Documents in the History of the U.S. Civil Space Program*, Volume I, Organizing for Exploration, NASA SP-4407 (Washington, DC: Government Printing Office, 1995), 559.
11. Memorandum from Danny Boggs to Edwin Harper, "NASA Space Shuttle Pricing," April 26, 1982, Box 83, Papers of Danny Boggs, RRL.
12. Memorandum from Edwin Harper to James Beggs, "Revenue Potential of Paying Passengers on Future Space Shuttle Flights," May 12, 1982 and letter from James Beggs to Edwin Harper, June 9, 1982, both in Box 82, Papers of Danny Boggs, RRL.
13. Letter from James Beggs to Representative Don Fuqua, October 7, 1982; NASA Press Release, "NASA Expands Payload Specialist Opportunities," October 22, 1982, both in Folder 8960, NHRC.

14. "Shuttle Customers will be Allowed to Fly Specialist with Payload," *Aerospace Daily*, October 1, 1982, Folder 8960, NHRC. Memorandum from Gilbert Rye to William Clark, "NASA Payload Specialist Policy," October 18, 1982, Box 91011, Papers of Gerald May, RRL. At the Reagan Library, Gil Rye's papers are filed in the boxes associated with his successor at the NSC, Col. Gerald May.
15. Interview with Alan Ladwig, May 6, 2016. Letter from James Beggs to Edwin Harper, June 9, 1982, Box 82, Papers of Danny Boggs, RRL.
16. Senior Interagency Group (Space) Space Launch Working Group, "Issue Paper on FY84 Budget Issues," undated but late November or early December 1982; IG (Space) Working Group Report on Space Launch Policy, December 2, 1962, both CIA-RDP83M00914RR000600030026-9, CREST.
17. Memorandum from E.A. Burkhalter, Director, Intelligence Community Staff, for Deputy Director of Central Intelligence, "SIG (Space): Fifth Orbiter Issue," December 9, 1982, CIA-RDP92B00181R001701610015-9, CREST; memorandum from Gilbert Rye to William Clark, "Shuttle Orbiter Production Capability," December 13, 1982, Files of Valerie Neal.
18. Craig Covault, "Budget for Fiscal 1984 Omits Fifth Orbiter," *AWST*, December 13, 1982, 16; memorandum from E.A. Burkhalter, Director, Intelligence Community Staff, for Deputy Director of Central Intelligence, "SIG (Space): Fifth Orbiter Issue," December 9, 1982, CIA-RDP92B00181R001701610015-9, CREST.
19. National Security Council, Senior Interagency Group (Space), "Issue Paper on the Space Transportation Systems (STS) Fifth Orbiter," undated but mid-December 1982, Files of Valerie Neal.
20. Memorandum from Gilbert Rye to William Clark, "Shuttle Orbiter Production Capability," December 13, 1982, Files of Valerie Neal.
21. Memorandum from William Clark to the President, "Shuttle Orbiter Production Capability," with attached draft NSDD, undated but December 14 or 15, 1982, and memorandum from Gil Rye to Bud (Robert McFarlane), December 13, 1982, Files of Valerie Neal.
22. Memorandum from David Stockman to Edwin Meese III, "NSDD on Space Shuttle," December 16, 1982, Box 5, Papers of Edwin Meese, RRL.
23. Memorandum from William Clark to Edwin Meese III and James Baker III, "Shuttle Orbiter Production Capability," December 17, 1982, Files of Valerie Neal.
24. Memorandum from Edwin Meese III and James Baker III, "NSDD on the Space Shuttle," December 22, 1982, Files of Valerie Neal.
25. Memorandum from Fred Khedouri to Craig Fuller, "NSDD on Space Shuttle," January 4, 1983 and memorandum from Craig Fuller to Edwin Meese III and James Baker III, "Shuttle Orbiter Availability," January 5, 1983, Files of Valerie Neal.
26. Memorandum from William Clark to the President, "Shuttle Orbiter Production Capability," with handwritten notation by RR (Ronald Reagan), January 10, 1983, Box 8, Outer Space Files, RRL. Memorandum from Craig Fuller to Ed Meese, "Space Shuttle Issues," January 27, 1983, Box 5, Papers of Edwin Meese, RRL.
27. Ibid.
28. Memorandum from John Poindexter to Ed Meese, February 2, 1983, Files of Valerie Neal. The NSDD-80 directive can be found at https://www.reaganlibrary.gov/sites/default/files/archives/reference/scanned-nsdds/nsdd80.pdf.

CHAPTER 7

The Next Logical Step

The decision with the most lasting impact on the U.S. space program made during the presidency of Ronald Reagan was likely his commitment to developing a space station. He announced during his January 25, 1984, State of the Union Address that he was directing the National Aeronautics and Space Administration (NASA) "to develop a permanently manned space station." This declaration was both the culmination of a decades-long quest by space advocates and the beginning of a program that would continue well into the twenty-first century. Because of its importance to Reagan's space legacy and because the process leading to the space station decision was a full example of how Ronald Reagan made major space policy decisions, this and the following three chapters give a detailed account of how that decision was made. A fifth chapter details the initiative to invite U.S. "friends and allies" to partner with the United States to develop and operate the space station.[1]

A Long-Time Ambition

A permanently occupied human outpost in Earth orbit had been a central element in planning for space development since even before government programs began in the late 1950s. Most famously, Wernher von Braun in the *Collier's* magazine issue of March 22, 1952, had described a wheel-shaped outpost in orbit 1075 miles above the Earth.[2]

Project Apollo's original goal as it was announced in summer 1960 was to develop a three-person spacecraft to fly in low Earth orbit as a precursor to a larger space station, as well as to carry a crew on a trip around the Moon. This Earth-orbiting step was bypassed as President John F. Kennedy decided in May 1961 to send the Apollo spacecraft directly to the Moon. But as Apollo moved toward that goal during the 1960s, NASA engineers and the U.S. aerospace industry continued to study space station designs and uses. An interim space

J. M. Logsdon, *Ronald Reagan and the Space Frontier*,
Palgrave Studies in the History of Science and Technology,
https://doi.org/10.1007/978-3-319-98962-4_7

station using spare Apollo hardware, later named Skylab, was part of NASA's early post-Apollo planning. Then, as the space agency began to focus on its post-Apollo activities in late 1968 and early 1969, NASA decided that some sort of permanently occupied space station would be its top priority human spaceflight program for the 1970s. President Richard Nixon in February 1969 chartered a top-level Space Task Group chaired by Vice President Spiro Agnew to provide him a "definitive recommendation on the direction which the U. S. space program should take in the post Apollo period." NASA Acting Administrator Tom Paine tried to preempt that review by proposing that the president make an immediate commitment to a space station as "the next major evolutionary step in man's experimentation, conquest, and use of space." Paine was not successful in this gambit. The Nixon White House was not interested in early approval of a major space initiative.

However, as the Space Task Group issued its report in September 1969, a 12-person space station launched by the Saturn V Moon rocket emerged as its central post-Apollo recommendation, to be followed by 50-person and then 100-person space bases, outposts on the Moon, and voyages to Mars in the 1980s. To be developed in parallel with the initial space station would be a reusable vehicle to carry crew, experiments, and supplies to and from the station—a space shuttle.

The Space Task Group's recommendations got a cool White House reception. Richard Nixon in March 1970 decreed that the post-Apollo space effort would not be "a series of separate leaps" and that "space expenditures must take their proper place within a rigorous system of national priorities." There would not be an Apollo-like follow on to Apollo. NASA was faced with a constrained budget outlook for the remaining years of the Nixon administration, and its leaders in summer 1970 decided that there would insufficient funds available for simultaneous development of a space station and a space shuttle. They further recognized that, while space station development was not financially viable without being accompanied by a lower-cost logistics system in the form of the space shuttle, there were rationales for developing a space shuttle that did not depend on the presence of a space station. They thus decided to develop the space station and the space shuttle in sequence rather than at the same time, and to make the space shuttle first in that sequence. NASA was not giving up on its space station aspiration, only deferring it until space shuttle development was completed, then planned for 1978 or 1979. This was recognized by at least some at the time of the shuttle decision; Dan Taft, the head of the Office of Management and Budget (OMB) unit overseeing the NASA budget, presciently commented in October 1971 "in a sense, a commitment to the shuttle is an implicit commitment to a subsequent space station program."[3]

NASA followed through on that strategy, beginning preliminary planning for a station as the shuttle follow-on during the Carter administration. Signaling that such planning would not bear fruit was a major reason for including the statements in the October 1978 Carter administration space policy statement that "our space policy will become more evolutionary rather than center around a single, massive engineering feat" and that "it is neither feasible nor necessary at this time to commit the US to a high-challenge, highly-visible space engineering initiative comparable to Apollo."[4] While it was clear that the Carter administration would not look in favor of a space station proposal, the belief persisted among NASA's space planners that a space station was the desirable next development after the space shuttle.

An Early Attempt at a Presidential Endorsement

It was not at all surprising, then, that James Beggs and Hans Mark, as they met soon after they had been selected as Ronald Reagan's top two NASA officials, would immediately agree that getting presidential approval for a space station program would be their highest priority in terms of new NASA activities. Both men were very familiar with past thinking with respect to the concept of a permanently crewed orbital outpost and both recognized that with the end of shuttle development a program of the scope of a space station would be needed to sustain the engineering and industrial base of NASA and its aerospace industry contractors. Assembling and then servicing a space station would provide a major mission for the shuttle, returning it to the original rationale for its existence. They gave little thought to alternatives to the space station as NASA's major development program for the 1980s, believing that gaining the experience in long-duration space operations that a station could provide was a necessary step toward any more ambitious objective such as a return to the Moon or journeys to Mars.

As a first step in NASA's strategy of convincing Reagan to grant early space station approval, Beggs identified top policy adviser Ed Meese as the most promising person to bring the space station idea before the president. Meese had attended the third launch of space shuttle *Columbia* on March 22, 1982, and had made very positive comments in an early-morning interview on the *Today* television show that day, saying that the Reagan administration had "a very deep commitment" to the space program, and that "there is so much to be gained from the space program we cannot afford, as a nation, not to support it fully." When asked specifically about potential administration approval of a space station, Meese was more cautious, saying, "we will do our best to support that at the appropriate time," and adding, "I think we have to go ahead with the space program. If we don't look to the future, we are being penny-wise and pound foolish."[5]

Beggs wrote to Meese on May 21, 1982, telling him that "it is timely for the Administration to decide on the next major goals in space, and for President Reagan to chart the course to be followed for the rest of this century." The White House had tentatively agreed that Ronald Reagan would be present on July 4 as the shuttle orbiter *Columbia* landed after its fourth flight; Beggs suggested that "the President's potential presence at the Shuttle landing at Edwards Air Force Base" would "provide an opportune time for announcing his plans." In Beggs's view, those plans "should embrace the development of a permanently manned space station." He offered four reasons for such a presidential commitment:

- Given European, Japanese, and Russian advances in space, "the United States [leadership] role in space is thus being vigorously challenged."
- "The Shuttle will occupy our development capabilities in manned systems for relatively little longer," and without a "major new challenge" those skills "will quickly erode."
- A space station could be operational by 1990 and thus is "a timely opportunity for a declared Presidential initiative."
- "The space station could also have major foreign policy advantages for the United States," with "Europe and Japan interested in participating" and likely to "contribute significant funding if they are given a significant role."[6]

The NASA campaign to gather early White House support for a space station also had a public dimension. On June 23, 1982, Beggs made a highly promoted speech in which he suggested that the "next logical step" in space development was to "establish a permanent manned presence in low-Earth orbit," and that this could be done "by developing a manned space station." The tag line "the next logical step" was to become closely associated in NASA communications with the space station proposal. This speech was part of a pro-space station campaign that presidential science adviser George "Jay" Keyworth found objectionable; he told the *Wall Street Journal* that Beggs's remarks were "plain totally premature" (Fig. 7.1). That NASA was orchestrating space station support from many quarters was clear to White House officials; OMB

Fig. 7.1 Presidential science adviser George "Jay" Keyworth, here seen talking with President Reagan, was a committed opponent of approving space station development. (Photo courtesy of the Reagan Presidential Library)

Associate Director for Energy and Natural Resources Fred Khedouri, who oversaw the part of OMB which reviewed the NASA budget, suggested to Director of Cabinet Affairs Craig Fuller that "those of you who wish to have the President get some benefit out of a space program initiative will never have the opportunity. NASA will have 'forced' the President—with the assistance of its congressional and corporate supporters—to accept its plans."[7]

A June 29 draft of the president's speech at the shuttle landing prepared by Dana Rohrabacher of the White House speechwriting office (later a California congressman) contained the sentence "we must look aggressively to the future by demonstrating the potential of the shuttle and establishing a more permanent presence in space." In reviewing the speech draft, OMB Deputy Director Ed Harper deleted that sentence. In subsequent days, there continued to be controversy over what, if anything, Reagan might say in his remarks with regard to a space station. What finally emerged in Reagan's July 4 speech was a compromise. The president included in his remarks the sentence that OMB had deleted, but did not specifically mention a space station or link the concept of permanency to human presence.[8]

The reinsertion into Reagan's remarks of the phrase about "a more permanent presence in space" was thus only a partial victory for NASA. The word "manned" was missing; having humans permanently in space was a central element of what NASA was proposing. This omission was purposeful. In a July 2 message back to the White House for Meese, Baker, Stockman, and Harper, who were not with the president in California, Darman and Fuller "for the record" noted:

> The president's remarks include a commitment to look toward "establishing a more permanent presence in space." This language was accepted by some concerned agencies [likely OMB and OSTP] only on an understanding that this does not necessarily imply or prejudge endorsement of a manned platform now being advocated by NASA. The manned platform proposal—and any other possible translation of the remark's general commitment—is to be subject to subsequent review and decision through ordinary policy and budgeting policies.[9]

NASA's hope to get early presidential approval for "the next logical step" had not succeeded. A space station decision would have to wait.

Preparing for a Space Station Decision

The Darman/Fuller message—that any decision to move forward with space station development would be "subject to subsequent review and decision through ordinary policy and budgeting policies"—reflected President Ronald Reagan's approach to making major decisions. Reagan might well be personally in favor of a particular course of action, in this case, approving NASA's space station proposal, but he seldom allowed his preference to be decisive until after having his staff carry out a full review of an issue and after hearing the range of views of his senior advisers on options for proceeding. In the case of the space station decision, all involved recognized that what was at stake was a multiyear, multibillion-dollar undertaking that could shape the character of the U.S. civilian space program for years to come. It was no surprise, then, that NASA's attempt to get early presidential endorsement of a space station had failed. There was no pressing external reason for such a commitment, and the hoped-for economic stimulus from the tax cuts that the Reagan administration had pushed through Congress in 1981 had yet to appear. With Congress resisting most of the budget reductions that the White House had proposed, this meant that the federal deficit was increasing at a rapid rate, making it very difficult to gain presidential approval for an expensive new space initiative. There was recognition, grudging on NASA's part, that a decision on the space station would almost certainly require a year or more of both policy and budget reviews before reaching the Oval Office. As those reviews went forward, NASA would be hard at work making its best case for why a space station was indeed "the next logical step."

NASA Builds Its Case

While Beggs and Mark had agreed on advocating a space station as NASA's next major program, they had not embraced a particular space station concept that would offer a specific set of capabilities or functions. Even though they had pushed for early approval of an orbital outpost, before NASA could realistically expect White House approval of its space station plans, it had to decide for itself what those plans were.

NASA in early 1982 created a group at NASA headquarters to support its space station planning; this step reflected the recognition that there had to be centralized coordination to bridge the varying views on the station inside NASA overall. Beggs brought back to NASA to head the coordinating group John Hodge, who had first come to NASA from Canada in 1959 to work on the Mercury program, and had stayed on to be involved in Gemini and Apollo. Hodge had left NASA in 1969 to work at the Department of Transportation, where he and Beggs collaborated on several projects. The planning group was formally established as the Space Station Task Force on May 20, 1982, with its charter being to develop "the programmatic aspects of the space station as they evolve, including mission analysis, requirements definition and program management." As its work progressed, the Space Station Task Force would also become NASA's primary marketing arm for the space station idea, mobilizing support within and without the government and representing NASA in interactions with working-level staff at the White House and the Executive Office of the President.

As the Task Force began work, Hodge made an important decision. The group would refrain from trying to design a space station, instead focusing on the missions it might perform and the requirements that could justify its development. Most members of the group were engineers with a tendency to want to begin their work by drawing blueprints of the facility they hoped to build, but Hodge laid down a "no pictures" guideline. Instead, the task force would focus its early work on identifying and analyzing the missions and requirements that would drive an ultimate facility design. To help in that effort, the Space Station Task Force in August 1982 issued small contracts to eight aerospace firms (basically all that existed) to "identify and analyze the scientific, commercial, national security and space operational missions that could be most efficiently conducted by a space station." The Department of Defense (DOD) added modest funds to help the contractors examine potential national security requirements.[10]

One of the leaders of the Space Station Task Force, Terry Finn, who had a background in political science and was responsible for the group's external relations, commented in a 1985 interview that "Jim Beggs and Hans Mark had a very lonely task … of selling the Space Station," adding that this was the case "because it wasn't as though there were lots of people around saying 'yeah, man, go.'" It was the job of the task force "to provide some kind of analysis so that when Beggs got up and said, 'We think we ought to build a space station,' and somebody said, 'What is a space station?' he had an answer. And to do a lot of nurturing and general support of Beggs and Hans Mark." Finn noted that the Space Station Task Force had three jobs—"to figure out exactly what a space station was and how much it costs and what it would do."[11]

An "Incremental Budget Strategy"[12]

NASA in mid-1982 had still held out the faint hope that it could get an implied administration commitment to the space station program by making the case to OMB for including modest funding to begin space station preliminary design efforts in President Reagan's Fiscal Year (FY) 1984 budget proposal, to be sent to Congress in January 1983. NASA, as it submitted its FY 1984 budget proposal to OMB in September 1982, gave higher priority to getting White House approval of the construction of a fifth space shuttle orbiter than it did getting additional funds to support the early space station work being coordinated by the Space Station Task Force. It asked for $200 million for the fifth orbiter and only $63 million for space station work. The task force had told Beggs that NASA would be ready to start preliminary design of a space station, Phase B in NASA's parlance, on October 1, 1983, the start of FY 1984, and the NASA budget request was based on enabling that action. Although Beggs in his letter submitting the NASA budget told OMB that a presidential decision to start building a station would not be required for "about two years," McCurdy notes the NASA hope that OMB approval of the $63 million request would "imply a commitment to proceed" with the station program. As he made his budget request, Beggs "was prepared to appeal the issue all the way to the President to get that commitment."

Beggs was following what McCurdy calls an "incremental budgeting" approach, "when an agency head initiates an important new program with a small increment [of funds] laid over last year's base." This approach did not get a positive reception. The staff in OMB reviewing the NASA request recommended to their boss, OMB Associate Director Khedouri, that NASA should receive no new funding for space station-related work, not only in FY 1984 but also in the following years. That outcome would effectively end NASA's hopes for getting a space station program approved during the first Reagan term. Learning of the OMB position, one member of the Space Station Task Force commented that "OMB came in and said, 'There isn't any way we are going to let you start Phase B' ... They were the dark days."

After receiving the initial OMB allowance in late November with no funds for space station work, Beggs filed two *reclama*, the term used for an appeal to higher levels in OMB and ultimately to the president that OMB's initial decisions be reversed or modified. The first appeal on November 29 indicated that NASA would accept $53 million rather than $63 million in space station funds. In his second appeal, filed on December 8, Beggs said that he would accept a $30 million allowance, but that if OMB continued to zero out station funding he would take the issue to the OMB director, the top-level Budget Review Board, and ultimately, if necessary, to the president. In response to this second appeal, Khedouri offered a compromise; NASA would be provided an additional $14 million for space station work in both FY 1984 and FY 1985, but there would be no commitment in either year to the station program.

Beggs made a judgment, "after more mature thought," not to appeal OMB's offer all the way to the president. He concluded that "I don't think the timing was right." NASA thus accepted the $14 million for FY 1984, while reserving its position for the following year. McCurdy comments that this decision was a short-term "defeat" and "a clear signal" that NASA's incremental budget strategy had failed in its first attempt. NASA "would not be able to get the space station proposal approved by pushing a small piece of it through the White House budget office."

Still, NASA did not abandon its incremental strategy, arguing that it was possible to buy the space station "by the yard" and asking for only a small initial budget to begin the station design effort. But it also recognized that it would have to find a way to get the station proposal before the president by following another path than through the budget process. By the end of 1982, the newly created Senior Interagency Group for Space [SIG (Space)] had begun to review the space station issue. Getting SIG (Space) support for the space station seemed the most promising approach toward eventual presidential approval. Commenting on his decision not to appeal the OMB allowance, James Beggs said: "I decided to keep my powder dry and flush this thing [the space station] up in a different way."

SIG (Space) Space Station Review Gets Started

As the SIG (Space) met for the first time on September 23, 1982, its members had before them a brief paper on a "Manned Space Station" that outlined the proposed approach to be taken in laying the basis for a Reagan administration decision regarding a commitment to space station development. The basic question to be addressed was, "What policy issues must be identified and resolved in order to establish the basis for an Administration decision on whether or not to proceed with development of a permanently-based, manned space station?" The paper noted that "the various policy issues surrounding such a program must be carefully examined because of the large multi-billion dollar investment required and because of the program's significant impact on space activities in the civil, national security and international sectors." The relevant policy issues were identified as follows:

1. How will a manned space station contribute to the maintenance of U.S. space leadership and to the other goals contained in our national space policy?
2. How will a manned space station best fulfill national and international requirements versus other means of satisfying them?
3. What are the national security and foreign policy implications of a manned space station?
4. What is the overall economic and social impact of the program?

The paper noted that a SIG (Space) Space Station Working Group would be formed to address these questions; it would begin its work immediately and would have over a year, until November 1983, to prepare its report.[13]

At the initial meeting of the secondary-level Interagency Group for Space on October 1, Hans Mark introduced the individual chosen to chair the Space Station Working Group, NASA's John Hodge, who was already heading NASA's internal Space Station Task Force. There had been a decision that NASA should chair the working group, and Hodge was the logical person to fill that role. Other members of the working group came from the Departments of State and Defense, the Intelligence Community, the National Oceanic and Atmospheric Administration of the Department of Commerce, and the Arms Control and Disarmament Agency and from the White House, OMB, and Office of Science and Technology Policy (OSTP).

Initial Working Group Activities

The SIG (Space) Space Station Working Group first met on October 21, 1982. The subsequent seven months were spent in bureaucratic debates regarding how the group would go about its work and, in particular, the character of its final report. This lengthy period of bureaucratic churning presaged the inability of the working group to prepare a report that could form the basis for presidential decision.

The working group decided to prepare "issue papers" responding to the questions raised in its charter; issues related to foreign policy were separated from national security issues, so there would be five papers. NASA provided four possible alternatives (called "scenarios") for future development:

1. *Space Shuttle and Unmanned Satellites*—basically continue the *status quo* of using the shuttle, with a one week stay time in orbit, to deploy satellites for various purpose and, when required, have shuttle crews interact with those satellites.
2. *Space Shuttle and Unmanned Platforms*—develop large platforms that would host instruments and payloads delivered and possibly retrieved by the shuttle, but would operate with only occasional astronaut intervention.
3. *Space Shuttle and an Evolutionary/Incrementally Developed Space Station*—develop a permanent crewed, multifunction and multidisciplinary facility at rate constrained by available financial resources. NASA would later call this approach "buying a space station by the yard."
4. *Space Shuttle and a Fully Functional Space Station*—develop a fully capable space station at the earliest possible time.

The working group would analyze each of the five issues in terms of each of the four hardware scenarios, identifying the pros and cons of pursuing a particular scenario with respect to each policy issue. This approach sowed the seeds of bureaucratic deadlock, in effect requiring 20 separate analyses.[14]

It is worth noting that at this point, the space station was being evaluated as a facility available for military and intelligence uses, not just one developed to meet civilian requirements. At a March 28, 1983, SIG (Space) meeting, Deputy Director of Central Intelligence John McMahon commented that the "Intelligence Community is actively examining potential intelligence requirements for a Space Station. We must be careful to insure that these requirements are treated with due regard to their sensitivity and classification." The DOD also had underway several internal reviews of military requirements that a Space Station could meet, as did NASA's eight mission analysis contractors.[15]

Gil Rye and the Space Station

Gil Rye had early on in his assignment as National Security Council (NSC) lead for space policy become convinced that NASA's space station was essential to maintaining U.S. leadership in space. Rye took several actions as the SIG (Space) review of the space station was underway to improve the chances that Ronald Reagan would approve the station proposal once it was formally presented to him. Rye convinced National Security Adviser Clark to complain to Ed Meese about science adviser Jay Keyworth's public opposition to a space station; Keyworth had called the station "an unfortunate step backwards" in congressional testimony on February 15, 1983, as the working group was still debating its terms of reference. In a March 1 memorandum, likely drafted by Rye, Clark told Meese that "there is no need for Jay" to use such a characterization; rather "the correct response is that the President recognizes the importance of the issue and has established a disciplined decision-making process to review the various aspects and present him with a recommendation." The memo noted that in 1982 OMB and OSTP had "criticized NASA for getting out ahead of the President on the Space Station. The White House Staff should not commit the same sin."

In mid-March 1983, Rye wrote a two-page memo to Clark to "sensitize" him to the importance of the space station decision, which would have a major impact on both "the future of the civil space program" and "the future of NASA." Rye suggested that if a space station were not approved, "a gradual downward trend in the civil space program would inevitably result which, combined with the increasing defense/intelligence use of space will only heighten the arguments that the Administration is 'militarizing space.'" This memorandum was written six days before Ronald Reagan, on March 23, 1983, announced what became known as the Strategic Defense Initiative; Rye was one of the few in the White House who knew of the president's intent.

Rye also suggested to Clark that "the absence of a Space Station program will fundamentally alter the character and scope of NASA activities … The absence of a major R&D program would significantly reduce NASA manning, fundamentally restructure the organization, and thus possibly result in the dismantlement of one of the nation's greatest assets." Rye added that "one final question that must be considered is whether we need a manned (as opposed to an unmanned) program simply because of the emotional impact on the American public, the pride it generates in our technology and know how and international perceptions of space leadership." He observed that "the nation needs heroes." Rye suggested to Clark that issues such as those he had laid out in the memo "are essential ingredients into the President's decision and we must prepare ourselves to present them to the President in the most objective manner possible."[16]

Rye had been detailed to the NSC from the Pentagon "to advance his career." He had been promoted from lieutenant colonel to colonel soon after his arrival at the White House. His tour of duty there was intended as a step on the way to more responsible Air Force assignments and to promotion to at least brigadier general. But Rye's advocacy of the space station and his cooperation with NASA in building the case for its approval would turn out to mean that he had "gone too far" from DOD's perspective. Mark in his account of the space station decision comments that "I doubt very much whether space station could have been pushed through the administration if it were not for the persistent and astute work of Gil Rye." Rye paid a high price for his advocacy; "his military career ended when he embraced the space station."[17]

Ronald Reagan Meets the Space Station

As another action related to his belief in the importance of getting presidential approval for the space station, Rye suggested to Clark that President Reagan be briefed on the space station review, so that Reagan could be "an active participant in the process." This to Rye was a way of alerting the president to a major issue that he would have to decide in coming months. Clark agreed, and the briefing was set for April 7. The briefer would be NASA's Beggs. Rye wrote Beggs in advance of the briefing, suggesting to the NASA head that "we need to stay away from an 'advocacy' pitch, instead informing the President on our game plan for advising him (via the SIG (Space)) on the Space Station issue."

The 20-minute briefing was held in the White House Situation Room. Attending were Vice President George Bush, Reagan senior advisers Baker and Clark, White House staffers Fuller and Rye, and NASA's Beggs. In his briefing, Beggs made sure that Reagan was aware that the Soviet Union already had a space station, but did not emphasize the national security aspects of the program. He instead emphasized a space station as key to "national leadership in commerce and technology," a theme that Reagan "liked very much." Reagan asked Beggs about the connection between a space station and a mission to Mars, and Beggs replied that it was a necessary "way station" for such a journey. Among Reagan's favorite reading as a young person had been Edgar Rice Burroughs's science fiction books about life on Mars, and his question at the briefing indicated his continuing interest in that topic.

Returning to NASA after the briefing, Beggs "was quite optimistic about the future." He reported that "the president was very interested in the general direction of the space program and the space station in particular" and that NASA "had some very strong support among members of the White House staff." Although Beggs had interacted with President Reagan in large settings on previous occasions, this was his first small group meeting with the president, coming almost two years after he had been nominated for the top NASA job.[18]

After the April 7 briefing, the NSC decided to incorporate the terms of reference for the space station study into a National Security Study Directive (NSSD) signed by the president, rather than just a study charter issued by SIG (Space). It had become clear at the March 28 SIG (Space) meeting that there likely would be significant opposition to the space station from the national security community, and having the space station study be a presidentially directed effort was a way to avoid the possibility of such opposition blocking the study from reaching the president. Rye later commented that "the President's signature on the NSSD vividly demonstrated the importance of the issue, and the bureaucracy took it seriously."

Ronald Reagan signed the two-page document, labeled NSSD 5-83, on April 11. As he did so, Reagan "seemed extremely interested and said that he would anxiously await the study results." The directive called for a study "to establish the basis for an Administration decision on whether or not to proceed with NASA development of a permanently based, manned Space Station." The study would be carried by the same SIG (Space) Space Station Working Group that had been meeting since October, now operating with presidentially approved terms of reference. The directive repeated the same five policy questions that had been first posed in the background paper for the first SIG (Space) meeting in September 1982 and listed the same four "example scenarios" that had appeared in NASA's input to the working group. The report of the working group would be "a summary paper that assesses the issues and identifies policy options." The study results would be presented to the SIG (Space) "not later than September 1983 prior to presentation to the President."[19] After seven months of debating how to proceed, a SIG (Space) space station study effort could finally begin. Another eight months of heated debate lay ahead before Ronald Reagan would decide whether to take "the next logical step."

Notes

1. There are already in being two detailed accounts of the space station decision, one by a participant in the choice and one by a well-respected scholar. Those accounts are by Hans Mark, *The Space Station: A Personal Journey* (Durham, NC: Duke University Press, 1987) and Howard McCurdy, *The Space Station Decision: Incremental Politics and Technological Choice* (Baltimore, MD: The Johns Hopkins University Press, 1990). The following chapters draw heavily on those two books, but supplement the narrative with documents and interviews not available to either author. Both Dr. Mark and Professor McCurdy have been very generous in providing me with material from their personal files, and for that they have my gratitude.
2. Wernher von Braun, "Crossing the Last Frontier," *Collier's*, March 22, 1952, www.rmastri.it/spacestuff/wernher-von-braun/colliers-articles-on-the-conquest-of-space-1952-1954/.
3. This account of post-Apollo planning is drawn from John M. Logsdon, *After Apollo? Richard Nixon and the American Space Program* (New York: Palgrave Macmillan, 2015). Taft's comment is on 216.
4. The Carter space policy statement can be found in John M. Logsdon et al., eds., *Exploring the Unknown: Selected Documents in the History of the U.S. Civil Space Program*, Volume I, Organizing for Exploration, NASA SP-4407, (Washington, DC: Government Printing Office, 1995), 576–578.
5. Ed Meese interview with Chris Wallace, *Today*, March 22, 1982. I am grateful to Howard McCurdy for sharing his copy of the *Today* show transcript with me.
6. Letter from James Beggs to Edwin Meese III, May 21, 1982, NHRC.
7. Memorandum from Fred Khedouri to Craig Fuller, "NASA Marketing Effort," June 24, 1982, with attached *Wall Street Journal* article, Folder 12772, NHRC.
8. Mark, *The Space Station*, 248; Memorandum from Edwin Harper to Aram Bakshian (chief Reagan speechwriter), undated, with attached speech draft, Box 83, Papers of Danny Boggs, RRL; Ronald Reagan: "Remarks at Edwards Air Force Base, California, on Completion of the Fourth Mission of the Space Shuttle Columbia," July 4, 1982. Online by Gerhard Peters and John T. Woolley, APP, http://www.presidency.ucsb.edu/ws/?pid=42704.
9. Telegram from Richard Darman and Craig Fuller to Ed Meese, Jim Baker, David Stockman, Ed Harper, July 2, 1982, Box 83, Papers of Danny Boggs, RRL.
10. See McCurdy, *The Space Station Decision*, 49–51, for a discussion of the origins of the Space Station Task Force and Chaps. 9 and 10 for a description of its organization and early work.
11. Interview of Terence T. Finn by NASA Historian Sylvia Fries, June 12, 1986, NHRC.

12. Unless otherwise noted, this account is based on Chap. 13 of McCurdy's *The Space Station Decision.*
13. "Manned Space Station," undated but September 1982, File 12905, NHRC.
14. Memorandum from Daniel Herman to Members of the Space Station Working Group, "SIG (Space) Space Station Working Group Meeting," January 21, 1982 (actually 1983), File 12905, NHRC; memorandum from Daniel Herman to Members of Space Station Working Group Members, "Record of SIG (Space) Space Station Working Group Meeting, January 28, 1983," January 31, 1982 (actually 1983), NSSD 13-82 File, Executive Secretariat, National Security Council, RRL.
15. Minutes, Senior Interagency Group for Space, March 28, 1983, CIA-RDP85M00364R001101630035-5. CREST.
16. Memorandum from William Clark to Edwin Meese, III, "Administration Policy on Space," March 1, 1983, NSSD 13-82 File, Executive Secretariat, National Security Council, RRL; memorandum from Gilbert Rye to William Clark, "Manned Space Station," March 17, 1983. This document was provided to the author by Valerie Neal of the Smithsonian National Air and Space Museum.
17. McCurdy, *The Space Station Decision*, 163, 165; Mark, *The Space Station*, 147.
18. Memorandum from Gil Rye to William Clark, "SIG (Space) Working Group on the Space Station," February 9, 1983, NSSD 13-82 File, Executive Secretariat, National Security Council, RRL. Letter from Gil Rye to James Beggs, March 14, 1983, in McCurdy, *The Space Station Decision*, 138. Mark, *The Space Station*, 165.
19. National Security Study Directive Number 5-83, "Space Station," April 11, 1983, Box 5, Papers of Edwin Meese, RRL. Letter from Gil Rye to Hans Mark, October 25, 1984, commenting on a draft of Mark's space station book. Quoted with permission from Gil Rye.

CHAPTER 8

Debates and Disagreement

There were many issues to be resolved before the question of whether to move forward with a space station could be ready for presidential decision. They included: whether the scientific community or national security community would support a space station program; what was the design of the space station the president would be asked to approve; how much that station would cost; and whether the U.S. space station program would be open to international participation. Those issues were extensively analyzed and debated during the April-August 1983 period. The Senior Interagency Group for Space [SIG (Space)] study process authorized by the April 1983 NSSD 5-83 (National Security Study Directive) proved highly bureaucratic, as interested agencies and White House staff offices had widely differing perspectives, and John Hodge's SIG (Space) Space Station Working Group report turned out to be poorly suited to defining policy options for presidential choice. There emerged deeply felt disagreements among senior Cabinet officials and White House staff on the wisdom of approving the space station proposal.

The SIG (Space) process was also only one of the routes that the space station issue would take to the president for decision. The Office of Management and Budget (OMB) was fully aware of the National Aeronautics and Space Administration's (NASA) hope to get station approval in fall 1983, with its Fiscal Year (FY)1985 budget including a start on detailed design of the space station that reflected that approval. As the SIG (Space) study moved forward, in parallel OMB was preparing for the possibility of station approval as it assessed the likely NASA budget for FY 1985 and beyond. The OMB director, David Stockman, his associate director Fred Khedouri, and most of the OMB staff dealing with the NASA budget opposed the station on both programmatic and budgetary grounds, but they also had to plan for possible presidential approval of the initiative.[1]

J. M. Logsdon, *Ronald Reagan and the Space Frontier*,
Palgrave Studies in the History of Science and Technology,
https://doi.org/10.1007/978-3-319-98962-4_8

Indeed, Ronald Reagan at several points during 1983 signaled that he was personally in favor of developing a space station. But Reagan's decision-making style meant that he did not impose his personal preferences early on in the policy-making process; issues were fully debated, at times overly so, before policy options were presented to the president for choice. This was certainly the case with respect to the space station decision.

SIG (Space) Space Station Working Group Struggles

Eight days after President Reagan signed NSSD 5-83, laying out the terms of reference for the space station study, John Hodge proposed to the members of the SIG (Space) Space Station Working Group a schedule for meeting the September 1983 due date for the working group's final product. NASA by the end of May would provide a "capabilities document" that would lay out in some detail alternative hardware scenarios for proceeding. The NASA capabilities document would be "a result of one year of developing mission requirements, two months of evaluating architectural options to fulfill those mission requirements, and one month of developing the cost data for a Space Station concept and its operations." A "background requirements document" would also be prepared by the end of May; based on the results of the NASA-funded mission studies, it would lay out the civilian, industry, and national security requirements that a space station could serve. Hodge planned a "final line-in-line-out" session on each issue paper and the draft working group report the week of July 11. That would be followed by two months for agency review, leading to presentations to the Interagency Group (Space) and SIG (Space) in September.[2]

Working Group Report and the National Space Strategy

Since the beginning of the National Space Strategy study initiated in December 1982, the future of the U.S. human spaceflight program had been seen as an important element of that broader study. As NASA and Department of Defense (DOD)/Intelligence Community inputs for the strategy study came into the National Security Council in late May 1983, National Security Adviser Judge Clark decided that the results of the space station study should be merged with these other submissions into a comprehensive space strategy report. A draft of that report was circulated for review in mid-July. The reaction to the draft was predominantly negative, with the national security community continuing its criticism that a single assessment of all U.S. space activities was artificial and unproductive.[3]

With respect to the space station portion of the report, the deputy director of the National Reconnaissance Office (NRO), J.D. "Jimmy" Hill, already a legendary figure inside the national security community for his persistence in advancing NRO interests, commented, "it is impossible to determine in any detailed fashion the schedule, cost or impact of any of the proposed space station options," and "the paper does not provide distinguishable options with the pros and cons to allow for a decision." Hill noted that "we were unable to obtain substantive data to enforce the NASA assertion that a permanently manned space station is the only way to maintain overall U.S. civil space leadership." He added that "the apparent method used by NASA to develop civil 'requirements' was to list any potential civil space activity (regardless of priority, requirements, cost or technical feasibility) as a civil sector mission for conduct on the space station." In addition, "the opportunity cost of a space station on other space programs is not discussed in any significant fashion. That cost is likely to be large."[4] Hill's criticisms were representative of the attitude of the national security community toward NASA's case in support of station approval.

The draft space strategy report was discussed at a meeting of the Interagency Group for Space on July 28. That meeting agreed that the attempt to combine the space strategy and the space station reports had not succeeded, and that two separate reports for the president should be prepared, one with respect to a decision on the space station, to be available by September, and the other on a National Space Strategy, to be available later in the fall.

The National Security Council's Gil Rye was tasked to prepare a new version of the space station report. It was clear from the reactions to the Space Station Working Group report that it could not be used as the basis of the new version. Rye asked NASA to assign someone not previously involved with the working group to work with him on a full-time basis to help prepare the new report. That new person was Margaret "Peggy" Finarelli, who had come to NASA in 1981 after working at the Central Intelligence Agency (CIA), the Arms Control and Disarmament Agency, and the White House Office of Science and Technology Policy. Finarelli was a tough-minded individual and would end up playing an important role in both the domestic and especially the international issues associated with the space station program. Hans Mark characterized Finarelli as NASA's "front line soldier" in the effort to gain support for the space station proposal, and commented that "she did her job with grace, intelligence, and good humor."[5] Other members of Rye's drafting group came from the DOD, CIA, and OMB. There was a sense of urgency in getting a new report drafted, since Rye had scheduled an August 10 meeting of SIG (Space); that meeting now would focus only on the space station issue. The report would also have to reflect the current state of play on a variety of issues that had gotten somewhat submerged as Hodge's working group had struggled with its analysis.

Space Station Issues as of Mid-1983

Those issues included how to deal with the message from the space science community that it did not see a need for a space station and, more important, the conclusion emerging out of the influential national security community that it had no requirements for a space station and would prefer that available funds be allocated to improving the capabilities and performance of the space shuttle. The question of whether international participation in the station would be encouraged also had to be addressed. After avoiding for over a year settling on a specific space station design, NASA would have to select its preferred design and indicate its likely cost. All of these issues had been bubbling up in the first half of 1983, and the report Rye, Finarelli, and their team was preparing had to state them in terms relevant to senior policy-makers and ultimately to Ronald Reagan.

Space Science Community Says "No Need"

The Space Science Board of the National Academy of Sciences, the top-level forum for the space science community, took the initial position that a space station would not significantly improve the quality of most scientific activities being proposed for the facility, saying that "the Board examined the set of specific missions proposed for implementation from the space station during the years 1991–2000." The Board judged that "few of these missions would acquire significant scientific or technical enhancement by virtue of being implemented from the space station." Given this reality, "and the adequacy of the present space transportation system for the purposes of space science, the Board sees no scientific need for this space station during the next twenty years."

The Space Science Board perspective reflected the view of traditional space scientists, accustomed to answering their questions through instruments launched on spacecraft dedicated to a specific scientific purpose; there were few scientists represented who had focused their research on the effects of very low gravity on physical and biological processes or on having human operators interact with their experiments in orbit. The space science community had also for many years viewed human spaceflight activity as a competitor for resources. Thus, it was not particularly surprising that there was no support for a station from the elite representatives of the space science community.

Beggs pushed back against the Board's initial views, with some success. In his subsequent testimony to Congress, Board chair Thomas Donahue gave a lukewarm endorsement, saying that "there are many ways in which the Space Station will enable crucially important space science investigations of the future."[6]

No National Security Requirements

NASA had hoped that significant requirements for a permanent facility in orbit would emerge from studies conducted by the national security community, making that community a supporter of developing a space station. That did not happen. Instead, the Intelligence Community and DOD turned out to be the primary opponents of the space station within the government.

On June 3, Charles Cook, a senior Air Force civilian official dealing with space issues, wrote to retired General Richard Stillwell, who as deputy undersecretary of defense for policy was the DOD representative on the Interagency Group (Space). He told Stilwell that "for over a year, the Air Force has been working with NASA examining the requirements for a manned space station." The service had reached the conclusion that "there are no currently identifiable mission requirements that could be uniquely satisfied by a manned space station. Further no requirements were found where a manned space station would provide a significant improvement over alternative methods of performing the given task." Cook noted that "the preferred DOD approach regarding a manned space activity is an evolutionary program ... DOD believes that man's role in space requires better understanding; this information can be obtained by using Space Shuttle capabilities and extending Space Shuttle flights." He added that "should a national decision be made to develop a space station based on other than military requirements, the DOD should continue to consider the space station for use by military programs." Cook also told Stillwell that "the United States must ensure that a viable Space Shuttle system is available to meet national launch needs, including those of national security. This Nation must be certain that adequate resources have been dedicated toward developing the Space Shuttle to fully utilize its designed capabilities and to ensure the operation of a Shuttle fleet that is adequate to meet national needs."[7]

Cook's memo encapsulated what was to be the DOD's position in following months—that there was no national security requirement that could justify developing a space station and that the human and financial resources potentially devoted to such development could be better applied to ensuring that the space shuttle had all the capabilities and operational characteristics needed for its national security missions. On June 20, Stillwell wrote to Bud McFarlane at the National Security Council. Most of his letter was taken *verbatim* from Cook's memorandum. Stillwell informed McFarlane that his letter represented "the current DOD statement on space station requirements for use in responding to NSSD 5-83."

International Participation[8]

From the start of its planning for a space station, international involvement was part of NASA's thinking. A bias toward international cooperation in its activities had been part of the NASA culture since the organization's inception. The space shuttle program had a heritage of such participation, with the European Space Agency (ESA) contributing the Spacelab laboratory and associated facilities and Canada, the remote manipulator arm.

When NASA convened the initial workshop on space station planning in November 1981, international involvement was a prominent agenda item, and the report of the workshop noted that "there appears to be substantial foreign interest in NASA's future plans for its manned space activities." However, "the subject of potential international participation in a U.S. space station program must be approached carefully and proceed under clear assumptions and guidelines. A fundamental ground rule should be that planning for a space station will be conducted as if the entire project is to be developed as a wholly U.S. effort. Planning should proceed, however, on the basis that it does not foreclose international cooperation."

As a follow on to the November 1981 meeting, Ken Pedersen, the head of NASA's Office of International Affairs, convened a January 1982 gathering of potential international partner space agencies to give them an early look at the agency's space station planning. Those partners during the preceding two decades had been critical of NASA for deciding by itself on the objectives and design of projects, and only then inviting foreign involvement on terms largely dictated by NASA. Pedersen's major point at the January meeting was that there would be a shift in NASA's approach; potential partner agencies would be invited to become involved at a very early stage in program definition, so that their inputs could help influence NASA's choices and they could understand options for their participation from the start.

The response to Pedersen's initiative was positive. At the end of a June 1982 meeting between Beggs and ESA head Eric Quistgaard, ESA announced that in parallel to NASA's study activities it would study the possibility of participating in the space station program. This announcement stimulated a Japanese decision that it too would examine such possibilities. Japan had missed the post-Apollo opportunity to participate in the space shuttle program, and did not want to repeat that experience. At the start of 1983, Canada initiated its own study of ways of participating.

Technology Transfer an Issue

However much international participation was attractive to NASA, by mid-1983 it had become clear that the potential for international involvement was not a strong selling point to other government agencies. From the start, Pedersen, Finarelli, and others had recognized that the possibility of technology transfer associated with such involvement would be of concern to the national security community and especially to Reagan administration appointees at DOD. However, they were also surprised to discover that the individuals within the Department of State overseeing the foreign policy aspects of science and technology were concerned about technology transfer and not enthusiastic about the potential of international cooperation in the space station program to serve broader foreign policy objectives. Recognizing that strong advocacy of international participation could jeopardize getting White House approval to go ahead with the station, NASA decided not to emphasize the international potentials of the program in the SIG (Space) process.

Even so, NASA publicly continued to discuss international participation as a rationale for station approval. NASA asked the American Institute of Aeronautics and Astronautics to convene a workshop in July 1983 to provide an overview of space station planning; the workshop was part of NASA's ongoing campaign to rally U.S. aerospace community support for a station. In the workshop's keynote address, Beggs noted that "the space station uniquely lends itself to international cooperation. If we can attract that international cooperation, then other nations will cooperating with us in the resources they spend, rather than competing with us."[9] Beggs did not specify what competitive space projects international cooperation would preempt, but there were worries that

Europe might develop some type of space shuttle as a means of acquiring its own means of carrying crew to orbit, thereby threatening the U.S. leadership position in human spaceflight. Japan was seen as a challenger to the U.S. position as the world's leading economic power based on its advanced technology, including in the space sector. By making the United States the leader in a large-scale undertaking that would require other countries to spend a significant share of their space budget on a U.S.-led project, the United States, Beggs was suggesting, could retain its overall leadership position in space. NASA had decided that the best way to convince Ronald Reagan to approve space station development was to cast it as essential to such leadership, and Beggs's remarks were calculated to emphasize that point.

Which Station to Propose?

The NASA Space Station Task Force during its first year of operation had assiduously avoided identifying a specific design for the space station, concentrating instead on identifying the missions a station could perform. But as the time for a presidential decision on the station neared, it was time to make that identification. Task force leader Hodge recognized that he needed "architectural details and he needed them fast. Without some preliminary architectural details, he could not calculate the cost of the space station. Without knowing the cost of the space station, he could not tell members of the White House staff and SIG (Space) what sort of program the money would buy."[10]

To undertake this narrowing of hardware options, the task force in April 1983 created a Concept Development Group, headed by an engineer from Marshall Space Flight Center named Luther Powell. At this point, Beggs and Hodge in their various briefings and other interactions had been talking about a space station that could sustain a permanent human presence in space, perform over 30 of the missions that had been identified by the task force and its contractors, and cost between \$4–6 billion. Beggs also wanted a station that could be developed incrementally, "by the yard." Powell and his group examined task force work to date, and soon focused on a space station design that consisted of a five-module core, including a habitat, two laboratory modules, a utility module, and a docking module for space shuttle logistic missions, all attached to a central truss; the facility could accommodate a crew of six to eight people. The core would orbit at a 28.5-degree inclination to the equator. It would be accompanied in that orbit by a platform that could provide accommodations for experiments not requiring frequent human intervention; it would be serviced by occasional astronaut visits from the core facility. Another platform would be in a polar orbit and not be accessible for astronaut visits, but still be considered part of the station complex. The initial operating capability of such a station design would be 1991. The cost of such a station, Powell estimated, would be between \$7–12 billion. The \$12 billion high-end estimate, Beggs and his associates judged, was not politically viable. Thus, Hodge mandated that the cost range should be indicated as between \$7–9.5 billion.[11]

An $8 Billion Space Station

There was increasing pressure on NASA during 1983 to provide an estimate of space station cost. The OMB in its preparation for decisions on NASA's FY 1985 budget was interested in the immediate and future budget impacts of a new start on the space station, should President Reagan, in spite of OMB's objections, approve such a step. In addition, the SIG (Space) report being prepared for the president would certainly have to discuss station cost.

McCurdy comments that "as the space station initiative moved toward the President's desk, the desire of White House officials for a good look at a specific number began to outweigh the efforts of NASA officials to avoid one … NASA executives began to lean toward an $8-billion mark." To James Beggs, the $8 billion number reflected "the mathematics of budgetary politics." His calculation was that a space station program costing $8 billion over the remaining years of the Reagan administration could fit into an overall NASA budget adjusted for inflation plus a "little bit of growth," and he thought he could convince Ronald Reagan that such growth was acceptable.[12]

Peggy Finarelli provided a breakdown of the $8 billion figure in a September 8, 1983, memorandum to OMB. Between FY 1985, the budget year then under discussion, and FY 1987, NASA would spend $630 million, defining in more detail the station it wanted to build. Between FY 1987 and FY 1991, it would spend $7.370 billion on station development. Finarelli provided the estimated cost of each element of the design NASA was proposing.[13]

The $8 billion cost was more a "political" figure than a careful cost estimate. There was recognition that the cost of the station would grow as additional capability was added. What NASA was proposing was a station that would cost $8 billion while Ronald Reagan was president, not $8 billion overall. The cost estimate was based on a set of assumptions and guidelines that were controversial even within NASA. Typical of how NASA dealt with activities coming out of different budget accounts, the cost of space shuttle flights to assemble the station and of early space station experiments were not included, nor were the NASA institutional costs associated with the program. The program "reserve" to allow for unanticipated cost growth was set at a lower percentage of program cost than normal. But once it had provided the $8 billion figure both in the SIG (Space) space station report and to OMB, NASA's leaders had to consistently state in the final stages leading to a presidential decision on the station that they were seeking approval for an $8 billion program.

SIG (Space) Space Station Report

A draft of the space station report prepared by Rye's group was ready by August 4. The report ran to 22 pages of single-spaced text. It identified three options for presidential choice. As Rye, Finarelli, and their associates prepared the report, they had received important policy guidance from NASA Administrator Beggs. He told them to "forget the politics [of attempting to gain national security community support] ... and develop our arguments for the space station with no military justification whatsoever." NASA would ask Ronald Reagan to approve "a completely civil station."[14]

The three options for presidential choice were:

1. Option 1—a presidential decision to "commit now to a permanently manned Space Station";
2. Option 2—a presidential decision to "commit now to an evolutionary development of expanded Space Transportation System capabilities and unmanned platforms"; this was the option that DOD seemed willing to support if the president did not decide to defer commitment.
3. Option 3—a presidential decision to "defer commitment to either Option 1 or Option 2 pending additional definition of requirements, costs and risks."

The report contained a two-page discussion of the Soviet space program based on a recently completed National Intelligence Estimate (NIE). As discussed below, President Reagan had been briefed on that assessment on July 21.[15] The estimate indicated that "the principal goals of the Soviet space program are to provide global support to Soviet military forces, enhance the worldwide influence and prestige of the Soviet Union, deny enemies the use of space in wartime, and contribute to the Soviet economy." The SIG (Space) report noted that "manned space activities are receiving increased interest in the Soviet space program. A comprehensive manned space program is underway," with the Soviet leadership announcing "the national objective of establishing a continually manned Space Station, which the Intelligence Community thinks will be achieved by about 1986." The report suggested that "a visible, highly-publicized, continuously manned Soviet Space Station will receive frequent worldwide attention." The report also briefly discussed the space programs of Europe, Canada, and Japan, commenting that "foreign space programs reflect awareness of growing and technically challenging international markets for space products and services" and that "other nations are beginning to share in the international space market that has been dominated by the U.S."

The final 12 pages of the report described in some detail the characteristics of and pro and con views on the two hardware options, Option 1 and Option 2, and devoted two pages to the pros and cons of Option 3, deferring a decision. With respect to Option 1, committing to space station development, the report noted that the station concept NASA was advocating would "require approximately $8 billion with FY 1985 costs of $225 million." Advocates of this option, the report noted, argue that "United States space leadership is being actively challenged by the Soviets." Not only would going ahead with a U.S. space station respond to that challenge; it would also enable the United States "to compete in 'races' yet to be defined, such as a manned lunar base or a manned Mars mission." In addition, "a Space Station is necessary to maintain real and perceived leadership in space," since a decision to move forward on the station "would be a reaffirmation to the world of technological superiority and space leadership."

Opponents of Option 1, said the report, "argue that a commitment now to a permanently manned civil Space Station would be premature and a serious error." Not only were NASA's cost estimates understated, but diversion of resources to a space station would "threaten NASA's ability to satisfy the national priority of making the STS fully operational and cost effective." Also, "the presidential commitment to an accelerated program for a manned Space Station could define a 'race' in which the United States would be widely perceived as having lost," since "the Soviets have a Space Station already on-orbit and plan to have a more advanced station in place well before any of the NASA program options."

The report described Option 2 as continuing "current Shuttle and Spacelab efforts," supplemented by "additional evolutionary development," such as a "Power Extension Package" that "would extend the Shuttle's on-orbit stay-time from seven to approximately twenty days." The option also envisioned developing "two man-tended platforms—one in a low inclination orbit and one in a polar orbit." The costs of pursuing this option were estimated at $1.3 billion with FY 1985 funding of $190 million required.

Advocates of this evolutionary approach argued that building upon space shuttle capabilities was a way "to understand the role of man in space and the value of permanent space facilities." Using shuttle and Spacelab as the basis for orbital operations would allow the United States to "explore the advantages of man in space for achieving civil and national security objectives"; it would also be possible to "evaluate man's role in satellite inspection, servicing and repair."

In opposing Option 2, the report suggested that such an approach did "not represent an effective use of USG [U.S. government] resources," since the "vast majority of existing and projected civil and commercial space needs are best satisfied by Option 1." If Option 2 were pursued, the report suggested, it could best be justified on national security requirements, implying that its funding would have to come from the national security budget. With a limit of some 20 days in orbit, it would "not improve on the U.S. understanding of man's permanent role in space." Also, Option 2, "because it will tie up Shuttles on-orbit for extended periods of time, may impact the ability of the Space Transportation System to meet its launch requirements." Thus, "a potential hidden cost in this option is the purchase of at least one additional Orbiter at approximately $1.5 billion."

Deferring a space station decision, Option 3, would require "that a Space Station decision be put in context with other major new space-related activities such as defense against ballistic missiles and future space system survivability enhancements." This comment was one of the few instances in the space station decision process in which the potential fiscal and manpower needs of the just-announced Strategic Defense Initiative were explicitly identified as competing with a positive decision on the space station. The report noted that "a premature decision to make a major commitment to a civil Space Station may preclude the ability to fund … potentially higher priority initiatives."

Reacting to these arguments, the report argued that NASA was "ready to begin the development of a Space Station." The space agency was "ready to respond to the President. Further study of this issue is not required. The nation has been in space for 25 years and it is this President's opportunity to take the next major step forward."[16]

As they prepared for the August 10 meeting to decide how, and indeed whether, to proceed, SIG (Space) members now had a summary report that highlighted the major disagreements among their representatives during the intensive interagency scrutiny of the space station proposal over the preceding four months. It was already clear that there were little grounds for agreement. If the United States was going to develop a space station, it would be Ronald Reagan himself who would make that decision.

Keyworth and Weinberger Make Their Views Known

During the preceding months, James Beggs had publicly communicated optimism that approval of the space station was on the not-too-distant horizon. He told the July AIAA symposium in July that he expected White House approval of the station initiative "within the next 6–12 months."[17] Beggs was chastised by the White House for this comment, since it appeared to presuppose a presidential decision that had not yet been made.

Questioning of Beggs's optimistic forecast came from two highly placed individuals. One was presidential science adviser George "Jay" Keyworth; the other was Secretary of Defense Caspar "Cap" Weinberger. While Keyworth was situated within the Executive Office of the President and thus able to make his views directly known to Ronald Reagan and his top advisers, he was not a Reagan insider and had somewhat limited influence; he thus chose a public path to communicate his perspective. By contrast, Weinberger was, according to Hans Mark, "the most brilliant member of President Reagan's cabinet." Weinberger was "experienced, articulate, and infinitely knowledgeable," and enjoyed Reagan's "complete confidence." In NASA's quest to gain presidential approval, he would be "a formidable opponent."[18]

Keyworth's "Curve"

Keyworth in February 1983 had drawn National Security Council ire when he had said in a congressional testimony that approving a space station would represent "an unfortunate step backwards." A few months later, Keyworth modified this view. In a June 27 speech to a technical conference, Keyworth started his remarks by saying, "I'm going to throw you a curve this morning." Instead of addressing the technical focus of the conference, aerospace propulsion, Keyworth addressed the past, present, and future of the U.S. space program, suggesting that "I sense that the United States is near an important decision point in determining the future direction of our space efforts." He wanted "to challenge the aerospace community to do some bold thinking about the future." Keyworth saw the need for "revolutionary new ventures in space exploration ... the kinds of initiatives that open new frontiers, develop new technologies, or recapture the sense of national unity that Apollo did." He asked: "How can we inject a new sense of vision into America's space exploration?"

Keyworth questioned the return on investment from a space station focused primarily on activities in low-earth orbit. He noted that "most advocates of a space station readily acknowledge that it is, in truth, only an intermediate step in a more ambitious long-range goal of exploring the nearby solar system. So if we're going to be asked to make an installment payment in the form of a space station, why don't we let the American people share the grand vision of the future of space—whether ... it's a manned lunar station, or even manned exploration of Mars?"[19]

Keyworth's "sharp new tack," with its expansive view of the future in space, was the result of "personal meditations on the national spirit and the space program ... I think the country would take a major thrust in space very seriously." In a 2017 interview Keyworth suggested that he was "always very enthusiastic about manned missions to Mars, with perhaps a lunar base along the way," but "was, am, and always will be violently opposed to a space station" as an end in itself. Keyworth's thinking may also have been influenced by his interactions with his mentor Edward Teller, the controversial nuclear scientist who had recommended Keyworth for the science adviser position. In a July 19, 1983, letter, Teller told Keyworth that he had given "some more thought" to their prior discussions of the "space lab"—Teller's name for a space station. Teller commented that as he understood the situation, multiple reasons were being put forward for developing a space station, including "preparation for a moon lab, preparations for landing a man on Mars, scientific and industrial research, as well as reasons of defense." To Teller, "the valid ones among these are preparation for a lunar colony, defense reasons insofar as they deal with ... defending our eyes in the sky, and those scientific and industrial researches which are connected with the absence of gravity. The latter, while important, may be overemphasized. The first, preparation for a lunar colony is not emphasized enough. I hope you ... put your effort on appropriate changes in emphasis."[20]

From NASA's perspective, linking space station approval directly to a plan for "exploring the nearby solar system" would be falling into a trap, since those opposed to setting a new exploration goal would also oppose the station as part of the planning to achieve that goal. NASA did not take up Keyworth's challenge; no missions beyond the year 2000, such as developing a lunar outpost or sending astronauts to Mars, were included in NASA's space station justifications. McCurdy observes that "given their view of the federal budget, NASA officials decided to stick with the incremental approach and press for a low-budget station," not one capable of supporting deep space missions.[21]

Weinberger Tells Reagan that Approving Space Station Would Be a Mistake

On July 21, 1983, President Reagan met with Secretary of Defense Weinberger, NASA Administrator Beggs, National Security Adviser Clark, and Rye to be briefed on a recently completed NIE on the Soviet space program. Such estimates represented the best judgments of all elements of the Intelligence Community on a particular topic. The NIE described "current Soviet space capabilities," identified "elements of the space program in various stages of development," and estimated "how these will affect future Soviet capabilities in space through the 1980s and into the 1990s." The NIE suggested that the Soviet space budget had doubled in the last three years and was the equivalent of $20 billion, compared to U.S. government space expenditures of $13 billion. It noted that "in the 1980s continuously manned Soviet space stations will provide the opportunity to gain international recognition as a leader in space." It noted that the Soviet Union would likely deploy a successor to the *Salyut* family of space stations by 1986, and that this "modular Soviet station, designed for crews of six to 12 persons, will probably be followed by a large space station capable of accommodating 12 to 20."[22]

On another front, President Reagan on August 3 lunched with a number of leaders from the private sector space community and heard their optimistic views on the economic potentials of space activity and the important role of a space station in enabling the realization of that potential. (This lunch is discussed in Chap. 13.) NASA hoped to use the argument that there were multibillion dollar industries that might emerge on the basis of research carried out aboard the space station as a major justification for station approval. Although the space station issue was not an explicit focus of the lunch, one press report on the gathering was headlined "Reagan Briefed on Space Station" and subtitled "corporate executives inform the President it is a critical element needed to stimulate commercialization of space." Reagan was quoted as saying to the luncheon attendees, "I want a space station, too. I have wanted one for a long time."[23]

Hans Mark in his book on the space station decision suggests that following the July 21 briefing and this lunch, Weinberger, fearing that the president was being swayed by station advocates before he heard the arguments of station skeptics, "escalated" the decision process by requesting another meeting with Reagan, nominally to provide the president with a "briefing from the Department of Defense which compares US and Soviet space capabilities," but also "to argue against going ahead with the space station." This seems not to be correct; Mark was being a bit paranoid. In fact, the second briefing had been requested already on July 26, as "a follow-on presentation to the President … required to provide the full military implications of Soviet space activities and a comparison of U.S. and Soviet space capabilities."[24]

The briefing was scheduled for August 8, two days before SIG (Space) was to meet to consider Gil Rye's space station report. Attending the 45-minute meeting were Weinberger, director of the CIA William Casey, Director of the Defense Intelligence Agency General James Williams, Chief of Naval Operations Admiral James Watkins, National Security Adviser Clark, and top Reagan advisers James Baker and Ed Meese. NASA's Beggs was out of town, so Mark attended the gathering; he remarks that given the high-level status of others in the White House situation room, he was "outgunned."

In his diary notes for August 8, Mark notes that, as Weinberger introduced the day's briefings, one by the Defense Intelligence Agency on the Soviet space program and one by Director of the NRO Edward "Pete" Aldridge on comparable U.S. national security capabilities, he made clear to President Reagan his strong opposition to going ahead with a space station. Weinberger's "essential argument" was "that it costs too much and will take resources away from badly needed national security programs"; the issue was "resources, not requirements." One element of Weinberger's opposition was the growing concern that the space shuttle would not be able to meet all national security needs, particularly in terms of weightlifting capability from its west coast launch site. Weinberger suggested that NASA focus its attention, best people, and funding on making the shuttle fully capable rather than on beginning a major new project. He also "made a great point of saying how important the President's strategic defense ("Star Wars") initiative was and that anything we do in space has to be related to that."

During the meeting, Reagan was "alert and attentive," and at the end of the meeting "asked a few questions that were perceptive and to the point." Reagan sensed that there was tension between NASA and the national security community. As was his style, Reagan broke the tension with an anecdote, saying that "he never thought he would see in reality what he had seen on the Hollywood stage forty years ago on the set next to the one he was using, where Buster Crabbe was playing the part of Flash Gordon."

Mark concluded his assessment of the situation by noting that "Weinberger is solidly against us because he looks upon the Space Station as a threat to the funding for his program. This is the real meaning of his statement that the Space Station would cost $18B." In Weinberger's view, it was possible to "spend the $18B better for national security." The weakness in Weinberger's case, thought Mark, was linked to an assessment in the Defense Intelligence Agency briefing with respect to Soviet space station plans. Mark thought that the United States could not "ignore the political consequences of having 20–100 Russians in orbit when we have no one." Writing in his diary after what he characterized as an "anti-Space Station briefing," Mark suggested that "we are in real trouble."[25]

SIG (Space) Deadlocked

As scheduled, SIG (Space) met on August 10 to consider the space station report Rye, Finarelli, and their associates had prepared. Only NASA, the Department of Commerce, and the Arms Control and Disarmament Agency supported Option 1, "commit now to a permanently manned space station." Cabinet secretary Craig Fuller, not formally a SIG (Space) member, who was by this point also a strong supporter of a space station, also spoke out in favor of Option 1. The representatives of the national security community from the DOD, the Joint Chiefs of Staff, and the Intelligence Community all favored Option 3, to defer a commitment, as did the State Department and the observers from OMB and the Office of Science and Technology Policy. There was no support for Option 2, a commitment to an upgraded, longer-duration space shuttle.

Rye provided a brief summary of the previous day's SIG (Space) meeting to President Reagan during the president's daily national security briefing on August 11. This briefing likely indicated to Reagan that the space station decision would come to him with divided opinions among his agency heads. Rye later suggested that at that point, "without making any verbal commitments, Reagan gave what Rye took to be a clear signal of interest in the space station initiative." Rye later wrote that it was following this briefing that "I knew for certain" that Reagan would approve the space station, once that issue was placed before him.[26]

As the August 4 draft of Rye's report for discussion at the SIG (Space) meeting was distributed, the National Security Council also requested written comments on the three options for presidential choice. The plan was to send those responses to the president as part of the space station decision package, along with Rye's report. DOD response was signed by Cap Weinberger rather than his deputy Paul Thayer, who was the DOD member of SIG (Space), once again emphasizing Weinberger's personal opposition to space station approval. The DOD "supports Option 3," he wrote, given the lack of identifiable national security requirements. He noted that "any proposed commitment to a permanently manned civil space station should be made in the light of potentially competing, defense-related space initiatives." Weinberger acknowledged that "emphasizing manned space flight is one possible way to maintain a perception of U.S. civil leadership in space." However, "a decision document for Presidential consideration should identify possible alternatives" for achieving that objective. Thus, "competitive alternatives should be reviewed before the permanently manned space station is reconsidered."[27]

The Intelligence Community response came over the signature of SIG (Space) member John McMahon, deputy director of Central Intelligence. The response echoed most of the themes in the DOD memo, noting that "the approximately 8 to 9 billion dollar cost estimate of the initial increment of the Space Station ... understates the total cost." McMahon suggested that the real cost of the station "could have an adverse impact on resources for the intelligence space program." Deferring commitment to a station would allow continued study of "potential national [i.e., intelligence] uses for it." McMahon concluded his response by noting that "a Space Station decision should be made only within the context of overall national needs and its direct contributions to national objectives. A premature decision to make a major commitment to a civil Space Station may preclude the ability to begin and fund potentially higher priority initiatives."[28]

The Intelligence Community also made its opposition known in a face-to-face meeting with Judge Clark on September 7. The meeting was attended not only by McMahon but also by Director of Central Intelligence Casey, who "pressed Clark to keep the President from committing himself" to the space station "without knowing the needs of other important programs." McMahon also suggested that "if the nation is to go with a manned space station, it should be done primarily for national security reasons, not for the benefit of a private pharmaceutical company," adding that "DOD, not NASA, should be the key actor." The "private pharmaceutical company" reference by McMahon was apparently linked to the planned tests aboard the space shuttle by Johnson & Johnson and McDonnell Douglas of a method of separating different substances in a low-gravity environment, a technique thought to have significant benefits in producing high-value drugs.[29]

The Department of State response was provided by William Schneider, Undersecretary for Security Assistance and Science and Technology. Schneider said that "the Department of State strongly supports Option 3." In addition to the points made in the DOD and Intelligence Community responses, Schneider insisted that "a study must be undertaken to form a basis for a decision on whether the U.S. should invite foreign participation in a manned space station and, if so, from whom and to what extent." Given the need for this and other studies, he suggested, "submission of a Presidential decision document on space station should be deferred to the FY 86 budget cycle time frame"; this would mean a year's delay in the station decision.

The response from the Office of Science and Technology Policy reflected the perspective that Keyworth had first enunciated in his June speech, saying, "capturing the vision, intent, and spirit of the President's support for space programs, requires a bold new space initiative. A U.S. lunar station or manned mission to Mars may be the kind of initiative that demonstrates our technological and scientific superiority … The proposal for a manned space station should be developed in this context." The office thus also supported Option 3 in the Rye report.[30]

There was some support for going ahead with the station. Secretary of Commerce Malcolm Baldridge signed the department's response, supporting Option 1 by saying that "this is the appropriate time for the United States to take the next major step in advancing the technology and utilization of space systems. There has been no major official guidance in this area since that of President Nixon. No initiative will reflect more favorably upon the United States and this Administration than the development of a permanently manned station in space." In fact, suggested Baldridge, his "chief concern" was that the options discussed at the August 10 SIG (Space) meeting were "too timid." If the United States was to "obtain the industrial stimulation and receive the benefits I believe to be possible," it should "move more aggressively." The response from the Arms Control and Disarmament Agency also supported Option 1.[31]

And, of course, NASA supported that option. In his written response, James Beggs made a strong argument for a positive decision on a space station, concluding by stating that "NASA is ready to respond. Further study of this issue is not required. The nation has been in space for 25 years and it is this President's opportunity to take the next major step forward."[32]

After the August 10 SIG (Space) meeting and agency responses to Rye's report were received, Rye prepared two alternate National Security Decision Directives for presidential decision. One directive reflected a presidential decision to commit to space station development; the alternate directed deferral of a station decision pending further study. Rye also prepared a cover memorandum to accompany the two directives; that memo told the president that "agency views indicate a lack of consensus on either of the two options" being offered. The memo noted that supporters of the space station "argue that a Space Station would counterbalance the visible Soviet challenge to U.S. space leadership, perform valuable civil functions, stimulate the commercialization of space, offset claims about the militarization of space, stimulate international cooperation, and serve as a centerpiece for a major revitalization of America high technology industry." Those suggesting deferral of a space station commitment "argue that the Space Station lacks sufficient analysis; is not based on either current national security requirements or adequately substantiated civil requirements; overstates the Soviet manned space challenge to U.S. civil leadership in space; involves the risk of major cost growth; and would divert significant resources from other potentially higher priority space programs."

Rye forwarded the alternate directives, the cover memo, and the written responses from SIG (Space) member agencies to Clark, noting his personal belief that "a Space Station is in the best overall interest of the Nation" and that "the President's commitment to the vigor of our manned space program will vividly demonstrate that he is a forward-looking leader who is determined to carry this Nation into the next decade as pre-eminent in space and technology."[33] Rye later suggested that by this time it was clear that "we would have to play out the rest of the process to bring the bureaucracy along" before presenting the space station proposal to the president for decision. Thus, Clark did not forward the alternate decision directives and accompanying material to the president. There was no space station decision made at this point. It would take a bit more than three additional months to make that choice.[34]

Notes

1. Interview with Jeffrey Struthers, in 1983 OMB Branch Chief for Space and Science, May 29, 2016.
2. Memorandum from Chairman, SIG (Space) Space Station Working Group to SIG (Space) Working Group Members, "SIG Activity for the Balance of This Year," April 19, 1983, Folder 12905, NHRC. NASA, "A NASA Capabilities Evaluation Document," Preliminary Draft, June 16, 1983, CIA-RDP92B00181 R001901730063-1, CREST.

3. Memorandum from Gil Rye to William Clark, "Schedule for the National Space Strategy," June 1, 1983, NSSD 13-82 File, Executive Secretariat, National Security Council; memorandum from William Clark to Members, Senior Interagency Group for Space, "Schedule for National Space Strategy," Box 30, Subject Files, Executive Secretariat, National Security Council, June 6, 1983, RRL.
4. Memorandum from J.D. Hill to the deputy director, Intelligence Community Staff, "NSSD 5-83 Space Station Report," July 25, 1983, CIA-RDP85 M00364R000400550062-3, CREST.
5. Hans Mark, The *Space Station: A Personal Journey* (Durham, NC: Duke University Press, 1987), 165, 172–173.
6. The Space Science Board's views are quoted in Howard McCurdy, *The Space Station Decision: Incremental Politics and Technological Choice* (Baltimore, MD: The Johns Hopkins University Press, 1990), 158.
7. Memorandum from Charles Cook to General Stillwell, "DOD Space Station Requirements," June 3, 1983, Box 1, Papers of Terence Finn, NARA.
8. Much is this section is drawn from John M. Logsdon, "Together in Orbit: The Origins of International Participation in the Space Station," NASA, Monographs in Aerospace History #11, November 1998, available at https://history.nasa.gov/monograph11.pdf. Unless otherwise noted, all quoted material in this section is drawn from this monograph.
9. James Beggs, "Keynote Address," July 18, 1983 in Mireille Gerard and Pamela Edwards, eds., *Space Station: Policy, Planning and Utilization* (New York: AIAA, 1983), 4.
10. McCurdy, *The Space Station Decision*, 152–153.
11. Ibid., 155.
12. Ibid., 171. Much of this section is drawn from McCurdy's study.
13. Memorandum from Peggy Finarelli to Bart Borrasca, "Space Station Funding," September 8, 1983, Box 1, Papers of Terence Finn, NARA.
14. McCurdy, *The Space Station Decision*, 169.
15. National Intelligence Estimate 11-1-83, "The Soviet Space Program," July 19, 1983, available at https://catalog.archives.gov/id/7327262.
16. National Security Council, "Space Station Report (DRAFT)," August 4, 1983, Box 5, Papers of Edwin Meese, RRL. Quoted passages are on 1, 3–5, 7, 10–12, 14, and 16–22.
17. Craig Covault, "NASA Chief Foresees Space Station Approval," *AWST*, July 25, 1983, 18.
18. Mark, *The Space Station*, 176.
19. Proposed Remarks of Dr. George A Keyworth, II to the 19th Joint Propulsion Conference, Seattle, Washington, June 27, 1983, Folder 1187, NHRC.
20. Mitchell Waldrop, "Keyworth Calls for Bold Push in Space," *Science*, July 8, 1983. Interview with Jay Keyworth, January 25, 2017. Letter from Edward Teller to George Keyworth, July 19, 1983, RAC 13, Papers of George A. Keyworth, RRL.
21. McCurdy, *The Space Station Decision*, 160–162.
22. National Intelligence Estimate 11-1-83, "The Soviet Space Program," July 19, 1983, 1, 10, 19, 21, available at https://catalog.archives.gov/id/7327262.
23. Craig Covault, "Reagan Briefed on Space Station," *AWST*, August 8, 1983, 16.

24. Mark, *The Space Station*, 173. Memorandum from Gil Rye to John Poindexter, "Briefing on Soviet and U.S. Space Capabilities," July 26, 1983 and memorandum from William Clark to Caspar Weinberger, "Space Briefing for the President," July 27, 1983, both in Box 30, Subject Files, Executive Secretariat, National Security Council, RRL.
25. Diary notes of Hans Mark, with attached "White House Briefing" summary, August 8, 1983. I am grateful to Dr. Mark for providing me with a copy of this material. See also McCurdy, *The Space Station Decision*, 166 for another description of the August 8 meeting.
26. McCurdy, *The Space Station Decision*, 181. Letter from Gil Rye to Hans Mark commenting on a draft of Mark's *The Space Station* book, October 25, 1984, cited with Rye's permission.
27. Memorandum from "Cap" to the Assistant to the President for National Security Affairs, "Defense Position on the SIG (Space) Station Issue," August 19, 1983 Files of Valerie Neal.
28. Memorandum from John McMahon to William Clark, "DCI Position of the SIG (Space) Space Station Issue," August 18, 1983, Box 5, Papers of Edwin Meese, RRL.
29. Name of author redacted, Memorandum for the Record, "Meeting with Judge Clark, 7 September 1983," September 8, 1983, CIA-RDP85M 00364R000601010070-0, CREST.
30. Memorandum from John Marcum to the Assistant to the President for National Security Affairs, "Comments on the Space Station Report," August 19, 1983, Box 5, Papers of Edwin Meese, RRL.
31. Letter from Malcolm Baldridge to William Clark, August 23, 1983, NHRC.
32. Letter from James Beggs to William Clark, August 18, 1983, NHRC.
33. Memorandum from Gil Rye to William Clark, "Space Station," August 25, 1983, with various attachments, Box 5, Papers of Edwin Meese, RRL.
34. Rye letter to Hans Mark, October 25, 1984.

CHAPTER 9

The Space Station Decision

With the majority of Senior Interagency Group for Space [SIG (Space)] agencies being opposed to going ahead with a space station, according to Hans Mark by the end of August he and Jim Beggs were "quite pessimistic" about getting a positive presidential decision. That pessimism increased in the following weeks. Mark suggests that "the last weeks of September … were the low point" in the space agency's hope to gain space station approval, as former National Aeronautics and Space Administration (NASA) administrator James Fletcher "came by for a visit and told us that we should think about backing away from pushing the space station project." Mark suggests that "the actual nadir of our fortunes" came at an October 4, 1983, staff meeting at which Beggs "wondered aloud whether Fletcher might not be right … We really were in the position of fighting an uphill battle, and there does come a time to cut losses." This may have been Beggs musing out loud, not a considered position. At any rate, reports Mark, "while we considered backing away, we finally decided to stick to our guns."[1]

Beggs was somewhat more optimistic than Mark; through his occasional interactions with President Reagan he had become convinced that Reagan wanted to move forward with a major space initiative in the civil sector, among other things as a counterbalance to the Strategic Defense Initiative (SDI) and the accompanying accusation that his administration was "militarizing space." Beggs continued between August and December 1983 to seek support for the station initiative from key individuals close to the president. For example, NASA asked Alan Shepard, the Project Mercury and Project Apollo astronaut who had been the first American in space and who by 1983 was a very successful businessman in Houston, to talk to his friend James Baker, White House chief of staff, about the space station. In his approach to managing White House activities, Baker seldom took an active role in advocating or opposing a particular policy choice, but behind the scenes he had become one of the president's most trusted policy advisers. Thus, his support for the station

J. M. Logsdon, *Ronald Reagan and the Space Frontier*,
Palgrave Studies in the History of Science and Technology,
https://doi.org/10.1007/978-3-319-98962-4_9

initiative could be extremely important. Shepard reported back that Baker was interested in the station proposal and wanted additional information about it. In an August 24 letter, Beggs provided Baker "with a number of papers which will familiarize you with our argumentation, which present Mac Baldridge's strong views on the program, and which provide some information on the history and the current state-of-play of this endeavor." Beggs even sent Baker "a statement that might be used by the President in announcing a decision to go ahead with the Space Station."[2]

There was a report in early October that the White House was becoming "increasingly receptive toward approval of a NASA space station because of the political benefit such a decision could bring just prior to an election year." Reagan's possible 1984 opponent was former astronaut John Glenn (D-OH), who had endorsed the station, and Reagan's political advisers did not want the president's lack of support for the civilian space program to become an election issue, should Glenn be the Democrat candidate. The same report noted: "Another factor is Reagan's enthusiasm for the U.S. space program. Corporate officials who have discussed space with U.S. presidents since Dwight D. Eisenhower say Reagan is more openly enthusiastic about the program than any other, including John F. Kennedy."[3]

Another development acting in NASA's favor was that the space shuttle, at least in the public's eye, had been a spectacular success in its initial two years of flights. This success had demonstrated to the White House and Congress both the political benefits of highly visible human spaceflight activity and NASA's capability to bring a challenging development program to a successful outcome.

The question facing NASA and its White House supporters such as Gil Rye and Craig Fuller in late 1983 was how, given the national security community's opposition to the undertaking and its majority position within the SIG (Space), to get the space station decision before the president in a positive context. It was unlikely that Reagan would render a decision before hearing what his top advisers thought about the station. In addition, NASA first had to navigate its way through some confusing signals from the White House.

Should NASA Stay the Course?

In the final months of 1983, there were two developments that muddied the situation regarding the fate of NASA's space station proposal. They raised the issue of whether, on one hand, the space agency should accept a less ambitious compromise approach to its future programs or, on the other hand, abandon its current focus on space station advocacy and seek presidential approval for a longer-term space goal that would require the station for its achievement.

OMB Compromise Proposed

Although in its discussions with NASA with respect to the space station Office of Management and Budget (OMB) had continued throughout 1983 to oppose the project, the budget office internally was preparing itself to deal with presidential approval of the undertaking, developing budget forecasts for Fiscal Year (FY) 1985 and beyond that included an allocation for the space station. The OMB permanent staff was getting "vibes" from its political leadership that Ronald Reagan was likely to approve some version of NASA's space station as a way of showcasing U.S. technological prowess and of maintaining space leadership. At some point early in the year Fred Khedouri, a political appointee as OMB Associate Director for Energy and Natural Resources, told the head of the OMB space and science branch, Jeff Struthers, "It's [the space station] going to happen." From OMB's perspective, the struggles during the SIG (Space) process were almost irrelevant. The actual space station decision, OMB believed, as it turned out correctly, would be made through the budget process, and OMB's assumption throughout 1983 was that the decision would ultimately turn out to be positive. Thus, OMB's goal was to limit the short-term budget of a positive station decision.[4]

Reflecting its organizational skepticism regarding the space station, early in the FY 1985 budget process OMB tried one maneuver to preempt a positive presidential decision to approve NASA's space station proposal. Meeting with NASA's Beggs on October 19, OMB's Khedouri offered to include in the NASA budget "a space platform that would initially be unmanned but that might be expanded later on to include a manned capability." Khedouri told Beggs that "we could afford to do something like this if the total runout cost could be kept to something like two billion dollars for the development of the platform." Beggs's response to Khedouri was "noncommittal"; Beggs judged that Khedouri "would not make such an offer unless he thought that there was a chance that NASA's proposal for a fully manned space station might succeed."[5]

Interpreting a Reagan Speech

NASA celebrated its 25th anniversary with an October 19 gala at the National Air and Space Museum. Ronald Reagan would attend, see a large-screen movie on early space shuttle flights, make some remarks, and cut an anniversary cake. Before the event NASA held out the slight hope that those remarks would include an indication of presidential support for the space station, but this was not to be. Ed Meese was quoted as saying, "NASA's birthday cake is not going to be in the shape of a space station."[6]

Reagan in his remarks did indicate how the space program fit into his broader concept of American exceptionalism, saying, "NASA's done so much to galvanize our spirit as a people, to reassure us of our greatness and of our potential." To NASA's dismay, much of his speech closely echoed the views that had been set forth by science adviser Keyworth several months earlier:

> We're not just concerned about the next logical step in space. We're planning an entire road, a "high road" if you will, that will provide us a vision of limitless hope and opportunity, that will spotlight the incredible potential waiting to be used for the betterment of humankind. On this 25th anniversary, I would challenge you at NASA and the rest of America's space community: Let us aim for goals that will

> carry us well into the next century. Let us demonstrate to friends and adversaries alike that America's mission in space will be a quest for mankind's highest aspiration—opportunity for individuals, cooperation among nations, and peace on Earth.
>
> The first 25 years of NASA opened a new era. Let us all rededicate ourselves today that NASA's next 25 years will ensure that this new chapter in history will be an American era.[7]

Reagan's speech, which included a rather dismissive characterization of the space station as only a step in a broader space effort, seemed to pose a threat to NASA's strategy of focusing its advocacy on the station and the shorter-term missions it would serve. McCurdy comments that the speech "placed NASA executives in the curious position of having to assume that Reagan was not speaking for himself." NASA soon discovered that the speech was the product of Keyworth's Office of Science and Technology Policy (OSTP). It was not surprising, then, that the speech reflected Keyworth's view of the need for an expansive space effort. What was not clear was whether it also represented Ronald Reagan's views.[8]

NASA decided not to interpret Reagan's remarks as a signal of the president's desire to postpone a space station decision while a more ambitious vision for the future in space was developed. Beggs was quoted as asking, "Is the country ready for another Apollo-type effort? I would like to think so, but I doubt it … We have to constrain those dreams to fit the kind of funding profile we are likely to get."[9] Thus, NASA's focus remained on getting a positive decision on the space station before the end of the year. The long process of getting the space station issue formally before the president was soon to reach its climax.

Finding the Path to the President

As the space station debate reached its final stage, there was a major change in the senior White House staff. National Security Adviser William Clark submitted his resignation on October 17, 1983; Clark had been engaged in a number of contentious battles with White House Chief of Staff James Baker and Secretary of State George Shultz, and had lost the confidence of senior adviser Michael Deaver and Ronald Reagan's wife Nancy, a behind-the-scenes powerful influence on the president. Reagan nominated Clark to replace James Watt as secretary of the interior, a position he assumed on November 18 after receiving Senate confirmation. Before Clark left the White House, there was a short-lived attempt to have Baker take Clark's job, with Deaver becoming chief of staff. Clark led the group within the White House that aborted this scheme.

Instead, Clark was replaced by his deputy, Robert "Bud" McFarlane, who had come from the State Department together with Clark in January 1982 and was a veteran of foreign policy and national security deliberations and actions. Unlike Clark, McFarlane was not a long-time Reagan associate and social friend, and his interaction with the president was on most days limited to the

15-minute national security briefing that was part of the Reagan morning routine. McFarlane would turn out to be a supporter of a strong space program. In his early weeks in office, as the station decision was coming to a head, McFarlane was dealing with the aftermath of the Soviet Union shooting down a Korean airliner on September 1, killing all 269 people aboard, and then an October 23 attack on a Marine barracks in Beirut, Lebanon, that killed 240 people. He left dealing with the space station issue to his subordinates, particularly Gil Rye.[10]

SIG (Space) in fall 1983 had also been the venue for selecting the lead agency to oversee the commercialization of expendable launch vehicles (see Chap. 12). As that issue was approaching a presidential decision, Gil Rye and Craig Fuller had agreed to switch the forum through which the lead agency question would be put before the president. Rather than employ SIG (Space), they decided, with Ed Meese's agreement, to convene the Cabinet Council on Commerce and Trade (CCCT), with Ronald Reagan in attendance, for a November 16, 1983, session to consider the lead agency issue. Commenting on this switch of venue, Rye explained that "Craig Fuller and I have worked out an arrangement whereby SIG (Space) issues which are principally civil or commercial in nature can be referred to the CCCT for resolution … The final forum for presentation to the President would then be defined by the nature of the issue."[11]

Rye and Fuller agreed on a similar switch with respect to the space station issue. This switch had a certain logic behind it, once Beggs had decided to advocate the space station as a civilian-only program rather than a facility serving both civilian and national security purposes. In addition, the switch to the Cabinet Council, according to Fuller, "allowed a little bit broader participation." Unlike the situation in SIG (Space), where the representatives of the national security community were in the majority, in the Cabinet Council they would be counterbalanced by representatives of domestic agencies. Rye later suggested that the Department of Defense (DOD), by stating that it had no requirements for a space station, made a "serious blunder," giving up its "leverage over the final decision." In addition, DOD "paranoia over being stuck with part of the [space station] bill (plus a 'NASA-bashing' attitude) related to NASA/Air Force disputes" with respect to the space shuttle "overcame their common sense."[12]

In scheduling the Cabinet Council meeting, Fuller recognized that Reagan "supported the space station conceptually," but wanted to hear the full range of pros and cons regarding the undertaking. Fuller did not want the station issue "to get lost due to internal rivalry," particularly between NASA and the national security community; the CCCT venue provided "a little more open process." At the end of an August 3 lunch with a dozen aerospace executives, Reagan had indicated that he favored going ahead with a space station as a facility to promote space commercialization, if only its cost could be accommodated in projected budgets. Ed Meese, who had attended the lunch, wanted to be sure representatives of this private sector interest were present as the president was briefed on the station, and thus readily agreed to the change of venue. The chair of the CCCT, Secretary of Commerce Malcolm Baldridge, had developed a strong interest in the economic potentials of space activity; he was actively seeking to make his department the focal point for promoting that potential, and had also given his strong support to the space station idea. Given all of these factors, Rye and Fuller recognized that by presenting the station issue through the CCCT, the "right people" would be sitting around the table to present a positive picture to the president.[13]

Fuller and Rye visited NASA on November 18. Mark comments that by that meeting "Fuller was an open and strong supporter of the space station program" and "was confident that we would eventually be able to persuade the president to go ahead with it." One action Mark took following the meeting was to ask NASA's Langley Research Center to build for use at the presidential briefing a high-fidelity scale model of the station design NASA was advocating.[14]

On November 22, 1983, Fuller sent a memorandum to the members of the CCCT, announcing a December 1 meeting of the council to discuss "whether or not to proceed with the NASA development of a permanently based, manned Space Station." He attached to the memo Rye's August SIG (Space) 22-page report on the space station, noting that it contained "three options for the President's consideration," but that the "Secretary of Commerce recommends that the President approve both Space Station and enhanced Shuttle capability (i.e., both Options 1 and 2.)" This added an Option 4 for presidential consideration. Fuller in his memorandum noted that "at the SIG (Space) meeting on August 10, the majority of members supported either Option 1 or 3"; he omitted the fact that the majority had supported Option 3, deferring a decision. Fuller noted that "the meeting will consist of a summary of the options by Gil Rye … and a presentation by NASA followed by a general discussion."[15]

One possible reason for scheduling the presidential briefing for December 1 was that on the prior day, November 30, Reagan was already scheduled to participate in a National Security Council meeting to discuss how to proceed with the Strategic Defense Initiative (SDI) that he had announced in eight months earlier. The trade weekly *Aviation Week & Space Technology* suggested that "an important element in the station decision could be which space-based missile defense development option is selected by President Reagan" at the November 30 meeting.[16]

Getting Ready

Terry Finn of the NASA Space Station Task Force was responsible for preparing the charts that Beggs would use in his briefing to the president. Rather than pull already-prepared charts from his large supply used in prior briefings, Finn decided to prepare all-new viewgraphs (transparent sheets of plastic upon which words and images could be printed). Finn in a 1985 interview noted that this presidential briefing "was very, very carefully constructed." It was "an effort to appeal to emotion and pride … to adventure, to leadership, and portrays the Space Station in a very traditional sense of American greatness."

An unresolved issue was whether to mention the potential for international participation in the space station. Given the skepticism of the national security community with respect to such participation and in particular its concern with respect to the station project being a channel for unwanted technology transfer, NASA decided to separate the issue of international involvement from the basic decision whether or not to go ahead with the station at all. Finn recalled that "we put together a couple of charts on international [participation] in the early editions of the briefing. And they didn't make the final twelve."[17]

Mark met with Rye on November 22 to discuss the December 1 session. The two discussed how to "deal with OMB." Rye's view was that the December 1 briefing would "not be a decision meeting but rather an information meeting for the president." He predicted that in its passback to NASA, OMB would "zero out the Space Station," a decision that NASA would then appeal. It was Rye's view that NASA would have to take the space station issue to the president as part of the budget appeal process, and that "the real decision will be made in this arena." By scheduling the Cabinet Council meeting, he and Fuller "were going through the motions for the benefit of the nay-saying 'ankle-biters' so they could say they had their day in court."[18] Rye was prescient; events over the next two weeks unfolded as he predicted.

Selling the Space Station

The December 1 briefing was held in the White House Cabinet Room. Finn commented that he, Mark, and John Hodge took the model to the White House, but had "trouble getting in," possibly because the Secret Service had not been alerted to the model's arrival. Finn continued: "We finally [did] get in, and we set it up and ... we practiced with the lights to make sure the view-graphs can be seen." The three placed the model "by the door where the President will come in. And Mark positions himself by the model." As the meeting participants gathered, Mark "would snag people and show them" the model. Finn's assessment was that if a "picture is worth a thousand words," a model "is worth ten thousand ... Whatever this model cost, it was worth it."[19]

Although the gathering had been convened as a meeting of the CCCT, the actual attendance list made it closer to a convening of the overall Reagan cabinet. Those present included Vice President George H.W. Bush, Ed Meese and James Baker from Reagan's inner circle, Secretary of Commerce Baldridge, the now-Secretary of the Interior Judge Clark, Secretary of Energy Donald Hodel, Attorney General William French Smith, Trade Representative William Brock, Deputy Secretary of Treasury R.T. McNamar, Deputy Secretary of Defense Paul Thayer, Undersecretary of Labor Barney Ford, Deputy Secretary of Transportation James Burnley, Deputy Director of Central Intelligence John McMahon, and William Niskanen from the Council of Economic Advisers. Among the dozen White House staff present were OMB director Stockman and his associate director Khedouri, science adviser Keyworth, Baker's deputy

Richard Darman, National Security Adviser McFarlane, Press Secretary Larry Speakes, and, of course, Fuller and Rye.

Although Gil Rye's space station report had been distributed in advance of the meeting, it is likely that few of the senior attendees had read it. Rye led off the session by summarizing the report and the options for presidential decision it had examined. Like Fuller in the memo calling the meeting, Rye did not mention that at the August 10 SIG (Space) meeting a majority had voted for Option 3, deferring a decision.

Beggs then presented NASA's 20-minute space station pitch. Hans Mark in his diary notes suggests that Beggs "unfortunately read from his notes too often, so that it [his briefing] did not seem to me to be as effective as it could have been." By contrast, another participant in the meeting described Beggs as "very eloquent. He stated it at just the right level, with vision, but not trying to commit to something too ambitious."[20]

Beggs started off by quoting President John Kennedy: "The capacity to dominate space is essential to the United States as a leading world power." He then cited President Richard Nixon as he announced approval of the space shuttle, speaking of "man's epic voyage into space—a voyage the United States of America has led and still shall lead." Beggs noted that the focus on leadership had been "reaffirmed" in the Reagan administration's July 1982 National Space Policy. A next chart showed NASA's budget, peaking during the Apollo program and then, after a rapid decline, essentially flat at a much lower level since the last missions to the Moon. Then an image of the space shuttle was projected, with the claim that it represented "leadership in space for the 1980's." Beggs noted that the shuttle "flies beautifully" and "has captured worldwide attention," reassuring "our Allies of America's technological strength." Beggs transitioned to an image of the space station NASA was proposing, saying, "What's needed now, what was originally envisioned, is a place to shuttle to." Below the station image was written: "A highly visible symbol of U.S. strength." Beggs elaborated on the various functions that a space station would enable, suggesting that it would "dominate the space environment for twenty years" and "stake out some options for the future, enabling a President in the years to come to embark the United States upon missions that transcend the boundaries of earth." This comment was intended as a defense against Keyworth's suggestion that a decision on the station should wait for the decision on more expansive space goals. The next chart played to Ronald Reagan's interest in space commercialization, with Beggs noting that "space is going to become a place of business" and that "a space station is going to make that possible." It was followed by an image of astronauts in the shuttle payload bay, with Beggs suggesting that "some people say you can do it all in space with robots. In fact, you must have man. He—and she—are the essential ingredients." Playing to the central theme of the briefing, Beggs added: "The presence of man is the key to leadership in space."

Although NASA had decided to downplay competition with the Soviet Union as a theme in its case for a space station, the briefing contained two images of the Soviet *Salyut* space station. Beggs suggested that he was worried about "what the Soviets are up to. What are they planning to fly in the late 1980s and the 1990s? Will they be successful in their plans to dominate space?" The next chart listed the three SIG (Space) options plus the fourth option proposed by Malcolm Baldridge, approving both a space station and a shuttle with expanded capabilities. Beggs noted that an extended duration shuttle orbiter might be a "worthwhile" project, but "hardly America's next step in space." Deferring a space station decision, he said, "derails new commercial endeavors in space"; "sends a signal to the American people that their space program is going to rest on its laurels"; and "means that, in the years ahead, that we are going to forfeit our hard won leadership in space." A final viewgraph in Beggs's presentation traced the history of NASA's accomplishments over the past quarter century. Beggs closed his remarks by saying, "Now, today, here in this room, we must look forward … The stakes are enormous: leadership in space for the next 25 years."[21]

Reagan interrupted Beggs at one point in his presentation, asking him how a space station figured in the possibility of establishing a base on the Moon. Beggs replied that a station was a necessary step on the way to the lunar surface.

Mark commented that Reagan "looked better than he did yesterday and he was more animated." As noted earlier, on the previous day, November 30, Reagan had participated in a difficult National Security Council meeting to review activity since his speech announcing his hope to develop a defense against ballistic missiles. In the interim, the SIG for Defense Policy had examined options for how to proceed and had recommended proceeding with research and development as quickly as technology would allow, but not to make any commitment to deploying a defensive system. Reagan's somewhat disappointed reaction to this recommendation was that the United States should "hope and pray" that it would be first to deploy a defensive system; otherwise "we can expect nuclear blackmail." Mark commented that "this meeting [the December 1 space station briefing] was more fun than the one on 'Star Wars.'" He added that "there is no doubt that the President is extremely sharp and quick on the uptake. He does not concern himself with details … but he is in full command of the big picture."[22]

After Beggs finished his presentation, the floor was opened for discussion. There are available two accounts from meeting participants of what transpired, one prepared by Hans Mark, writing in telegraphic style, and the other by Deputy Director of Central Intelligence John McMahon. The two accounts complement one another.

Mark: Keyworth says man in space is not new … Russians have substantial manned presence … Russians have a station, why should we catch up? Argues that we should make the goal of going to the moon right away. Wants to optimize program toward that goal. Wants to have a "space summit" of the best people to study going to the moon. Wants to broaden outlook.

McMahon: George Keyworth commented that the MSS [manned space station] is not new and he feels we shouldn't emphasize a simple way station but rather go for a bold step and optimize for a goal which would mean something in space, advocating a moon space station. He readily admitted, however, that he didn't know what we would do on the moon or how we would get there but suggested over the next six months we study it to determine those answers.

Mark: Baldridge says that options are too narrow. Says we should include commercialization and try to put together a goal for the State of the Union message … Baldridge commented that we have not heard yet from the private sector and that there is some talk that several companies may form a consortium to build the MSS and lease it back to the U.S. [government.] This generated a few smiles and smirks around the table.

McMahon: Baldrige then commented to the President that the options we considered were too narrow and probably we ought to undertake a study into the commercial and civil uses of space.

Mark: Thayer [says] man in space is attractive. DOD viewpoint is that there is no significant requirement for men in space. Concerned about effect that station would have on shuttle. Says it has to work. Says that shuttle is short of meeting operational commitment. Says cost of $8B is optimistic. Will cost another $18 to $20B to get it into operation. Should not follow Russians. The Russians need men in space because they don't have the technology to do things automated. Says he cannot put it high on the priority list … Says that if we had the money it would be a good thing to have. Real problem is that it will bleed off effort from the shuttle.

McMahon: Thayer then spoke of the impact that the MSS would have on diverting NASA's attention away from their problems with the shuttle and optimizing that for the purpose for which it was designed. He also felt that the MSS would not be a $8 billion program but rather a $20 billion one … If we had the money maybe we could take the chance and do a thing like the MSS but there is a shortage of funds to do essential items and we can't visualize the MSS in any way recouping its investment. He said because of his own uncertainties on the future demand for dollars he proposed not scheduling the MSS now but deferring it for six months to a year.

Mark: Stockman concerned about cost. All other things being equal we ought to do this. Not unreasonable to spend $10–$12B per year [on NASA]. Says all things are not equal. Says we are bankrupt. Deficit is $208B—unacceptable. Congress has not helped. Congress will not help and in an election year especially. Says this is a crisis. $600B deficits in 3 years. Says that the environment is wrong to start anything. We will run up a deficit of over one trillion in the 1980's. Must win election first and get Congress under control before we make commitment that will cost $20B to $30B before it is finished. Must have defense buildup but nothing else. [Mark commented: "an eloquent speech."]

McMahon: Stockman then spoke, saying that the MSS was excessive in costs when resources are scarce. He felt that it would cost at least $10 to $15 billion and that amount would push the budget to fracture. He noted that the $3.5 trillion program puts us out perilously close to fiscal bankruptcy … He said that you, Mr. President, have got to get hold of Congress and force them to do what they ought to do.

Mark: Brock strongly supports the economic argument for going into space. Must look at the quality of govt. expenditures. Says that we must make investments.

McMahon: Bill Brock, U.S. Trade Representative, then spoke. He claimed that we really haven't spent any dollars [in space] in the past 20 years because every dollar spent in space has been returned to us 20 times over in economic growth and flow of dollars to the treasury. McNamar interrupted him by saying that the Treasury has yet to see its first dollar from space and if it's true that we can get 20 to 1 for an investment like the MSS then he and [Secretary of the Treasury] Don Regan are going to leave Treasury and set up a company to do that … Brock then gave an emotional talk on how dollars spent on welfare are wasted, we never see them and even noted how the seed corn that we gave the Indians which would have increased their harvest tenfold was eaten instead of planted.

Mark: Fuller [comments that] process is finished—don't need any more. Makes the private sector case. Definition and certainty are important. Partners want certainty. Private sector will not build platform … We are not talking about much money. Have the opportunity to make an investment that we want to make.

McMahon: Craig Fuller then commented about he must represent U.S. commercial interests with whom he has been in touch and he feels that if the President makes this commitment that U.S. industry is with him and that he would even have our European partners cheering us on. Whereupon the meeting ended and we all withdrew.

As the meeting broke up, according to Hans Mark, "the President, Smith, Clark, Thayer and others came by to look at the model."[23]

During the meeting, Ronald Reagan commented that "he didn't plan to make a decision today, but just wanted to hear everyone out." There had been during the meeting one light-hearted exchange, also typical of how Reagan dealt with a contentious issue. As Treasury's McNamar was protesting that there had been no economic return on space spending, Attorney General William French Smith interrupted him, saying, "I'll bet the Comptroller of Ferdinand and Isabella made exactly the same speech when Columbus proposed to them that he sail west to reach the far east." Reagan quickly reacted, commenting, "Yes, Bill, you're right, but you remember that the Comptroller won the argument: Isabella had to pawn her jewels to pay for Columbus' trip. What I want to know is, who has the jewels around here?" According to Mark, "There was general laughter at this quip and from then on, there was definitely a more relaxed atmosphere in the room."

Mark "was disappointed at the outcome of the meeting ... there were more people against us than for us." Mark told Beggs: "I thought we did not lose anything at the meeting but that we also did not gain much." Beggs replied: "You're wrong. I could tell the president was with us. He winked at me a couple of times while the other people were making their critical remarks."[24]

Beggs was correct. In his diary entry for December 1, Ronald Reagan wrote: "Cabinet Council on 'space' and where we go. The issue is whether to move on a program for a permanent manned space station. I'm for it as I think most everyone is but the question is funding such a new course in face of our deficits."[25] The space station decision had come down to whether the funding for the program could be accommodated in an acceptable overall NASA budget, not only for FY 1985, but for the following years of what Reagan and his associates hoped would be a second term in the White House.

Decision Made, Protested, Reaffirmed

There were in the first Reagan term two paths to presidential approval of a policy initiative. If the initiative required a policy decision with minimal budget implications, it most often would reach the president through one of the Cabinet Councils or, in the case of national security issues, an SIG or the National Security Council itself. If the initiative had significant budget as well as policy implications, the path to president's desk usually ran though the annual budget process. Although SIG (Space) had spent the better part of a year debating the space station issue, it would be through this second path that Ronald Reagan would give final approval to the space station program. This was not a surprise. Both NASA and those handling the station policy issue in the White House, particularly Gil Rye and Craig Fuller, had for several months been saying that the final decision on whether to proceed with the station would come as the FY 1985 budget for NASA was finalized.

Space Station Approved

As NASA submitted its FY 1985 budget request to OMB in September 1983, it had asked for $235 million to support space station Phase B detailed design efforts, and forecast that it would need an additional $460 million in the next two fiscal years to complete that phase of the program. NASA, playing a bit of a budget game, did not include any money in its five-year budget forecast for the Phase C space station development phase that would follow the design efforts, even though that phase would have to start in FY 1987 if the station were to begin operation in 1991, as NASA was promising. The OMB staff estimated that over $6 billion in development funding would be needed by FY 1989, the last Reagan administration budget request if the president were reelected; this was a need that NASA hoped to downplay by not mentioning it in its budget request.

A crucial step in the annual budgeting process is the OMB "Director's Review," at which the OMB professional staff presents its evaluation of an agency's budget request to the OMB director and his top assistants, with its recommendations for the OMB position. After this review, the OMB leadership makes an initial decision on an agency budget allowance and communicates it to the agency in a "passback." In its November Director's Review presentation with respect to the NASA budget, the OMB staff pointed out that "the need for increased funding in 1985 to meet approved commitments along with the desire of NASA to initiate a Space Station and expand sharply all of its major R&D programs leads to a requested $0.9 billion increase [in FY 1985] over 1984 and to dramatic growth in outyear costs." Key questions for decision were: "Should a commitment to an initial manned Space Station be approved for 1985? If not, what are the appropriate funding levels and constraints for related activities?" The material for the Director's Review included a seven-page "issue paper" on the space station. That paper pointed out that NASA's budget proposal was only for an "initial stripped down version" of a space station costing $8 billion, and that NASA acknowledged that the cost of "a fully developed Space Station may run as high as $28 billion." Also, the $8 billion figure "does not include any operating costs … or the cost of science and applications to be performed on the Space Station." The staff was skeptical of NASA's cost estimates, since "history shows that NASA estimates of total costs for large projects are a factor of 2–10 too low"; also, there was "major uncertainty as to whether a Space Station would be the most suitable means for meeting the future needs of most space user groups." Noting that NASA was considering international participation in the station, the staff suggested that "NASA may in fact give away our technological leadership" through such participation. Another consideration was that "other possibly higher priority alternatives exist on two fronts: extended duration shuttle missions and unmanned space platforms; and the President's DABM [Defense against Ballistic Missiles] initiative, which may require construction of heavy lift launch vehicles and large unmanned platforms."

Given all these considerations, the staff recommendation was to "defer commitment" but "consider establishing a long term goal to establish permanent space facilities." If this recommendation were accepted, OMB would "provide a small amount of money to explore alternatives to the manned Space Station for consideration in future budgets."[26]

This recommendation was accepted by OMB leadership, and in its passback to NASA, OMB included only $6 million for continued low-level space studies. This was essentially a holding decision; OMB knew that NASA would protest the low budget allocation, and that the actual decision on whether or not to approve the space station would be made either by the president's senior advisers or by Ronald Reagan himself in the final stages of formulating the NASA FY 1985 budget.

NASA in late November did indeed protest the OMB budget offer, beginning the appeal process that would eventually reach the president. The final stop on that path was the White House Budget Review Board, consisting of James Baker, Ed Meese, and David Stockman. The board meeting on the NASA FY 1985 budget was held on December 2, the day after the station briefing to the president. Beggs made his presentation on the overall NASA budget to the three top White House aides, and there was a lengthy discussion of whether NASA's cost estimates for its core program (without the space station) were accurate. But NASA's budget request for the space station, by now reduced through negotiations between NASA and OMB to $150 million, was not addressed; rather, NASA was informed that the space station decision "would require a special meeting with the president before a final decision could be made." That meeting was scheduled for Monday, December 5.

On December 3, Meese, Baker, and Stockman sent a decision memorandum to the president that outlined four alternatives for the NASA budget.

- Alternative I would fund NASA for FY 1985–1989 at the original target that OMB had provided some months earlier, curtailing some ongoing projects and not allowing initiation of any new projects during the next five years;
- Alternative II would provide an increase above the original target sufficient to maintain all current projects and to allow some new starts other than the space station; this alternative would add $4.97 billion to the NASA five-year budget;
- Alternative III would augment the budget proposed in alternative II by funding the full NASA request for the space station; this alternative would add $12.2 billion to the NASA five-year budget;
- Alternative IV would provide NASA with a fixed five-year funding allowance that would yield 1 percent real growth and allow a start on the space station; this alternative would add $10.4 billion to the NASA budget.

Meese, Baker, and Stockman advised the president to "approve Alternative II if you decide to defer approval of the Space Station" and to "approve Alternative IV if you decide to give your approval to starting the Space Station in FY1985."[27]

NASA in fact did not have to wait for the outcome of the December 5 meeting to learn that Reagan had approved the space station and thus that funds for the program would be included in its FY 1985 budget. In response to the memo from the members of the Budget Review Board, Reagan indicated that he would approve the station, but asked them not to discuss that decision, with the exception of communicating it to the NASA leadership. Beggs met with OMB's Khedouri on Saturday, December 3; he was told that "the $150 million we had requested would be in the president's budget and that this would be confirmed the following Monday (December 5, 1983) at a formal decision meeting with the president." Rye comments that "it must be one the great ironies … that Beggs would learn of the Space Station decision from Fred Khedouri (one of the 'ankle-biters'). Craig [Fuller] and I were sworn to secrecy."[28]

NASA's meeting with Reagan was scheduled for 11:00 a.m. on the morning of December 5. It took place in a very positive context from NASA's perspective. Just before that meeting, Reagan, speaking from the Oval Office, had engaged in a three-way televised conversation with German Chancellor Helmut Kohl and astronauts aboard space shuttle *Columbia*. The shuttle had been launched on November 28 for a ten-day mission; it was the first shuttle flight to carry in its payload bay the Spacelab pressurized laboratory, a European contribution to the shuttle program. Germany had been the lead funder of Spacelab development, and aboard the mission was the first German to fly on the shuttle, scientist Ulf Merbold. (In fact, Merbold was the first non-American to fly on a NASA mission.) In his remarks during the conversation, Reagan noted: "This mission is also a shining example of international cooperation at its best … Building on that good will, this is the first time a citizen from another country has joined one of our space missions as a member of the crew. It is an exciting first."[29]

Going almost directly from his orbital conversation to his meeting on the space station certainly reinforced the president's positive attitude toward NASA and the space program. At the Cabinet Room meeting (Fig. 9.1), Meese, Rye, and Fuller "sat in Beggs corner," while Khedouri backed up Stockman. The budget director tried to get NASA to agree to develop the space station without an increase in NASA's overall budget in future years. Beggs insisted that NASA's budget had suffered over the previous decade, and that an "appropriate increase" was needed. Stockman finally was willing to offer Alternative IV in the December 3 decision memorandum, a 1 percent annual increase above inflation. Beggs accepted that offer, and Ronald Reagan said: "Done." With that, the space station decision was made. Writing in his diary that evening, Reagan noted: "A Budget appeal meeting on N.A.S.A.'s request. I think we're OK there & can still start to plan a space station." Reagan instructed those in the December 5 meeting not to communicate his choice to others; he wanted to announce it himself, and already had a plan for how to make that announcement.[30]

Decision Protested

With Reagan's directive not to publicize his positive decision on the space station, other senior members of his administration apparently believed that they still had time to weigh in against such a choice. On December 9, Director of

Fig. 9.1 Ronald Reagan gave final approval of NASA's space station program at a December 5, 1983, budget review meeting in the White House Cabinet Room. Sitting to the president's right is Edwin Meese. Across the table (l to r) are OMB Director David Stockman, Secretary of the Treasury Donald Regan, an unidentified person, NASA Administrator James Beggs, Cabinet Secretary Craig Fuller (standing), Gil Rye (against wall), NASA Comptroller C. "Tommy" Newman, and Deputy National Security Adviser Admiral John Poindexter. (Photograph courtesy of Reagan Presidential Library)

Central Intelligence William Casey, a long-time Reagan adviser, wrote to Meese, noting the lack of agreement at the December 1 meeting and saying, "I believe it is evident to everyone that without a national consensus it is questionable whether or not the nation can maintain the high level of commitment necessary to sustain, over the long run, major parallel efforts such as strategic forces and intelligence capabilities modernization, defense against ballistic missiles, and a bold new civil space initiative." Casey said that he wanted to "continue to stress that before the U.S. embarks on major new space initiatives we owe the President the complete menu of options which will permit him to address the civil program within the context of our national security needs."[31]

In an undated memorandum to Meese, possibly coordinated with the communication from Casey, Secretary of Defense Caspar Weinberger reiterated his opposition to going ahead with the station. He noted that the December 1 Cabinet Council meeting "underscored the fact that there is little support in government, the scientific community, or the private sector for NASA's proposal to initiate development of a manned space station." He argued that "US commitment to a manned space station would be better characterized as a major engineering effort rather than a bold new space initiative" and that "undertaking an engineering project of this magnitude would divert NASA expertise and resources from the task of pressing the frontiers of technology along lines more likely to produce economic and scientific benefits for the nation." He thought that "one should not further stress our resource constrained environment with any initiative that does not promise extraordinary return. The manned space station does not meet this criterion." Weinberger supported "the thrust of Jay Keyworth's proposal for a 'space summit' to assemble the best talent in the industrial and scientific communities and to chart bold new directions for the civil space program," since such a gathering "would provide the President with a range of options to weigh in the dual context of national security and space leadership." Given these arguments, Weinberger told Meese that he would "strongly recommend deferring commitment to the NASA space station proposal."[32]

Concerned that these protests might cause President Reagan to reverse his December 5 decision in favor of the space station, and likely to give a sense of closure to the SIG (Space) process regarding the space station, Gil Rye drafted a "Memorandum of Decision" for the president to sign recording his approval of station development. That three-page memo stated that "the United States will establish a permanent manned presence in space through development of a manned Space Station." Rye also drafted a cover memorandum to be signed by Meese and McFarlane, forwarding the decision memo to Reagan. The cover memo said: "While we certainly understand Dave Stockman's concern about growing deficits and Paul Thayer's assessment [expressed at the December 1 Cabinet Room meeting] that there are no current military requirements for a Space Station, we believe that the interests of the nation would be best served with your immediate approval of the program." The memo added: "We believe it is unnecessary and probably politically unwise for you to commit now to a goal of returning to the moon … Your approval of a Space Station program would preserve the option for more ambitious goals sometime in the future, perhaps during your second term when deficits are under better control." The memo concluded by recommending "public announcement of your decision during your State of the Union Address." National Security Adviser McFarlane signed the memorandum, but Meese did not, and thus it was not forwarded to the president.[33]

The strong protests by Casey and Weinberger, two of his most senior advisers, did not deter Ronald Reagan. He had made his decision to approve the space station, and by early December preparations were already underway to announce that decision. It was thus through the combination of inserting funding for space station design in the NASA FY 1985 budget and, more important, the upcoming high-profile announcement of space station development by the president himself that his approval would become public.

Notes

1. Hans Mark, *The Space Station: A Personal Journey* (Durham, NC: Duke University Press, 1987), 179–180.
2. Letter from James Beggs to James A. Baker, III, August 24, 1983, Folder 12776, NHRC.
3. Craig Covault, "Defense Dept. Studies Aid Space Station," *AWST*, October 3, 1983, 19.
4. Interviews with Jeff Struthers, May 29, 2016, and Norine Noonan, September 23, 2016. Noonan in 1983 was an OMB budget examiner working under Struthers.
5. Mark, *The Space Station*, 181–182.
6. Meese's quote is in Michael A. G. Michaud, *Reaching for the High Frontier: The American Pro-Space Movement, 1972–1984* (New York: Prager Publishers, 1986), 281.
7. Ronald Reagan: "Remarks at the 25th Anniversary Celebration of the National Aeronautics and Space Administration," October 19, 1983. Online by Gerhard Peters and John T. Woolley, APP, http://www.presidency.ucsb.edu/ws/?pid=40662.
8. Howard McCurdy, *The Space Station Decision: Incremental Politics and Technological Change* (Baltimore, MD: The Johns Hopkins Press, 1990), 181.
9. Craig Covault, "NASA Answers Planning Challenge," *AWST*, October 31, 1983, 22.
10. For an evaluation of Clark's tenure as national security adviser, see John P. Burke, *Honest Broker? The National Security Adviser and Presidential Decision Making* (College Station, TX: Texas A&M Press, 2009).
11. Memorandum from Gil Rye to Robert McFarlane, "Membership on the Senior Interagency Group for Space," November 30, 1983, Files of Valerie Neal.
12. Letter from Gil Rye to Hans Mark commenting on a draft of Mark's *The Space Station* book, October 25, 1984, cited with Rye's permission.
13. Interviews with Craig Fuller, December 14, 2016; with Edwin Meese, October 12, 2016; and Gil Rye, May 20, 2016.
14. Mark, *The Space Station*, 184.
15. Cabinet Affairs Staffing Memorandum, "CCCT Meeting on the U.S. Space Station," November 22, 1983, with attached memorandum from Craig Fuller to Members of the Cabinet Council on Commerce and Trade, "Space Station," undated, CIA-RDP85M00363R001002300012-7, CREST.
16. "Washington Roundup," *Aviation Week & Space Technology*, November 28, 1983, 17.
17. Interviews of Terence Finn by Sylvia Fries, June 12 and July 10, 1985, NHRC.
18. Handwritten note from Hans Mark to James Beggs, "In Strictest Confidence," November 23, 1983. I am grateful to Dr. Mark for providing me a copy of this note. Rye letter to Hans Mark, October 25, 1984.
19. Interview of Terence Finn by Sylvia Fries, June 12, 1985, NHRC.
20. Hans Mark, "Diary Notes for December 1, 1983," provided to the author by Dr. Mark; M. Mitchell Waldrop, "The Selling of the Space Station," *Science*, February 24, 1984.
21. NASA, "Revised Talking Points for the Space Station Presentation to the President and the Cabinet Council," with attached presentation charts, November 30, 1983, File 12766, NHRC.

22. Hans Mark, "Diary Notes for December 1." Minutes of the November 30 meeting on the Strategic Defense Initiative can be found at http://www.thereaganfiles.com/19831130-nsc-96-sdi.pdf.
23. This account of the December 1 meeting is based on Hans Mark's notes on "White House Meeting," dated December 1, 1983, and John McMahon, Memorandum for the Record, "Summary of Cabinet Council on Trade and Commerce Meeting," December 1, 1983, CIA-RDP86B00885R000300430043-0, CREST. Dr. Mark provided a copy of his notes to the author.
24. Mark, *The Space Station*, 186–187.
25. Hans Mark, "Diary Notes for December 1, 1983"; Douglas Brinkley, ed. *The Reagan Diaries* (New York: HarperCollins, 2007), 201.
26. Director's Review material is from Office of Management and Budget, "National Aeronautics and Space Administration, 1985 Budget," Overview and Space Station sections, undated but November 1983.
27. Memorandum from Edwin Meese III, James A. Baker III, and David A. Stockman to the President, "NASA Budget for FY 1985–FY 1989," December 3, 1983, Box 9, Agency Files, Executive Secretariat, National Security Council, RRL.
28. Hans Mark, "Diary Notes for December 2, 1983," provided to the author by Dr. Mark. Mark, *The Space Station*, 190. Rye letter to Hans Mark, October 25, 1984.
29. Ronald Reagan: "Remarks During a Conference Call With Chancellor Helmut Kohl of the Federal Republic of Germany and Crewmembers of the Space Shuttle Columbia," December 5, 1983. Online by Peters and Woolley, APP, http://www.presidency.ucsb.edu/ws/?pid=40834.
30. McCurdy, *The Space Station Decision*, 185; Brinkley, ed. *The Reagan Diaries*, 295.
31. Memorandum from William Casey to Edwin Meese, "Cabinet Council Meeting of 1 December 1983," December 9, 1983, CIA-RDP92B00181R001901730053-2, CREST.
32. Memorandum, unsigned but from Caspar Weinberger, to Edwin Meese, "Manned Space Station," undated, CIA-RDP92B00181R001901730069-5, CREST.
33. Memorandum of Decision, "Development of Manned Space Station," undated; memorandum from Edwin Meese III and Robert McFarlane to the President, "Space Station Decision," December 10, 1983; memorandum from Fred Khedouri to Craig Fuller, "Space Station Decision Memorandum," December 12, 1983, all in Box 7, Outer Space Files, RRL.

CHAPTER 10

"Follow Our Dreams to Distant Stars"

Article 2, Section 3 of the U.S. Constitution requires the president "from time to time give to the Congress Information of the State of the Union." This constitutional requirement has since early in the twentieth century been met by an annual speech delivered by the president to a joint session of the Congress, from 1934 on in the early weeks of the year. That speech is very carefully prepared, since it gives a president the opportunity to highlight his achievements and to set out his agenda for the coming year and beyond. Ronald Reagan viewed his major speeches as a primary means of communicating his priorities, and thus gave particular attention to the content of his annual State of the Union address.

Space as a "Great Goal"

Reagan and his advisers had at some point in 1983 decided to include in his 1984 State of the Union speech a significant discussion of the future of the U.S. space program. This decision reflected administration hopes to have completed the national space strategy by the time of the speech; that strategy would provide the foundation for what the president might say. By late 1983, administration initiatives regarding both the space station decision and space commercialization had reached the point where the president could personally announce them, even if the overall strategy had not been finalized.

Thus, a December 8, 1983, draft outline of the 1984 State of the Union speech (only three days after Ronald Reagan had given his formal approval to the space station) already said, "let us move boldly forward to develop America's newest frontier," by announcing the "goal of a permanent manned presence in space within a decade." The parallel with President Kennedy's 1961 speech announcing his decision to send Americans to the Moon were very much in mind; the outline noted: "(Recall Kennedy/Apollo spirit.)" A January 5, 1984,

J. M. Logsdon, *Ronald Reagan and the Space Frontier*,
Palgrave Studies in the History of Science and Technology,
https://doi.org/10.1007/978-3-319-98962-4_10

draft of the speech contained the passage "I am asking the Congress to move ahead with [SPACE INITIATIVE] not only for its immediate benefits but because it is a symbol that we as a Nation welcome the future, its challenge, its opportunities, its chance for greatness." A January 11 draft of the speech added: "Our progress in space—taking giant steps for all mankind—is a tribute to American teamwork and excellence ... We are first, we are the best, and we are so because we are free." That draft also included for the first time a mention of space commercialization, saying, "Space holds enormous potential for commerce today." A later-in-the-day January 11 draft, marked "Master," said: "Our second great goal is to build on America's pioneer spirit and develop our next frontier," and identified that frontier as "space." The draft also said: "America has always been greatest when we dared to be great. We can reach for greatness again. We can follow our dreams to distant stars."[1]

The January 11 "Master" draft also contained a handwritten notation that "language on space cooperation to come." The issue of possible international participation in the space station program had been set aside in the run up to the December 1 briefing to the president and had not been a focus of the decision to include funds indicating space station approval in the president's FY 1985 National Aeronautics and Space Administration (NASA) budget proposal. But by January 11, preparations were well underway to enable such participation.

Adding the International Element[2]

From 1981 on, international participation in any space station the United States might develop had been part of the thinking of those working within NASA to promote the station initiative. European countries, working through the European Space Agency (ESA), and Canada had made significant contributions to the space shuttle program. Japan had regretted not being ready to contribute to the space shuttle, and its space industry and space agency were determined not to miss the space station opportunity. Europe, Canada, and Japan during 1982–1983 had carried out studies of potential space station contributions, even as they had been cautioned by NASA that neither space station approval nor an invitation to participate could be guaranteed.

While the leadership of the State Department in August 1983 had recommended against presidential approval of the space station, some in the department had a different view. In a memorandum drafted by Tom Niles, a staff member of State's European Bureau, the head of that Bureau, Richard Burt, and the head of the Bureau for East Asian and Pacific Affairs, Paul Wolfowitz, had pointed out that "there are significant foreign policy advantages to be gained from it [station approval] in terms of our relations with the major European countries, Canada, and Japan." They acknowledged that "some technology transfer would be involved," but added that "we are persuaded by past experience that this can be managed in such a way as to minimize any dangers involved." Burt and Wolfowitz also suggested that "if the President decides to go ahead with the Space Station Project, international participation in it could be a useful contribution to the 1984 London Summit." At that gathering, President Reagan would be interacting with the leaders of the six major U.S. allies. These perspectives, ignored in August, became influential as the space station announcement neared.[3]

The idea of making a call for increased international cooperation, not only with U.S. allies but also with the Soviet Union, a feature of the State of the Union speech was also attractive to Bud McFarlane, Reagan's new national security adviser. On December 15, 1983, McFarlane discussed with his chief of staff Robert Kimmitt "whether the President should call for closer international cooperation in space, including with the Soviet Union" and asked Gil Rye for an "assessment of pros and cons … of such a call vis-à-vis the Soviet Union. How would they respond publicly? Is there the possibility they can turn this to their advantage rhetorically or strategically?"[4]

The next day, Rye responded to McFarlane's questions. He mentioned "the tremendous success of the Reagan/Kohl/Spacelab TV conference" of December 5; he told McFarlane: "Of course, the Space Station would provide the most opportune and visible platform for such an initiative." Rye suggested that "in the State of the Union Address the President could not only announce approval of the Space Station, but also solicit all nations to join with us in a peaceful and cooperative venture that will benefit all mankind." He added that the president "could especially call upon the Soviet Union to join us in this endeavor." Rye continued: "The President could indicate that he is sending Jim Beggs to various nations to solicit their cooperation and work out the details. To continue the scenario further, the London Summit in May 1984 could be used as the occasion to cement these details with the Europeans in a very visible way." As it would turn out, with the important exception of asking the Soviet Union to participate in the space station, Rye's suggestions described the path actually pursued.

Rye told McFarlane: "Over the past year, I have had several discussions with Jim Beggs and others on the potential for international cooperation … I believe the initiation of an International Space Station program would have an overwhelmingly positive effect both domestically and internationally." With respect to Soviet participation, Rye thought that "we win … whether they accept or reject the offer—it has tremendous propaganda appeal."

Rye offered some arguments for and against making the space station an international undertaking. Among the "pros" were:

- It would diffuse arguments about an arms race in space;
- Because of the above, Congressional approval of the Strategic Defense Initiative would be significantly improved;
- Domestically (and politically) it would diminish the President's "saber-rattling" image in the eyes of some;
- European and Japanese contributions would divert resources away from space programs designed to compete with the U.S. in the commercial marketplace.

Arguments against international participation included:

- There would be inevitable technology transfer in a cooperative venture. Our competitive edge would be reduced;
- Management of the program would be more complex.

Rye recommended to McFarlane that he "discuss this subject with the President at the earliest opportunity and solicit his approval for a major announcement of an International Space Station Program in his State of the Union Address."[5]

National Security Council Soviet expert Jack Matlock also assessed McFarlane's idea of inviting the Soviet Union to participate in the space station; he characterized such cooperation as "a powerful political gesture, indicating our willingness to consider limited cooperation in an important and highly visible area of activity." If a decision to propose cooperation was made, Matlock suggested, "the question [should] be discussed with the Soviets privately in advance of inclusion in the State of the Union message. If we spring it on them in a public statement, the Soviets will conclude that it is merely a propaganda gesture."[6]

McFarlane brought Rye and Matlock with him on January 4, 1984, for his daily national security briefing to Ronald Reagan. They summarized for the president the proposal that Reagan use the already-planned State of the Union address announcement of his space station decision to invite both allies and the Soviet Union to participate in the project. At the bottom of Rye's December 16 memo, the notation "Done 1/4" was added, indicating that the topic had been discussed with the president. Reagan indicated that he was supportive of the cooperative initiative. According to Hans Mark, at some point, Reagan called NASA's Beggs to ask, "Can we do this thing internationally?" Beggs responded: "Of course."[7]

The proposal to invite the Soviet Union to participate in the space station was short-lived. Opposition from hard-liners vis-à-vis the Soviet Union in the White House, the Department of State, and the Department of Defense torpedoed the National Security Council initiative. Beggs later recalled: "I would have brought the Russians in … had I the permission to do so. But I didn't … We were still in the Cold War."[8]

Cooperative possibilities continued to be discussed in early January, leading to the insertion of the "place holder" in the January 11 State of the Union speech draft, saying, "language on space cooperation to come." Peggy Finarelli at NASA and Niles at the State Department were most active in these talks, which led to a January 18, 1984, meeting convened by McFarlane. Attending the meeting were McFarlane's deputy, Admiral John Poindexter, NASA Administrator Beggs, Undersecretary of State for Political Affairs Lawrence Eagleburger, Undersecretary of Defense for Policy Fred Ikle, and CIA Deputy Director Robert Gates. This high-level group fleshed out Reagan's January 4 decision in principle to solicit international participation in the space station. They also agreed that Beggs, acting as the Reagan's personal emissary, would travel to key foreign capitals to personally extend the presidential invitation for space station cooperation. Unlike the extended and contentious process that had led to the December 1 briefing to the president, there were no interagency meetings or policy papers devoted to developing the cooperative proposal, nor any formal assessment of the benefits and risks associated with such international cooperation.

The text of the invitation as it was to appear in the State of the Union address was hurriedly drafted on the evening of January 18 and approved by the meeting participants the next day. The draft insert said:

> This nation long ago committed itself to work with all nations in the peaceful exploration and use of space. This new program [the space station] will allow us to deepen that commitment. We want our friends and allies to help us meet our ambitious goals and to share the benefits of this peaceful venture. I have therefore asked NASA's Administrator, James Beggs, to visit capitals around the world so he can explore opportunities for constructive participation in this program.[9]

Potential international partners in the space station program were given advance notice that the president would be inviting them to participate. On January 24, the day before he made the State of the Union speech, Ronald Reagan sent a personal letter to Chancellor Helmut Kohl of the Federal Republic of Germany, President Francois Mitterrand of France, Prime Minister Margaret Thatcher of the United Kingdom, Prime Minister Bettino Craxi of Italy, Prime Minister Yasuhiro Nakasone of Japan, and Prime Minister Pierre Trudeau of Canada. The message was straightforward:

> During my State of the Union address this Wednesday, January 25, I will be announcing the United States' intention to proceed with development of a manned Space Station program. It is my hope that we can work together on this project. To develop this cooperative effort I have asked James M. Beggs, the Administrator of the National Aeronautics and Space Administration (NASA), to act as my personal emissary and meet with senior officials of your government in the near future.[10]

"We Can Reach for Greatness Again"

There was one last-minute problem with the State of the Union speech, at least from NASA's perspective. When the text of the speech arrived at NASA on the afternoon of January 25, just hours before Reagan was to deliver it, NASA discovered that it called for a "permanent space station," not a "permanently manned space station." This left open the possibility, in NASA's somewhat paranoid view, of interpreting Reagan's words as approving an outpost permanently in orbit, but with a crew aboard only on a part-time basis. This of course was not what NASA had in mind. Finarelli called Rye, insisting that the wording of the speech text needed to be changed to say "permanently manned." Rye told her that at that late hour it was very unlikely to be possible, but in fact he was able to get the change made.[11]

Just after 9:00 p.m. on January 25, 1984, President Ronald Reagan began his State of the Union address before a joint session of Congress, with most of his Cabinet, his top military commanders, and the justices of the Supreme Court also in the House chamber. The president would stand for reelection later in 1984, and Reagan intended his address to set out the main themes of his reelection campaign. He thus began by telling the audience and the country that "America is back, standing tall, looking to the eighties with courage, confidence, and hope." A few minutes later, in perhaps the most memorable line in the speech, Reagan declared, "America is too great for small dreams." He then suggested: "Let us unite tonight behind four great goals to keep America free, secure, and at peace in the eighties together."

After discussion of his first great goal, "steady economic growth," Reagan indicated that "our second great goal is to build on America's pioneer spirit" by developing "our next frontier: space." Then came the words that the space community had been hoping to hear:

> Our progress in space—taking giant steps for all mankind—is a tribute to American teamwork and excellence. Our finest minds in government, industry, and academia have all pulled together. And we can be proud to say: We are first; we are the best; and we are so because we're free.
>
> America has always been greatest when we dared to be great. We can reach for greatness again. We can follow our dreams to distant stars, living and working in space for peaceful, economic, and scientific gain. Tonight, I am directing NASA to develop a permanently manned space station and to do it within a decade.
>
> A space station will permit quantum leaps in our research in science, communications, in metals, and in lifesaving medicines which could be manufactured only in space. We want our friends to help us meet these challenges and share in their benefits. NASA will invite other countries to participate so we can strengthen peace, build prosperity, and expand freedom for all who share our goals.[12]

Gathered in a meeting room at a Holiday Inn a few blocks from the Capitol building where the president was speaking were members of NASA's Space Station Task Force. McCurdy recounts that they were joined "by interested contractors, international contacts, congressional staffers, White House aides, and allies from other agencies with an interest in the initiative." (This author was also in the room.) As Reagan spoke these words, the crowd erupted in cheers. Some in the room had been waiting two decades to hear a president say these lines. The next morning, *The New York Times*, somewhat overstating what the president had said, headlined its page one story on Reagan's announcement "President Backs U.S. Space Station as Next Key Goal," but then added, "Permanently Manned Base for Colonizing Planets is Seen in $20 Billion Program."[13]

Reagan also briefly mentioned space commercialization in his address, saying, "Just as the oceans opened up a new world for clipper ships and Yankee traders, space holds enormous potential for commerce today … We'll soon implement a number of executive initiatives, develop proposals to ease regulatory constraints, and, with NASA's help, promote private sector investment in space." Three days later, in his regular Saturday morning radio broadcast, Reagan expanded on his State of the Union remarks, and placed them in the context of the emerging national space strategy. He told his audience he wanted to lay out "how we can advance America's leadership in space through the end of this century and well into the next, and how, by reaching for exciting goals in space, we'll serve the cause of peace and create a better life for all of us here on Earth." He added: "Our approach to space has three elements ... The first is a commitment to build a permanently manned space station to be in orbit around the Earth within a decade. It will be a base for many kinds of scientific, commercial, and industrial activities and a steppingstone for further goals." A second element of the strategy was "international cooperation," which "has long been a guiding principle of the United States space program." Reagan added "the third goal of our space strategy will be to encourage American industry to move quickly and decisively into space. Obstacles to private sector space activities will be removed, and we'll take appropriate steps to spur private enterprise in space." He concluded his remarks by saying: "Our space program has done so much to bring us together because it gives us the opportunity to be the kind of nation we want to be, the kind of nation we must always be—dreaming, daring, and creating."[14]

Conclusion

With his January 25, 1984, announcement, Ronald Reagan formally initiated a space station program that, together with the space shuttle approved by President Richard Nixon just over 12 years earlier, would dominate U.S. human space flight activities well into the twenty-first century. The space station program would have a tumultuous history, with many redesigns; the possibility of cancellation; finally, after the end of the Cold War and the breakup of the Soviet Union, adding Russia as a partner; and from 2000 on, continuous astronaut occupancy. In reflecting on how he convinced Reagan to approve the space station, James Beggs remembered: "I told him what it would do. I gave him a number of presentations on the potential of the space station, what we could do, the potential of commercial activities, the potential for long-term research." Beggs added "I pointed out, which is an argument he liked, that you'll be able to see it with the naked eye ... Knowing that he was an actor, I quoted from [Shakespeare's] *Julius Caesar*: 'There is a tide in the affairs of men which, taken at the flood, leads on to fortune. Omitted, all the voyages of our life are spent in shallows and in miseries. On such a broad course we are now engaged and we must take the current when it serves or lose our ventures.' He liked that." Beggs observed that Reagan "was not a hard sell. His staff was a very hard sell, but he was not."[15] From the first time he was exposed to the space station concept, Ronald Reagan had seen it as an important step in space leadership. He was willing to approve the station in the face of determined opposition from some of his most senior advisers. He might not have intervened as the station decision process unfolded, but in the end his preference prevailed. Ronald Reagan's approval of the space station was a leadership decision.

Notes

1. The December 8 11:30 a.m. draft and the January 11 6:00 p.m. draft, marked "Master," are in Box 129, Speech Drafts, Speechwriting Office; the January 5 draft and the January 11 3:30 p.m. draft are in OA 11688, Papers of Bruce Chapman, RRL.
2. For a discussion of the origins of international participation, see John M. Logsdon, "Together in Orbit: The Origins of International Participation in the Space Station," NASA, Monographs in Aerospace History #11, November 1998, available at https://history.nasa.gov/monograph11.pdf.
3. Memorandum from William Schneider to Robert Kimmitt, National Security Council, "Department of State Position on the Space Station," August 18, 1983; Memorandum from Richard Burt and Paul Wolfowitz to Under Secretary William Schneider, "SIG (Space) Discussion of Space Station Project," August 16, 1983, both in Box 5, Papers of Edwin Meese, RRL.
4. Memorandum from Robert Kimmitt to Gil Rye, "Space Cooperation with the Soviets," December 15, 1983, Box 99, Subject Files, Space Policy (11/21/83-1/15/84), Executive Secretariat, National Security Council, RRL.
5. Memorandum from Gil Rye to Robert McFarlane, "International Cooperation in Space," December 16, 1983, Box 99, Subject Files, Space Policy (11/21/83-1/15/84), Executive Secretariat, National Security Council, RRL.
6. Memorandum from Jack Matlock to Robert McFarlane, "Space Cooperation with the Soviets," December 22, 1983, Box 99, Subject Files, Space Policy (11/21/83-1/15/84), Executive Secretariat, National Security Council, RRL.
7. Interview of Hans Mark by Rebecca Wright for International Space Station Oral History Project, July 8, 2015, NHRC; Mark repeated this story in a May 24, 2016, interview with the author.
8. Memorandum from Jack Matlock to Robert McFarlane, "Space Cooperation with Russia," December 22, 1983, Space Policy Files (11/21/83-1/15/84), Executive Secretariat, National Security Council, RRL. Interview of James Beggs by Kevin Rusnak, March 7, 2002, NHRC.
9. "Eyes Only" memorandum from J.M. Poindexter to Larry Eagleburger, Fred Ikle, Bob Gates, and Jim Beggs, "Space Initiative," January 18, 1984, Space Policy Files (1/16/84-2/16/84), Executive Secretariat, National Security Council, RRL.
10. The Reagan message is quoted in John Logsdon, *Together in Orbit*, 20.
11. Interview with Peggy Finarelli, July 10, 2017.
12. Ronald Reagan: "Address Before a Joint Session of the Congress on the State of the Union," January 25, 1984. Online by Gerhard Peters and John T. Woolley, APP, http://www.presidency.ucsb.edu/ws/?pid=40205.
13. Howard McCurdy, *The Space Station Decision*, (Baltimore, MD: The Johns Hopkins University Press, 1990), 189–190. Philip Boffey, "President Backs U.S. Space Station as Next Key Goal," *The New York Times*, January 26, 1984, A1.
14. Ronald Reagan: "Radio Address to the Nation on the Space Program," January 28, 1984. Online by Peters and Woolley, APP, http://www.presidency.ucsb.edu/ws/?pid=40349.
15. Interview of James Beggs by Kevin Rusnak, March 7, 2002, NHRC.

CHAPTER 11

Together in Orbit: Round One

Ronald Reagan in his January 28, 1984, radio address, three days after announcing his space station decision, discussed that decision in terms of the White House attempts to craft a national space strategy. He added that a second element of that strategy was "international cooperation ... Just as our friends were asked to join us in the shuttle program, our friends and allies will be invited to join with us in the space station project."[1]

Inviting other countries to participate in the space station was only one of the steps toward enhanced international cooperation pursued by the Reagan administration in the 1983–1985 period. Invitations were extended by the White House to a number of countries to fly one of their citizens aboard the shuttle. Perhaps most intriguingly, there were preliminary discussions with the People's Republic of China about flying a Chinese-developed experiment, accompanied by a Chinese payload specialist, on the vehicle. Although the White House in 1982 had decided to let lapse the U.S.-U.S.S.R. space cooperation agreement first signed in 1972 and renewed in 1977, and in the final stages of the space station decision process had rejected the possibility of inviting the Soviet Union to participate along with U.S. "friends and allies" in the station, it did in January 1984 float the idea of a joint U.S.-Soviet space rescue demonstration. Although the Soviet rebuffed that idea, the White House decided to repeat it later in the year, after President Reagan on October 30, 1984, signed a congressional joint resolution calling for cooperative East-West ventures in space, including the joint rescue demonstration. The Reagan administration was trying to engage all spacefaring countries with the U.S. space program, thereby solidifying U.S. space leadership.

J. M. Logsdon, *Ronald Reagan and the Space Frontier*,
Palgrave Studies in the History of Science and Technology,
https://doi.org/10.1007/978-3-319-98962-4_11

Extending the Space Station Invitation[2]

Between January and June 1984, international participation in the U.S. space station went from a last-minute insertion in Ronald Reagan's State of the Union address to a significant U.S. foreign policy initiative to be discussed among Reagan and the leaders of the other six members of the Group of Seven—the leading economies of the world—as they met in London for their annual economic summit. It was the U.S. hope to get at the early June summit top-level political agreement, at least in principle, to participate in the U.S. space station. The summit would be attended by the heads of government of Canada, France, Germany, Japan, Italy, and the United Kingdom—precisely the "friends and allies" that the United States wanted to engage in space station cooperation.

Exploring Partner Interest

The decision to link space station cooperation to the London summit lent a sense of urgency to learning how potential partners would react to Ronald Reagan's invitation. That linkage was not part of the January 18 discussion that had led to the decision to include the cooperative invitation in the State of the Union speech, but it had quickly surfaced in the days following the January 25 address.

The connection had already been suggested in an August 1983 memorandum from the heads of the State Department Bureaus of European and East Asian and Pacific Affairs, saying that "if the President decides to go ahead with the Space Station Project, international participation in it could be a useful contribution to the 1984 London Summit." The drafter of that memorandum was Tom Niles of the European Bureau. Also, Gil Rye had in December suggested to National Security Adviser Bud McFarlane that if the president did indeed want to invite other nations to form a space station partnership, "the London Summit in May 1984 could be used as the occasion to cement these details ... in a very visible way."[3]

As he was made aware of the January 18 decision to invite international participation in the space station as part of the State of the Union speech, Niles quickly added that possibility to the already-underway preparations for the June summit.

> Having seen this proposal, my colleagues and I in the State Department who were responsible for the Department's participation in planning for the Summit concluded that this was an appropriate initiative. We based this conclusion on the obvious need for initiatives in connection with the Summit, the fact that the Summit participants were the obvious choices to join with us in the space station, and the reality that kicking a proposal of the magnitude of the space station up to the Head of State/Government level, through the Summit process, is often the best way to get a decision.[4]

The day before he made the public invitation to cooperate in developing the space station, Ronald Reagan sent a personal communication to other leaders alerting them to the forthcoming offer and telling them that he would be dispatching National Aeronautics and Space Administration (NASA) Administrator James Beggs, acting as his "personal emissary," to their countries to discuss the offer. After a month of discussion following Reagan's State of the Union speech, the approved terms of reference for the Beggs trip were issued by National Security Advisor Bud McFarlane on February 25. They specified the following:

- In his discussions with foreign officials, the administrator should
  - Explain NASA's current plans for development of a permanently manned space station, with emphasis on expected capabilities, modular design, anticipated availability, and relationship to the president's overall civil and commercial space program.
  - Assess the extent of foreign interest in program participation. This assessment should include the level of overall interest, the expected benefits to be achieved, and the foreign resource commitments that might be forthcoming.
- During the discussions with foreign officials, the Administrator should avoid making specific commitments regarding international cooperation until other U.S. government agencies have had the opportunity to review the implications.
- NASA will then take the lead, with the support of the Department of State, in preparing a report on approaches to international cooperation on the Space Station program. The report will outline various objectives and approaches for international cooperation, including those related to technology transfer and insuring consistency with overall U.S. foreign policy. The report will be presented to Senior Interagency Group for Space [SIG (Space)] members for review and comment.
- This report will be completed by April 2, 1984, and will be presented to the president in order that he may define the framework for international collaboration on the Space Station program which can be formally proposed and agreed upon at the London Summit in June 1984.[5]

Beggs Tours the World

The original plans for the Beggs trip called for the use of commercial airlines. Vice President George Bush, who had offered quiet support for the international initiative all along, suggested to the NASA administrator that he request the use of one of the Air Force planes available to the White House; Bush indicated that he would support such a request. Accordingly, on February 19, Beggs wrote to White House Chief of Staff James Baker requesting the use of a government airplane, arguing that it was "justified and appropriate" because of "the President's direct instruction, the extremely tight timetable, and the importance which space station has assumed here and abroad as a central feature of this Administration's leadership program." The plane was provided by the White House, and Beggs and an entourage that included Rye from the National Security Council (NSC) staff, Phil Culbertson, John Hodge, Ken Pedersen, Peggy Finarelli, and Lyn Wigbels from NASA, and Mark Platt and Michael Michalik from the State Department, left Washington on March 3. There were no representatives of the national security community in the delegation; there had been pressure from that community to at least include one Department of State official from the part of the department concerned about technology transfer, but that did not happen. The delegation first traveled to London, then Bonn, Rome, and Paris, and flew directly from Paris to Tokyo, returning to Washington on March 13. After a few days home, the group visited Ottawa on March 19. At every stop, Beggs and his delegation met with space officials and with the highest ranking nonspace officials available. Beggs formally reiterated Reagan's invitation to consider participation in the U.S. space station program, and he responded to questions and concerns.

An issue in almost every discussion was the character and cost of the contribution which NASA was hoping for. Beggs had asked Pedersen, NASA's director of international affairs, for an estimate of what a reasonable expectation might be. Pedersen's response noted that European countries had contributed approximately 12 percent of the costs of developing the space shuttle and associated capabilities, and that it was "reasonable to expect similar percentage contributions from these countries to Space Station." He noted that the German estimate for a potential European space station contribution was $1.5 billion and that Canada was considering a station contribution that "would cost roughly the same" as the $100 million Canada had spent on the space shuttle remote manipulator system. Pedersen thought that "it is probably not realistic" to expect Japan's contribution to be half that of Europe, but he noted that the pressurized module that Japan was considering "would cost Japan at least $500 million to develop given their current lack of related experience."

Another issue discussed at every stop on the trip was possible national security uses of the space station. Pedersen had prepared talking points for Beggs and his delegation, pointing out that the stated position of the Department of Defense (DOD) was that it "had no requirements for a Space Station." Thus President Reagan had approved "a civil Space Station to be funded entirely out of NASA's budget." However, "like the Shuttle, the Space Station will be available for [all] users." Thus, "if there are any national security users … they will surely be able to arrange to use the facility to do work in space." Pedersen added "in any event, international cooperation and use by our national security community are not necessarily incompatible. We've handled both well in the Shuttle program."[6]

Upon his return from Europe and Japan, Beggs wrote to Secretary of State George Shultz, summarizing his assessment of the trip to date. He told Shultz:

> The reaction so far to the President's call for international cooperation has been both strongly positive and openly appreciative. It has been positive in the sense that our principal Allies are moving quickly, or have already moved, to make

> political decisions to participate. And their reactions clearly show appreciation for the major foreign policy benefits that will flow from open and collaborative cooperation on such a bold, visible and imaginative project. I heard nothing but praise and admiration for the President's foresight and leadership in making this decision.

Beggs judged that Italy, Germany, and Japan had already made the political decision, at least in principle, to participate and that France was also likely, after tough bargaining, to be involved. French President Francois Mitterrand in February 1984 had proposed that Europe develop its own small space station for military purposes, but that proposal had gotten a cold reception from other European countries. When Mitterrand met with Beggs, who described the French president as "that old curmudgeon," he told the NASA chief that "it appears that [a European station] will not happen so we will have to be with you on this."[7] The reception in Great Britain had been the coolest on the trip; that country historically had been skeptical of the value of humans in space. It seemed that European cooperation would be organized through the European Space Agency (ESA), rather than on the basis of bilateral relationships between the United States and specific European countries. The uncertainty of an upcoming national election made it impossible for Canada to indicate its commitment to cooperation.

Reporting on the Beggs Trip

The terms of reference for the Beggs trip included a requirement for a post-trip report setting out "various objectives and approaches for international cooperation, including those related to technology transfer and insuring consistency with overall U.S. foreign policy." By insisting on such a report and by requiring it to be reviewed by all members of SIG (Space), officials in those agencies who had been bypassed in the leadership decision to include the cooperative invitation in the State of the Union address would have an opportunity to insert their perspectives before the report was presented to President Reagan.

A first draft of the report was sent to members of the Interagency Group for Space [IG (Space)] for review on March 22. Agency reactions to the draft were so numerous and varied that the NSC soon abandoned the April 2 deadline for completing the report and instead formed an interagency drafting team to incorporate agency comments into a second draft. The goal was now to have the report in final form by May 4. Beggs in an April 6 letter to McFarlane noted that "there's much work still to be done within the bureaucracy if we're going to implement the President's decision for an international Space Station."[8]

The second draft of the report was sent to IG (Space) member agencies on April 12. It identified two points of disagreement. NASA and the Office of Management and Budget (OMB) disagreed with respect to how to handle potential international contributions to the "core" $8 billion station. Noting that some of the suggested international contributions discussed during the Beggs trip might replace U.S. elements of the station design that had been presented to the president, OMB, as usual protective of the public purse, suggested that "to the extent that foreign participation in core elements is realized, the monetary value of such foreign contributions will be considered as offsets to the $8B U.S. program." NASA rejected the idea that foreign contributions to the station core were acceptable, arguing that "international government-to-government cooperation in the Space Station will take place on elements which are additive to the core capability of the U.S. Space Station." NASA's Hans Mark commented that "asking foreigners to help us save money is not how the world's leader in space goes about satisfying its requirements."[9]

The other area of disagreement was between NASA and the national security community; the latter continued its concern regarding the possibility of unwanted technology transfer. The Intelligence Community suggested that "the Senior Interagency Group on the Transfer of Strategic Technology (SIG(TT)) will be the responsible interagency body to protect against unwarranted technology transfer." NASA did not want a body on which it was not represented to have this responsibility, and suggested that the report say that "NASA recognizes that it is fully responsible for protecting against adverse technology transfer," that "NASA will undertake to consult on a timely basis with other agencies expert in this area," and that "SIG (Space) will review as necessary policy issues that might arise in these consultations."

The report noted that "a Summit declaration regarding the international nature of the Space Station project would be a highly visible demonstration of U.S. space leadership and would underscore the unifying effects of high technology cooperation in the alliance." It recommended that those in the United States preparing for the summit "should work to extract the most committing Summit language possible from potential Space Station partners."[10]

Space Station Cooperation and the London Summit

Having the invitation to participate in the space station come from the U.S. president and be extended to other heads of government had changed the stakes. On January 30, 1984, Ronald Reagan approved the suggestion that the invitation to participate in the space station should become an item for discussion at the forthcoming G-7 summit in London. This added even more political urgency to the invitation. The preceding two years of discussions at the space agency level, and the biases toward collaboration that had emerged from those discussions, had been transformed into an issue high on the political agenda.

No ally wanted to be in a position, without compelling reasons, to refuse President Reagan's public invitation. On the other hand, all three potential partners—Japan, Canada, and major European countries acting through ESA—were in the midst of their own internal debates over the future direction of their space efforts. Accepting the U.S. invitation, even in principle, implied that a significant share of their space budgets over the coming decade would have to be channeled into a partnership with the United States. Whatever their leanings toward accepting Reagan's invitation, in few of those potential partners had there yet been enough discussion to make their leaders ready to make an early political commitment to collaboration of that character and scope.

Preparing for the annual economic summit meeting among the leaders of the United States, the United Kingdom, France, Germany, Italy, Japan, and Canada—the "Group of 7"—was more or less a year-round process, gaining in intensity as the date of the meeting approached. The preparatory process was led in each summit country by a senior official designated as a "Sherpa," so called after the Nepalese ethnic group who traditionally have served as the porters who carry the heavy equipment required by Himalayan mountain climbers. Summit Sherpas "bear the burden" of managing summit preparations for their respective countries. The U.S. Sherpa for the London summit was W. Allen Wallis, undersecretary of state for economic affairs. Wallis was quick to follow up Reagan's approval. At a February 17–19 meeting of summit Sherpas from all G-7 countries, Wallis proposed that the summit leaders approve a statement that they "agree in principle to cooperate in the development of an international Space Station, demonstrating that free nations will continue to use outer space for peaceful purpose and for the benefit of all mankind." At an April 2–4 Sherpa meeting the United Kingdom proposed a much weaker summit statement; it would only "note ... the proposals of the United States for a manned Space Station and to pursue follow-up discussions as appropriate." The United States responded by proposing alternative language that deleted the "agree in principle" phrase deferred for a year the deadline for indicating an intention to participate; the new language indicated that "the United States agreed to report progress on Space Station development to the next Summit," scheduled for Bonn, Germany, in spring 1985.

Beggs wrote to Wallis after his March trip, noting that he had come to understand during his trip that: "The Summit declaration is ... extremely important to NASA's counterpart technical agencies in these other countries. To them it represents the political underpinnings necessary to proceed—analogous to the President's State of the Union guidance to us." The character of European participation was particularly uncertain. Wallis made arrangements to send a second, smaller delegation to Europe to respond to concerns of potential partners and to answer additional questions about the space station program. That delegation would meet with both summit Sherpas from European nations and representatives from those nations' space agencies, to make sure that the two were in communication. The delegation was led by Rye representing the White House, and included Finarelli and Robert Freitag from NASA and Platt from the State Department. Freitag had been one of the leaders of the internal NASA Space Station Task Force and was an enthusiastic advocate of international cooperation.

This group left Washington on April 25 and visited London, Bonn, Rome, and Paris before returning on May 5. In its meetings, the delegation emphasized that the U.S. interest was in "additive capabilities" that would "enable achievement of more ambitious goals," and that the United States was "not seeking resource commitment at the London Summit—only agreement in principle" to establish a political foundation for a "positive approach" to cooperation. They noted that "President Reagan will undoubtedly want to discuss participation with any interested Heads of State" in London, thereby putting pressure on those they met with to prepare their leaders to be ready to respond to Reagan. Another message was that a summit "in principle" commitment to space station cooperation would be "non-binding and contingent upon our ability to negotiate mutually agreeable arrangements."[11]

Preparations in Washington for the summit picked up their pace in May. At a May 14 meeting attended by President Reagan, Beggs reported on the results of his March trip. In a background paper prepared for the president in advance of that meeting, Reagan was told that "at the space/technical agency levels, there was universal enthusiasm for cooperation," but "at the political level, the reactions ranged from guarded to already committed." In particular, as they had met with the Rye delegation "the French, Germans and British all said they could make no commitments in time for a Summit announcement." The paper suggested to Reagan that "at London, you should reiterate the invitation, note that participation in the design stage need imply no commitment yet to build hardware and indicate that we will proceed in any case with those who wish to participate."[12]

Discussions with the summit nations continued throughout the month, but were not successful in moving toward the U.S. objective of including in the summit communique notice of an agreement in principle on the part of other governments to participate in the U.S. space station. Rye told McFarlane on May 25 that while "we are still trying to strengthen the language" in the summit statement, we are "still quite far from a firm commitment to participate in the program. Indeed a majority of the summit partners (with the possible exception of Japan and Italy) have indicated their reluctance to commit resources to the program."[13]

This reluctance did not stem from any fundamental objection to the U.S. cooperative invitation. Canadian uncertainty about making a commitment was primarily due to the fact that Prime Minister Pierre Trudeau had announced his retirement in late February 1984, effective on June 30, and until his successor was elected, the lame duck government was not in a position to make new agreements. European reluctance stemmed from the unsettled condition of European space policy, with France pushing for as much European autonomy, under French leadership, as was feasible, and Germany favoring a policy based on substantial transatlantic cooperation. Until Europe got its own space house in order, agreement on the character of its response to the U.S. invitation would have to wait. In addition, from the European perspective there were a number of issues to be worked out, such as the cost of participation, access rights to the station, and intellectual property rights, before an agreement, even in principle, was possible. European leaders were quite aware that one U.S. objective in inviting Europe and Japan to participate in the station program was to tie those countries' space spending to a cooperative project under U.S. leadership, rather than have it go to projects intended to compete with the United States, as had been the case with *Ariane*.[14]

Creating a Photo Opportunity

McFarlane had asked Rye whether it was wise to mount a significant public relations campaign during the summit meeting linked to space station cooperation as a way of "reflecting US leadership." Rye thought not, since such a campaign would suggest that "the U.S. was overplaying the agreement." Rye suggested that from a public relations standpoint "the most significant product we could hope for out of the Summit is a photograph of all heads of state surrounding the scale model of the Space Station. Such a photograph would visibly demonstrate that all leaders are united in planning for the future of the free world (under U.S. leadership)." Rye suggested that either McFarlane or the keeper of Reagan's image, Mike Deaver, call the British Sherpa, Sir Robert Armstrong, and "lay out our desires, stressing the political value to the President of the photograph of the Heads of State around the model." Rye told McFarlane: "I will plan to bring to London the large crate containing the Space Station model (for which NASA paid $40,000!)."[15]

Ronald Reagan had his final presummit briefing on May 29. Space station cooperation was an agenda item, with Rye reviewing both U.S. objectives and tactics with respect to the invitation and the current positions of other heads of state. Reagan departed Washington for Europe on June 1, spending four days in Ireland before arriving in London on the afternoon of June 4. He visited Queen Elizabeth and Prime Minister Margaret Thatcher on June 5; on June 6, the 60th anniversary of D-Day, he helicoptered across the English Channel to the invasion beaches in Normandy, where he delivered a memorable speech celebrating the heroics of American Army Rangers at Pointe du Hoc. Reagan returned to London that evening to prepare for the Economic Summit, which would begin the next day.

Before the summit meeting opened, Reagan was reminded by a briefing memorandum from McFarlane that "detailed discussions" with respect to potential space station participation had taken place in advance of the summit, and that "all of the Summit leaders expect you to raise this subject during both your bilateral and multilateral discussions." McFarlane told the president "we have reason to believe Nakasone and Craxi, in particular, are prepared to give quite positive responses to your invitation," but that "other leaders will probably be less forthcoming but will continue to study the proposal, reserving formal commitment until their internal decision processes have been completed." McFarlane provided a set of talking points for Reagan's bilateral discussions:

- Objective: Lay a technological base to move us into next century
- Space station symbolizes our faith in future and commitment to work together over the long term toward shared objectives
- The United States will develop space station by the early 1990s
- We welcome our friends and allies and intend to proceed with whomever wants to join
- Understand you have been considering the invitation
- Propose communique welcome invitation and agree to examine how to respond.

The British organizers of the summit had agreed to the idea of a photograph of the summit leaders gathered around the space station model. Rye had accompanied the model, which was the same one that had been used during the December 1, 1983, Cabinet Room briefing to the president, on a cargo flight to London also carrying the limousines the president would use during his trip. McFarlane in his June 7 memo told Reagan that "Mike Deaver is attempting to arrange an opportunity where all of the Heads of State could surround the model for a photograph for use by the media." He added: "Such a photo could visibly transmit the forward-looking objectives we think are so appropriate for this Summit. Obviously, such an image could have tremendous domestic impact as well."[16]

As expected, the declaration issued at the conclusion of the summit reflected the inability of the United States to achieve its original objective of getting the summit countries to agree in principle to participate in the space station program. It said only that:

> We believe that manned space stations are the kind of programme that provides a stimulus for technological development leading to strengthened economies and improved quality of life. Such stations are being launched in the framework of national or international programmes. In that context each of our countries will consider carefully the generous and thoughtful invitation received from the President of the United States to participate in the development of such a station by the United States. We welcome the intention of the United States to report at the next Summit on international participation in their programme.[17]

This statement was seen as "a sterling example of pious generalities," but it accurately reflected the state of affairs. It endorsed the station concept and thanked Ronald Reagan for his invitation. The requirement of a report on station cooperation on the agenda for the 1985 summit was particularly significant. It was intended to encourage speedy decision-making, because any delays or breakdowns in discussions over acceptance of Reagan's invitation would have to be reported back to the summit leaders at their next get-together. Although more time would be needed to find ways in which the U.S. invitation and the separate space goals and ambitions of Europe, Japan, and Canada could be combined in ways acceptable to all partners, there was now a deadline to provide a focus for deliberations around the world.[18]

Fig. 11.1 President Reagan, U.K. Prime Minister Margaret Thatcher, and Japanese Prime Minister Yasuhiro Nakasone, plus several other European officials, listen as Gil Rye explains details of the planned space station during the June 9 lunch break of the London Economic Summit. (Photograph courtesy of Reagan Presidential Library)

The United States also was not successful in its hope to get a picture of all seven heads of state clustered around the space station model. At the end of the morning session on June 9, the summit participants adjourned for prelunch refreshments. As they came back from drinks, they found the station model sitting in front of a flower-bedecked fireplace. British Prime Minister Margaret Thatcher stood next to Ronald Reagan and asked a series of questions. After a few minutes Rye stepped in to help Reagan field Thatcher's queries. There were a number of photographs taken, but none included all seven heads of state. Even so, photographs of Reagan, Thatcher, and Japanese Prime Minister Nakasone, in addition to other less senior officials, found wide distribution (Fig. 11.1).

Invitation Accepted, Negotiations Begin

Europe, Japan, and Canada over the year following the London Economic Summit decided to accept Ronald Reagan's invitation to participate in the U.S. space station program, at least through the two-year detailed design phase. Final agreement to participate would only come in 1988, as the time to begin developing station hardware arrived. Before that agreement, there would be difficult negotiations over the terms and conditions of foreign participation (see Chap. 22). Those negotiations were in the future as NASA in the first half of 1985 signed memorandums of understanding (MOUs) with the ESA, Japan, and Canada for design phase collaboration. The MOU with ESA was the last to be signed, at the Paris Air Show on June 3, 1985.

As the Group of Seven met in Bonn, Germany, on May 2–4, 1985, they were able to declare (with European participation already assured):

> We welcome the positive responses of the member states of the European Space Agency, Canada and Japan to the invitation of the president of the United States to cooperate in the manned space station program on the basis of a genuine partnership and a fair and appropriate exchange of information, experience and technologies. Discussions on intergovernmental cooperation in development and utilization of permanently manned space stations will begin promptly.

The inclusion of the phrase "genuine partnership" in the summit declaration reflected the views of European countries, which felt that they had been treated almost as industrial subcontractors in their cooperation with NASA during the space shuttle program. European space leaders believed that their space programs had advanced to the point that Europe deserved recognition as a well-qualified partner in space station development; they also believed that the terms of the space shuttle partnership had not been equitable and were determined to strike a better deal with respect to the payoffs to their industries from their contribution to the space station. The Bonn summit declaration thus also stated: "We also welcome the conclusions of the E.S.A. Council on the need for Europe to maintain and expand its autonomous capability in space activity, and on the long-term European space plan and its objectives."[19]

The United States had hoped at the Bonn summit to get an even broader commitment to future cooperation. National Security Adviser McFarlane had proposed to President Reagan in December 1984 that the United States should attempt to get its allies, when they met for the 1985 summit, to reach "agreement on a set of long-range goals in space." He noted that such agreement "would serve U.S. interests by: (1) demonstrating, in a highly visible and forward looking way, the unity of purpose that underpins the western alliance; (2) diverting foreign resources away from competitive space projects … (3) enabling the achievement of goals that would not otherwise be possible…: and (4) making SDI even more palatable to the American public, Congress and foreign governments and people through emphasis on the peaceful uses of outer space." However, discussions among the allies in the months prior to the Bonn summit did not lead to agreement on joint pursuit of broad long-range space goals. Indeed, there was no such agreement internal to the United States.[20]

With the acceptance by Europe, Japan, and Canada of Reagan's space station invitation, the largest-ever peacetime advanced technology cooperative project was begun. That cooperation would persist as the negotiations on the terms and conditions of cooperation veered close to collapse; as the space station went through multiple redesigns, delays, and budget overruns; as the U.S. Congress and U.S. President Bill Clinton considered cancelling the project; as Europe lowered its ambitions to develop comprehensive European space capabilities; as the original partners accepted the 1993 Russian suggestion, after the breakup of the Soviet Union, that the Russian human spaceflight program be merged with the U.S.-led space station; and as there was a lengthy stand down of space shuttle flights after the 2003 *Columbia* accident, with some suggesting that the shuttle be permanently grounded before it carried the partners' contributions into orbit.

With, finally, the 2011 completion of assembling what by then had been for many years called the International Space Station, Ronald Reagan's 1984 vision of the United States and its "friends and allies" to "follow our dreams to distant stars, living and working in space for peaceful economic and scientific gain" became reality. Like the space station itself, the international partnership that brought it into being is a major element of the Reagan space legacy.

Other International Initiatives

Offering other countries an opportunity to partner with the United States in developing and operating a space station was by far the most far-reaching initiative to enhance international involvement in the U.S. space program. But there were several other paths pursued in the 1982–1985 period.

Flying Foreign Astronauts

As part of the agreement with ESA to contribute the Spacelab pressurized laboratory and other elements to the space shuttle program, NASA recognized that ESA would want to fly European astronauts on missions carrying the Spacelab. An ESA astronaut program was begun in 1978 with the selection of four individuals from West Germany, Switzerland, the Netherlands, and Italy. One of those four, West German Ulf Merbold, became the first non-American to fly on the shuttle during the initial Spacelab mission, which took place from November 28 to December 8, 1983.

When the White House queried NASA's Beggs in 1982 about the possibility of flying fare-paying passengers on the shuttle, Beggs had suggested that "the interests of the United States might be better served by using these opportunities in support of foreign policy, as is the USSR practice in recent years using their Space Station to host foreign nationals." By 1982, the Soviet Union had flown several "guest cosmonauts" for short stays on the Soviet *Salyut* space station. Beggs's suggestion was pursued. An Italian/American working group established when Italian Prime Minister Giovanni Spadolini visited Washington in late 1982 recommended that the United States, as part of a joint effort to develop a tethered satellite system, invite an Italian astronaut to be aboard the shuttle when the experiment was flown, then planned for 1987. In the memorandum to President Reagan recommending that he issue such an invitation, National Security Adviser William "Judge" Clark had reminded the president that "similar offers of cooperation have been made with the Brazilian and Japanese governments." Reagan accepted Clark's recommendation, and on

May 12, 1983, in a letter to the new Italian prime minister Amintore Fanfani, he said that "the planned joint development of a Tethered Satellite System … provides a unique and highly visible opportunity to demonstrate our close cooperation in space … It is with pleasure and a sense of adventure that I invite Italy to nominate Italian candidates" to fly aboard the tethered satellite mission.[21]

In connection with his March 1984 trip to Europe to discuss participation in the space station program with potential partners, Beggs on February 28 wrote to McFarlane, Clark's replacement as national security adviser, suggesting that "it would be greatly appreciated if during my visit to London I extend a formal invitation to the British government to nominate an astronaut to fly with us … in the near future. It would be particularly fitting if I could do this on behalf of the President." McFarlane on March 1 forwarded for Reagan to sign a letter to British Prime Minister Margaret Thatcher extending such an invitation. He noted as a supporting rationale for the invitation that "the British have recently booked two spaces on the Shuttle for their Skynet satellite despite strong pressure to utilize a French-made launch vehicle." The plan was to have Beggs hand carry the invitation as he left for Europe and offer it to Prime Minister Thatcher at a March 5 meeting. It turned out that Mrs. Thatcher was not in London during the Beggs visit; it is not clear whether Beggs left the letter with the prime minister's staff for later delivery. In any event, the United Kingdom decided not to accept the initial U.S. invitation to participate in the space station program. (It decided to join the program only in the 1990s.) No British citizen who was not also jointly a U.S. citizen ever went into space aboard the shuttle.

In his March 1 memorandum to President Reagan regarding the letter to Prime Minister Thatcher, McFarlane mentioned that "plans are underway for a similar invitation to President Mitterrand during his visit in late March." The French and American presidents met on March 22–23. Space cooperation did not come up on the first day; upon learning this, Rye told McFarlane: "It is important that sometime during tomorrow's session, the President invite the French to fly an astronaut on the Shuttle and highlight the importance of cooperation on the Space Station program." He provided talking points for Reagan to use with Mitterrand. They suggested that the president say, "I would be pleased if you would accept my offer to nominate a French astronaut to fly on an upcoming shuttle mission" and that "this would be particularly fitting in the light of the on-going space cooperation between our two countries, including a partnership on the manned space station program." Reagan used these talking points as the first order of business when the two presidents and their associates met over breakfast on March 23. Mitterrand responded to the invitation by saying, "Space is a grand adventure"; he "readily accepted the President's offer," suggesting that "cooperation on space efforts was a very promising venture." Subsequent to the Reagan-Mitterrand meeting, Patrick Baudry on a June 1985 mission became the first Frenchman to fly aboard the space shuttle.[22]

Even a Chinese on the Space Shuttle?

Ronald Reagan visited China between April 26 and May 1, 1984. This was the first visit to China by a U.S. president since Richard Nixon's February 1972 path-breaking trip. China had launched its first satellite in April 1970 and in the following years had developed a family of Long March expendable launch vehicles and had considered beginning a human spaceflight program, but its space activities in the early 1980s were still modest in character. In preparation for the Reagan visit, science adviser Jay Keyworth's top assistant for space, Richard Johnson, who had replaced Vic Reis in late 1983, and Pete Smith, a staff person from NASA's Office of International Affairs, traveled to Beijing in March to discuss a possible space cooperation initiative as part of the U.S--China summit outcome. That initiative was focused on flying a Chinese science experiment on the space shuttle, accompanied by a Chinese payload specialist, with the flight to take place in the 1988–1989 time frame.

Johnson was told that China was very interested in the proposed space initiative, but could not respond in time for an announcement to be made at the time of the summit meeting. Instead, the United States and China set up an expert group to flesh out the proposal. That group met once in 1985 and again in January 1986; its discussions stayed at the technical level, with little political visibility. The group agreed on flying a plasma physics experiment in the shuttle's payload bay. In addition to operating that experiment, a Chinese payload specialist would participate in several life sciences tests. In parallel with the expert group discussions, the two sides began negotiating a framework agreement on space cooperation to provide the political and policy context for implementing the initiative. Agreement on the essence of that framework had been reached by January 1986.[23]

The January 1986 *Challenger* accident put a temporary hold on further U.S-China discussions, but China during 1986 and 1987 continued to push the United States for a flight opportunity. During 1987 meetings on science cooperation, NASA was "repeatedly reminded of President Reagan's 1984 offer" and asked for "continued U.S. consideration of accommodating a Chinese Payload Specialist on the Shuttle." Given the limited payload specialist slots available in the post-Challenger period, the flight opportunity was never reoffered.[24] One can only wonder about what path subsequent U.S.-Chinese space cooperation might have taken, if the 1984 initiative had come to fruition.

Cooperating with the Soviet Union

On July 17, 1975, an U.S. Apollo Command and Service Module docked with a Soviet *Soyuz* spacecraft in what NASA formally designated the Apollo-*Soyuz* Test Project. The undertaking was both a symbolic demonstration of U.S.-Soviet *détente* and an initial step toward developing a space rescue capability. The project was carried out under a U.S.-Soviet space cooperation framework agreement that President Richard Nixon and Soviet Premier Alexei Kosygin had signed at a summit meeting in May 1972. That agreement was renewed in May 1977; at that point, it presciently contemplated a rendezvous between the U.S. space shuttle and a Soviet space station, followed at some future time by U.S.-Soviet cooperation in developing a next-generation station. Following the Soviet incursion into Afghanistan in 1979, the Carter administration decided not to implement the 1977 agreement, and U.S.-Soviet space cooperation was reduced to a few low-level projects. The cooperation envisioned in the 1977 agreement did eventually occur, but not until the 1990s after the collapse of the Soviet Union.[25]

The 1977 agreement came up for renewal in 1982. In a 2002 interview, NASA's James Beggs suggested that he had recommended renewal to the White House. He thought that the agreement was "fairly innocuous." It governed only "the terms and conditions" under which the United States and the Soviet Union could cooperate. But, said Beggs, "the White House didn't want to renew it." Beggs went to see McFarlane at the NSC, saying, "We really ought to do this. I'd rather have an agreement where we can have some window into what they're doing." McFarlane apparently agreed, but told Beggs, "I can't persuade anybody to do it." McFarlane's boss at that time, National Security Adviser Judge Clark, was perhaps even more strongly anti-communist than Ronald Reagan. Reagan had not yet used publicly the phrase "evil empire" to describe the Soviet Union—that would come in 1983—but since his time in Hollywood he had been strong and persistent in his negative attitude toward the U.S. Communist party and its Soviet patron. A more immediate factor in the decision not to renew the space cooperation agreement was as a reaction to the Soviet support of the December 1981 imposition of martial law in Poland as a way of suppressing the Solidarity anti-communist movement.[26]

Matsunaga Initiatives

There was early countervailing pressure on the Reagan administration to work together with the Soviet Union in space. In July 1982, Spark Matsunaga (D-HI), the junior senator from Hawaii, wrote to Reagan, enclosing an article he had published in *The Washington Post* titled "Find Peace with Russia in Space." There had been reports circulating that the DOD was considering basing weapons on an orbiting platform; Matsunaga proposed instead making "the first orbiting space station a weapons-free international project involving the United States and the Soviet Union, as well as other interested nations having a space capability." In his letter, he suggested that "space [cooperation] offers the best opportunity" for reducing tensions with Soviet Union "without surrendering any aspect of our national security."

The Matsunaga article was the result of a suggestion to the senator from a young Hawaii-based consultant, Harvey Meyerson. Meyerson met with Matsunaga, sharing his concern about an armed space station and space weaponization overall, and Matsunaga told him "to see what I could do." In the next several years, it would be Meyerson, working through Matsunaga, who became the architect of a series of space cooperation proposals that culminated with a 1985 initiative embraced by the Reagan administration—that 1992 be designated as the "International Space Year."[27]

A Reagan Cooperative Initiative

National Security Adviser McFarlane in December 1983 had raised the possibility of inviting not only U.S. allies but also the Soviet Union to participate in the space station program that Reagan would announce in his January State of the Union Address. That possibility was short-lived. As an alternative to space station cooperation, a joint U.S.-Soviet demonstration of space rescue capability was then suggested. In this effort, a space shuttle would dock with the Soviet *Salyut* space station and the two crews would simulate a space rescue. This proposal was seen as a follow-on to the 1975 Apollo-*Soyuz* Test Project and had been mentioned as the U.S./Soviet space cooperation agreement had been renewed in 1977. Thus, the original January 18 insert into the text of the State of the Union Address that invited international participation in the space station had also said:

> We also recognize the importance that the Soviet Union attaches to its manned space flight program. Tonight, I extend an offer to that Nation to join with us in a manned flight project that can achieve mutual objectives and demonstrate a spirit of goodwill. More specifically, I propose that the United States and the Soviet Union join in a simulated space rescue mission using both the U.S. Space Shuttle and the Soviet *Salyut* space station. Such a project could demonstrate a capability that will be important to both of us as we develop a more permanent presence in space. Successful cooperation in this area might also open the door to more ambitious efforts in the future.

This proposal for a joint U.S./Soviet rescue mission, like the idea of inviting the Soviet Union to participate in the space station, had a short lifetime. Soviet officials were contacted through diplomatic channels and asked whether they would accept such an invitation. There was no immediate response, and with the January 25 date of the State of the Union address less than a week away, the rescue mission proposal was not inserted into the speech text. The Soviet Union finally provided an informal response in March, linking any discussion of a space rescue proposal to a treaty banning anti-satellite weapons. This linkage was not acceptable to the United States, and the proposal was not pursued. It was not forgotten, however. In a June 27, 1984, speech on U.S.-Soviet exchanges in which he listed a number of areas of possible enhanced cooperation, Reagan noted: "We've proposed a joint simulated space rescue mission in which astronauts and cosmonauts would carry out a combined exercise in space to develop techniques to rescue people from malfunctions in space vehicles."[28]

Senate Joint Resolution 236

After two years of effort by Senator Matsunaga, both houses of the Congress on October 12, 1984, approved Senate Joint Resolution 236, "Cooperative East-West Ventures in Space." The resolution had three operative clauses, calling upon the Reagan administration to:

1. Endeavor, at the earliest practicable date, to renew the 1972–1977 agreement between the United States and the Soviet Union on space cooperation for peaceful purposes;
2. Continue energetically to gain Soviet agreement to the recent U.S. proposal for a joint simulated space rescue mission; and
3. Initiate talks with the Government of the Soviet Union, and with other governments interested in space activities, to explore further opportunities for cooperative East-West ventures in space including cooperative ventures in such areas as space medicine and space biology, planetary science, and manned and unmanned space exploration.

The resolution also had a number of "whereas" clauses linking space cooperation to preventing a potential arms race in space and saying that "the opportunities offered by space for prodigious achievements in virtually every field of human endeavor, leading ultimately to the colonization of space in the cause of advancing human civilization, would probably be lost irretrievably were space to be made into yet another East-West battleground."[29]

There was a hurried interagency review examining whether President Reagan should sign the resolution. Only the DOD recommended against that action, saying that while it implied that "cooperative programs could moderate Soviet behavior, the historical record suggests otherwise." In addition, DOD was concerned that "cooperative East-West space projects would result in the undesirable transfer of critical technology" and would "reward the Soviets with prestige and technical benefits" despite negative Soviet behavior. The State Department and the NSC recommended that the president "sign the enrolled resolution and issue a signing statement which addresses its pitfalls." President Reagan chose this second course. He endorsed the resolution on October 30; his signing statement said: "I find portions of the language contained in the preamble to the Joint Resolution very speculative," but that "we are prepared to work with the Soviets on cooperation in space in programs which are mutually beneficial and productive."[30]

In the following weeks, McFarlane asked Reagan whether he wanted to renew the offer of the joint rescue mission to the Soviet Union. Reagan responded positively, but added in a handwritten note, "I've approved but with the assumption that these projects will not provide the Soviets with 'High Tech' information they do not have & which can be used for mil. [military] purposes." McFarlane then wrote to Secretary of State George Shultz, saying that "the President believes the Strategic Stability Talks in Geneva next month

[January 1985] would be an opportune occasion to renew our offer to the Soviet Union for a joint U.S./Soviet mission to develop space rescue techniques." McFarlane added that "although this mission is not an arms control issue per se, discussion on this subject could hopefully assist in diverting attention from divisive issues."[31]

On January 8–9, 1985, Shultz met in Geneva with his Soviet counterpart, Andrei Gromyko. The focus of their meeting was establishing a framework for resuming arms control talks, but Shultz did renew the offer of a joint space rescue mission. Once again, the Soviet Union did not accept the offer. The Soviet Union at this point linked civilian space cooperation with the United States to abandoning the U.S. Strategic Defense Initiative program and other military space efforts, a linkage that the Reagan administration strongly resisted. The prospect for increased U.S.-Soviet space cooperation would have to wait.

Notes

1. Ronald Reagan: "Radio Address to the Nation on the Space Program," January 28, 1984. Online by Gerhard Peters and John T. Woolley, APP, http://www.presidency.ucsb.edu/ws/?pid=40349.
2. Much of the material in this chapter is based on John M. Logsdon, *Together in Orbit: The Origins of International Participation in the Space Station*, NASA Monographs in Aerospace History #11, November 1998, available at https://history.nasa.gov/monograph11.pdf. Citations to primary material included in this monograph will not be repeated here.
3. Memorandum from Richard Burt and Paul Wolfowitz to Undersecretary William Schneider, "SIG (Space) Discussion of Space Station Project," August 16, 1983, Box 5, Papers of Edwin Meese; Memorandum from Gil Rye to Robert McFarlane, "International Cooperation in Space," December 16, 1983, Box 99, Subject Files, Space Policy (11/21/83-1/15/84), Executive Secretariat, National Security Council, both at RRL.
4. Letter from Thomas Niles to the author, May 1, 1990.
5. Letter from Robert McFarlane to James Beggs, February 25, 1984, with attached Terms of Reference, CIA-RDP92B00181R001901730047-9, CREST.
6. Memorandum from director of international affairs to administrator, "Military Involvement in the Space Station," February 28, 1984, File 12905, NHRC.
7. United Press International, "Mitterrand Proposes Station in Space for Western Europe," *NYT*, February 9, 1984, available at http://www.nytimes.com/1984/02/09/us/mitterrand-proposes-station-in-space-for-western-europe.html. "Washington Roundup," *AWST*, April 30, 1984, 13. Interview of James Beggs by Kevin Rusnak, March 7, 2002, NHRC.
8. Letter from James Beggs to Robert McFarlane, April 6, 1984, Box 30, Subject Files, Executive Secretariat, National Security Council, RRL.
9. Memorandum from Fred Khedouri to John Poindexter, "International Cooperation on the Space Station Program," April 18, 1984, and letter from Hans Mark to John Poindexter, April 17, 1984, both available at https://fas.org/irp/offdocs/nsdd/nsdd-144.htm.

10. Memorandum from John Poindexter to members of the Interagency Group for Space, "International Cooperation in the Space Station Program – Second Draft," with attached draft report, April 12, 1984, CIA-RDP92B00181R001901730041-5, CREST.
11. Peggy Finarelli, "Briefing Book for Trip to Meet with Summit Sherpas," provided to author by Ms. Finarelli.
12. Agenda for "Summit White House Group Meeting," with attached background paper for the president, May 11, 1984, and memorandum from Michael Deaver and Robert McFarlane for the President, "Meeting on Preparations for the London Economic Summit, May 14, 1984," undated, both in Box 4, Papers of Edwin Meese, RRL.
13. Memorandum from Gil Rye to Robert McFarlane, "Space Station and the Summit," May 25, 1984, Box 30, Subject Files, Executive Secretariat, National Security Council, RRL.
14. "Europeans Hesitate on U.S. Space Station Plan," *AWST*, June 4, 1984, 24.
15. Memorandum from Gil Rye to Robert McFarlane, "Space Station and the Summit," May 25, 1984, Box 30, Subject Files, Executive Secretariat, National Security Council, RRL.
16. Memorandum from Robert McFarlane to the president, "Space Station and the Summit," June 7, 1984, Box 30, Subject Files, Executive Secretariat, National Security Council, RRL.
17. Ronald Reagan: "London Economic Summit Conference Declaration," June 9, 1984. Online by Gerhard Peters and John T. Woolley, APP, http://www.presidency.ucsb.edu/ws/?pid=40025.
18. William Gregory, "Cooperation on Space Station," AWST, July 9, 1984, 9.
19. Ronald Reagan: "Bonn Economic Summit Declaration on Sustained Growth and Higher Employment," May 4, 1985. Online by Peters and Woolley, APP, http://www.presidency.ucsb.edu/ws/?pid=38586.
20. Memorandum from Robert McFarlane to the President, "International Space Initiatives," December 15, 1984, Box 31, Subject Files, Executive Secretariat, National Security Council, RRL.
21. Letter from James Beggs to Edwin Harper, June 9, 1982, Box 82, Papers of Danny Boggs, RRL. Memorandum from William Clark to the President, "Invitation for an Italian Astronaut," May 12, 1983, and letter from Ronald Reagan to Prime Minister Amintore Fanfani, May 12, 1983, File 12766, NHRC.
22. Memorandum from Gil Rye to Robert McFarlane, "Space Initiative with France," March 22, 1984, Box 30, Subject Files, Executive Secretariat, National Security Council, RRL; memorandum of Conversation, "Summary of President's Working Breakfast Meeting with French President Francois Mitterrand," March 23, 1984, found at http://www.thereaganfiles.com/19840323-hos.pdf.
23. Information on the U.S.-China interactions comes from an undated viewgraph presentation, "Overview of Objectives," RAC 14, Papers of George Keyworth, RRL. Also, memorandum from director of international affairs (Dick Barnes) to administrator (James Fletcher), "Space Cooperation with China," May 29, 1986, and Richard Barnes, "Cooperation with China in Space Science," July 28, 1986, both in Box 6, Papers of James Fletcher, NARA.
24. Peter Smith, "Cooperation with China in Space Science," July 28, 1987, and letter from Richard Barnes to associate deputy administrator-policy, July 28, 1987, both in Box 6, Papers of James Fletcher, NARA.

25. For background on early U.S.-Soviet space cooperation, including texts of the 1972 and 1977 agreements, see John M. Logsdon with Dwayne Day and Roger Launius, *Exploring the Unknown: Selected Documents in the History of the U.S. Civil Space Program*, NASA SP-4407, Volume II, External Relationships (Washington, DC: Government Printing Office, 1996), Chap. 1.
26. Interview of James Beggs by Kevin Rusnak, March 7, 2002, NHRC. On the Soviet Union and Poland, see Mark Kramer, "The Rise and Fall of Solidarity," *The New York Times*, December 11, 2011, www.nytimes.com/2011/12/13/opinion/the-rise-and-fall-of-solidarity.html.
27. Letter from Spark Matsunaga to the president, July 14, 1982, with attached "Find Peace with Russia in Space," July 4, 1982, Box 6, Outer Space Files, RRL. Personal communication from Harvey Meyerson to the author, December 20, 2015.
28. "Eyes Only" memorandum from J.M. Poindexter to Larry Eagleburger, Fred Ikle, Bob Gates, and Jim Beggs, "Space Initiative," January 18, 1984, Space Policy Files (1/16/84-2/16/84), Executive Secretariat, National Security Council, RRL. Ronald Reagan: "Remarks to Participants in the Conference on United States-Soviet Exchanges," June 27, 1984. Online by Peters and Woolley, APP, http://www.presidency.ucsb.edu/ws/?pid=40102.
29. Personal communication from Harvey Meyerson to the author. The text of the resolution can be found at http://www.princeton.edu/~ota/disk2/1985/8533/853310.PDF.
30. Memorandum from David Stockman for the President, "Enrolled Resolution S.J. Res. 236 – Cooperative East-West Ventures in Space," October 26, 1984, Box 3, Outer Space Files, and Office of the Press Secretary, The White House, "Statement by the President," October 30, 1984, RAC 14, Papers of George Keyworth, both in RRL.
31. Memorandum from Robert McFarlane to the president, "International Space Initiatives," December 15, 1984, Box 31, Subject Files, Executive Secretariat, National Security Council, RRL. Memorandum from Robert McFarlane to George Shultz, "International Space Initiatives," December 24, 1984, CIA-RDP86M00886R001000070002-5, CREST.

CHAPTER 12

Space Commercialization

In his January 28, 1984, radio address, Ronald Reagan spelled out his concept of a national space strategy. In addition to the commitment to a space station and to enhanced international cooperation, a third pillar of the space strategy would be "to encourage American industry to move quickly and decisively into space ... We'll take appropriate steps to spur private enterprise in space."[1]

One of Ronald Reagan's core convictions was that the American free enterprise system was a powerful force for ensuring both societal and individual well-being, and it was thus important for the government not to interfere with the workings of the market other than to provide a supportive policy and regulatory environment within which it could operate. The administration had brought with it to Washington a firm belief in the principle that the U.S. government should not carry out activities that could just as well, if not better, be private sector responsibilities. That perspective had been reflected in the July 1982 statement of National Space Policy, which set as one of its six major space goals expanding "United States private-sector investment and involvement in civil space and space-related activities." To achieve that goal, the policy directed government agencies to "provide a climate conducive to expanded private sector investment and involvement in civil space activities"; private sector space activities would be "authorized and supervised or regulated by the government" only "to the extent required by treaty or national security."[2] This was the first time a government role in encouraging commercial space activity had been called out in national space policy, and clearly reflected Reagan administration's intent to make space commercialization a major emphasis in its policies and programs.

There was follow through on this intent. As the administration in December 1982 launched the effort to craft a national space strategy, one area of strategic focus was identifying "new areas of private sector investment in space which the Administration should stimulate." Routing various forms of telecommunications

J. M. Logsdon, *Ronald Reagan and the Space Frontier*,
Palgrave Studies in the History of Science and Technology,
https://doi.org/10.1007/978-3-319-98962-4_12

through satellites had become a thriving business in the preceding two decades; what was sought were similar economic opportunities in other space sectors. During the following two years, first in parallel to the debates over whether to approve a space station and then on its own momentum, space commercialization was a major area of White House and interagency attention.

Administration activities in support of "expanded private sector investment and involvement in civil space activities" had several dimensions. One was to transfer to private sector ownership and management all or part of the existing government programs observing the Earth from orbit. A second was both to allow U.S. industry to assume ownership and operation of the existing expendable launch vehicles (ELVs) that were due to be phased out as the space shuttle became the primary launch vehicle for all government payloads and to encourage entrepreneurial firms interested in developing new space launch capabilities. A third area of emphasis was creating an overall policy and regulatory framework that would encourage and facilitate commercial space efforts. A final area of activity was urging a reluctant National Aeronautics and Space Administration (NASA) to partner with the private sector to explore the potentials of new, profit-generating activities carried out in orbit, first aboard the space shuttle and eventually on the space station.

The first two of these areas are discussed below, with the latter two areas covered in the following chapter. Taken together, they reflect one of the major Reagan administration space legacies—the opening up of the space domain to a variety of profit-oriented efforts. While those efforts were very slow to develop in the eight years of the Reagan presidency, administration actions during those years laid the policy foundation for what in recent years has become an increasingly dynamic business sector.

A Failed Attempt at Privatization

Beginning even before space commercialization became a stated administration policy objective, the White House in the first months of 1981 began an effort to apply to the government's civilian activities related to observing the Earth's surface and atmosphere from space its principle of supporting increased private sector engagement in space. This effort would ultimately fail. Market realities proved to be a more powerful force than free market ideology.

In the late 1960s, the NASA had promoted the idea that there were substantial benefits beyond weather forecasting from observing the Earth from space, and that the agency should carry out a research effort to demonstrate the payoffs from what it labeled "remote sensing." NASA in 1972 launched the first Earth Resources Technology Satellite; a second satellite followed in 1975, as the name of the program and its satellites was changed to Landsat. Two more second-generation remote sensing satellites were approved for subsequent launch; they were designated Landsat-D1 and Landsat-D2. NASA not only funded the development and operation of the satellites. In the hope of demonstrating significant economic and social benefits from diverse applications of Landsat-gathered data and derived information, it also provided financial support for research efforts carried out by a wide variety of potential users.[3]

The Carter administration in 1979 had set as a policy goal "the eventual operation by the private sector of our civil land remote sensing systems." The plan at that time was for a decade-long transition to private operation, concluding by the end of the 1980s; during that transition, the National Oceanic and Atmospheric Administration (NOAA) of the Department of Commerce would be responsible for managing the government's civilian remote sensing operations, taking over that role from NASA.

As it entered the White House in January 1981, the Reagan administration hoped to speed up this transition by five or more years. The Office of Management and Budget (OMB) in mid-1981 turned to the recently established Cabinet Council system to identify the best approach to organizing such a transition. There were five Cabinet Councils, each dealing with a specific area of government activities and each composed of the Cabinet members with responsibilities in that area. Ed Harper, OMB's deputy director, in a July 13, 1981, memorandum noted that "with the revisions to the 1982 Budget the Administration explicitly stated its intention to hand-off operational responsibilities for land remote sensing to the private sector in the mid-1980's or sooner, if possible." Harper requested that a Cabinet Council working group be set up to address the question: "What is the best mechanism to implement the current policy of transfer of civil land remote sensing systems (LANDSAT) to the private sector as soon as possible?" Harper also posed a second question for review: "Should the Administration consider simultaneously private sector transfer of both civil weather and land remote sensing systems?" This transfer would be an example of privatization of government assets, although it was often referred to as commercialization. In commercialization, the private sector invests in creating new capabilities; in privatization, the private sector takes over assets resulting from government investment.

As it became clear that the Reagan administration was planning to privatize Landsat on an accelerated basis, the possibility of also privatizing the well-established civilian meteorological satellites had been suggested by the Communications Satellite Corporation (Comsat). That corporation had been established in 1962 to be the U.S. participant in a global telecommunication satellite network. By the late 1970s, Comsat was looking to diversify its lines of business beyond satellite telecommunications. Taking over operation of the government's civilian Earth observation activities seemed a promising opportunity.[4]

Cabinet Council on Commerce and Trade Considers Privatization

The task of answering Harper's remote sensing questions was assigned to the Cabinet Council on Commerce and Trade (CCCT). The CCCT chair was Secretary of Commerce Malcolm Baldridge. Baldridge, it would turn out, was as much a believer in the efficacy of the free market as was Ronald Reagan. He also soon became an enthusiast with respect to the economic potentials of space.

Differences in point of view, particularly at the working level, between the CCCT representatives from government agencies, who saw issues from the perspectives of their organizations, and those members coming from the White House, more closely aligned to Ronald Reagan's antigovernment, market-oriented ideology, quickly became the central dynamic of the slow-moving CCCT review. At an initial December 1981 meeting, there was basic agreement that transferring Landsat was indeed desirable. The sticking point was the level of guaranteed government financial support of the costs of operating the Landsat system, in addition to government data purchases, that would be needed to keep it viable until a meaningful private market for Landsat data was established. At this point in time, the government was the primary customer for Landsat products.

Most executive agencies participating in the review warned that "the small data market today, combined with competition from subsidized foreign systems, poses too large a business risk with too small a return on investment for a commercially viable venture." They feared that "a hasty transfer could result in the demise of the United States program," thereby ceding the remote sensing sector to emerging French and Japanese competitors. In addition, the Central Intelligence Agency (CIA) pointed out that any transfer of Landsat to private sector operation had to take into account the fact that the Intelligence Community was using Landsat data for various national security purposes; CIA suggested that "the continuity of Landsat data collection provides the CIA with the timely, large area, multi-spectral data necessary for the preparation of grain estimates … assessments that in the past provided early warning of major events affecting world grain markets and US policy."[5]

The possible transfer of weather satellites to private sector operation proved to be a particularly thorny issue. Most agencies supported a go-slow approach; the CIA noted that "as the largest user of weather data on a worldwide basis for agricultural, transportation, and other economic, military, and intelligence activities, the US would be particularly affected by … precipitous actions to transfer the civil weather systems." Various White House officials pushed for a faster approach, suggesting that a delay in the transfer "may deny the Administration the opportunity to achieve this significant private sector transfer before the next Presidential election," and also "could result in the loss of interest in commercializing civil weather satellites expressed by one or two firms."[6]

The next CCCT meeting on the remote sensing question took place only on April 20, 1982. Participants at this meeting decided that weather satellites should remain under government control. This decision was based on "national security concerns, the need to meet U.S. commitments for international exchange of weather data, and a belief that a transfer would result in a controlled monopoly requiring government support and direction."[7]

Reversing the Decision on Weather Satellites

The next step in the decision process would normally have been for Secretary Baldridge to forward the CCCT recommendation to retain weather satellites under government control to President Reagan for his approval. Baldridge did not take this step, instead delaying the decision process for a number of months. According to one participant in the process, "the principal reason for the hold appears to be the arguments that COMSAT had presented directly to Secretary Baldridge shortly after the CCCT meeting, to the effect that COMSAT can accommodate all the cited concerns. These arguments have apparently led the Secretary to seek out additional information before a final administration decision is reached."[8]

On September 10, 1982, the Commerce Department solicited private sector views on the wisdom of simultaneous privatization of Landsat and weather satellites. The department's request for information noted that "it is the policy of this Administration to seek commercialization of Governmental activities which are not uniquely Governmental in nature since private enterprise is the primary source of our national economic strength."[9]

There were 14 responses to this request, ranging from comprehensive submissions from Comsat and several other firms to a postcard from a professor at a California university. Comsat "submitted the only proposal advocating total commercialization of civil remote sensing." The responses were reviewed by both an interagency panel and an external industry panel. Both panels expressed strong skepticism regarding both the desirability and the viability of a simultaneous transfer.[10]

Reagan Decides

These skeptical comments did not slow what one participant in the CCCT process characterized as the "considerable momentum" of the Department of Commerce in pushing for a positive recommendation to President Reagan regarding the privatization of both Landsat and weather satellites. The clear warning that privatization at a minimum would be unlikely to save the government money and could also pose a variety of risks was ignored by Baldridge and his associates. It took until February 1983 for a decision memorandum signed by Baldridge to get to the president's desk. The memo gave Reagan two choices. One option was to "continue the budget policy of funding operational remote sensing systems by the Government to a close nominally by 1988, or sooner if private industry is willing to take it over, and retain the civil weather satellites under Government control." The other option was "transfer to the private sector, via competitive means, the current operational civil weather and land satellites. Separate bids would be permitted for the land or weather satellites, or a firm could elect to submit a single bid for all." The decision memorandum noted that this second option was "unanimously supported by the Cabinet Council on Commerce and Trade." As was his usual practice when presented with a recommendation having the unanimous support of the concerned agencies and White House staff, Reagan approved the transfer option.[11]

Reactions to Reagan Decision

The White House announced the Reagan decision on March 8, 1983. It was the proposed transfer of the weather satellites that grabbed the most attention. Commenting on the proposal, consumer advocate Ralph Nader characterized it as a "rip-off of the American tax payer." One *Washington Post* columnist suggested that "the Reagan administration is hereby awarded the Nobel Prize for chutzpah. What other president would have the nerve to suggest selling the government's five weather and land-resource satellites to a private company at a big loss, and guarantee its profits for 15 years?" He added, "What we see here is the ultimate idiocy of the hard-core antigovernment philosophy of the Reagan administration – the theory that the less government the better, and the best government is none at all."[12]

Congress heard vocal opposition on the part of the agriculture and farming communities to transferring weather satellites to private management. This opposition and a general skepticism regarding the wisdom of that transfer in November 1983 led to the House of Representatives passing by a 377–28 majority a "sense of the Congress" resolution that "it is not appropriate at this time to transfer ownership or management of any civil meteorological satellite system and associated ground system equipment to the private sector." The Senate concurred in the resolution. The idea of privatizing weather satellites was dead.[13]

Privatization Fails

The request for proposals for the privatization of Landsat was finally issued in early 1984. Once the decision had been made to retain weather satellites under government control, Comsat did not submit a bid. The Department of Commerce narrowed the competition to two finalists, Eastman Kodak and a new entity named EOSAT, a partnership between Hughes Aircraft and RCA, the builders of the Landsats. Then OMB's David Stockman put limits on the annual federal subsidy that would be made available to the Landsat contractor, and Eastman Kodak withdrew its bid, leaving EOSAT by default as the winner of the competition. The final contract between EOSAT and the government was signed in 1985.

From the start of the debate over privatization, those close to the program had warned that the market for Landsat-type data was very unlikely to be large enough to support a successful private venture that included financing, developing, launching, and operating subsequent land remote sensing satellites. These warnings turned out to be prophetic. EOSAT was not able to generate enough revenue to pay for additional satellites. By the early 1990s, the administration of President George H.W. Bush, in order to maintain continuity of Landsat data, decided to return total control and funding of the Landsat system to government auspices.[14]

Reality Versus Ideology

The failed effort to privatize the U.S. civilian Earth observation satellites reflected a policy-making process during which the Reagan administration's desire to reduce government spending and particularly to apply its free market ideology to a space venture had prevailed over clear warnings that what was being proposed was neither politically viable nor economically feasible. The lengthy interagency process had enabled all views on the remote sensing commercialization issue to be identified; the failure was a reflection of what happens when there is a mismatch between reality and ideology, and that mismatch is ignored as a policy decision is made.

Commercializing Space Launch

The Reagan administration early in its tenure had also signaled that it was in principle open to allowing the private sector to take over the ELVs that were to be phased out as the space shuttle became the exclusive launcher for government payloads. Those ELVs had been developed with government funding and used government-owned launch facilities for their operations. The government had also been by far the primary launch customer, with only an occasional launch of a communication satellite for a commercial customer. The manufacturers of those vehicles—McDonnell Douglas for *Delta*, General Dynamics Convair for *Atlas*, and Martin Marietta for *Titan*—and other entities were by 1982 assessing whether in the absence of government launch contracts a successful commercial business using these vehicles to launch payloads for paying customers could be created. To be successful, private operators of commercialized ELVs would have to compete not only with the European Arianespace for launch contracts, but also with NASA-provided launch services using the space shuttle. NASA at this point had no intention of withdrawing the shuttle from competition for commercial launch contracts; the fees from launching such payloads were an important part of space shuttle economics. A commercial ELV industry in the United States to be viable would thus have to compete with launch services subsidized by both the U.S. and European governments.

In addition, there were aspiring new entries into the space launch business. Several of a new generation of space entrepreneurs who became active in the first years of the Reagan administration judged that their best opportunity for entering the space sector was developing new launch vehicles. They quickly discovered that if they were to create a rocket launching business, they would require a variety of permissions and licenses from the Federal government. They were interested in ensuring that any regulatory regime dealing with privately conducted space launches would not pose significant barriers to the business success of their startup ventures.

Let's Build a Rocket!

One of the new entrants into the space launch arena was Space Services Incorporated of America (SSI), a company created in 1980 by a prominent Houston real estate developer David Hannah. Hannah had during the 1970s become interested in the concept of large-scale space settlements, but by 1980 had redirected his ambitions to lowering the cost of access to space, becoming convinced that less expensive space launch was the key to private space development.[15]

As the Reagan administration was taking office, SSI contracted with a 31-year-old Californian, Gary Hudson. Hudson had come up with the concept of a low-cost liquid-fueled rocket he called *Percheron*, after a breed of hard-working horses. He convinced Hannah and other investors to provide him with $1.2 million for an early demonstration flight, to be launched only a few miles high from an island off of the coast of Texas. However, on August 5, 1981, *Percheron* exploded on its launch pad during a static firing. Hudson and Hannah soon ended their business relationship.

SSI Not Discouraged

Shortly after this failure, Hannah was advised by Chris Kraft, the director of NASA's Johnson Space Center, and NASA "chief designer" Max Faget him to concentrate on using solid fuel for the kind of small rocket he was trying to create. Hannah hired one of the original Mercury astronauts, Donald "Deke" Slayton, who was just retiring from NASA, to be the president of SSI. Recognizing that the company would need a presence in Washington, he also hired a young space activist named Charles Chafer, who was at that point working at Georgetown University's Institute for the Social Science Study of Space; Chafer would become the first Washington representative of an entrepreneurial space venture.

SSI took a more conservative path toward developing its next rocket, named *Conestoga* after the covered wagons that had in the nineteenth century transported settlers of the American West. *Conestoga* would be a relatively low-powered launcher, intended only to carry small payloads into space. The firm decided it wanted to develop the booster using the solid fuel motor of the second stage of a *Minuteman* Intercontinental Ballistic Missile. The company first approached the Department of Defense (DOD) about purchasing a surplus rocket motor; DOD referred SSI to NASA, which had acquired several such motors for use in its sounding rockets. NASA could not legally sell one of the motors to SSI, but its top lawyer, Neil Hosenball, came up with a creative arrangement. NASA would "lease" the motor to SSI, with the provision that SSI would have to pay the motor's cost if it were not returned to NASA in original condition. This was an unlikely outcome, given that SSI intended to fire the motor during a test flight of *Conestoga*.

Hannah soon realized that he would have to gain government approvals for the *Conestoga* launch. In the fall of 1981, he approached his friend from

Houston, Vice President George H.W. Bush, for advice on how to proceed. Bush referred Hannah to Reagan science adviser Jay Keyworth, who offered White House support to SSI; Keyworth was impressed by the company's entrepreneurial approach to space. Following up on his visit, Hannah on October 22, 1981, sent a three-page memorandum to Keyworth, presenting SSI's "views on the proper role of the Federal government in regulating private enterprise ventures operating in outer space." He told Keyworth that "in numerous meetings with representatives of the Federal government, Space Services has received very positive responses to the concept of private enterprise in outer space and to this company's efforts as the initial risk taker." Hannah hoped for a presidential statement in support of "this and future ventures." He noted that "appropriate review and approval procedures undoubtedly can be developed for private space transportation ventures." He observed that "there is no lead regulatory authority which authorizes and supervises private space transportation systems. As a result, Space Services must file numerous individual requests for permission for each launch."[16]

Hannah's complaints about the fragmented responsibility for approving a private launch were on target. Between January and September 1982, SSI spent $250,000 in legal fees to get the permissions required for the first launch of *Conestoga*. Eleven government agencies had to issue licenses or permits. Among them were: an export control license, since the launch of a payload beyond the limit of national sovereignty was deemed an "export"; a Federal Communications Commission (FCC) license to use the radio frequencies needed to control a launch; and Federal Aviation Administration (FAA) clearance to transit airspace during launch and reentry. Other considerations that had to be addressed were making sure that payloads were not subject to the International Trade in Armaments Regulations and how to deal with the international treaty obligation that the U.S. government was liable for any damages resulting from a space launch from U.S. territory.

With all the necessary permissions in hand, SSI was ready for a first test launch of *Conestoga*. On September 9, 1982, *Conestoga I* lifted off from Matagorda Island on the Texas coast on a near-perfect suborbital flight, becoming the first privately funded American launch vehicle. While the launch was "a symbolic turning point for space commercialization and the new entrepreneurs," it was not followed by a series of revenue-producing missions. The entrepreneurial space launch sector proved very slow to develop, primarily because there were few buyers for the launch services it was offering.[17]

Starstruck

While SSI took the initial lead in representing entrepreneurial interest in the policy and regulatory context for commercial launches, it was not alone in seeking a supportive framework. Four young California space enthusiasts, James Bennett, Philip Salin, Gayle Pergamit, and Bevin McKinney, in March 1981 formed Arc Technologies, with the intent of eventually developing a booster powerful enough to launch a communication satellite to geostationary orbit. They spent much of 1981 raising capital in Silicon Valley, which Bennett described as "a large pool of capital with independent perspective and judgment." Their approach to entry in the space business was infused by the entrepreneurial, often libertarian, spirit characteristic of Silicon Valley technological ventures. Arc Technologies' largest investor was the first chief executive officer of Apple Computers, Michael Scott, who became president of the new company in December 1982. The firm was renamed Starstruck. The company's initial focus was developing a small sea-launched rocket named *Dolphin* which would employ an innovative hybrid propulsion rocket motor.

Starstruck, like SSI, decided to hire a Washington representative. It chose another young space activist, Courtney Stadd, who came from a background similar to that of Chafer. Stadd had been Chafer's predecessor working at Georgetown University, and had left that position to take over day-to-day management of a space advocacy group named the National Space Institute. That group had been founded by Wernher von Braun after he had retired from NASA; it drew its support from both space industry firms and individual members. Both Chafer and Stadd soon became expert in navigating the halls of the White House, NASA, other member agencies of the Space Launch Policy Working Group, and congressional members and committees in involved in space matters. Stadd in particular would play an important role in Reagan administration space policy debates. But like SSI, Starstruck after a few years stalled in its attempt to build a business using its new launch vehicle.

Endorsing ELV Commercialization

Both for Chafer and Stadd and the entrepreneurial ventures they represented and for the more established representatives of major space companies, the initial focus of attention during 1983 were the conclusions and recommendations coming out of the deliberations of a working group on space launch of the Senior Interagency Group for Space [SIG (Space)]. That group was examining the question of whether it was in the national interest to commercialize existing ELVs, even as the government prepared to turn to the space shuttle as its primary means of access to space. The group's answer to this question came in an April 13, 1983, report; it was an enthusiastic "yes." The report concluded that "a US commercial ELV capability would benefit both the USG [U.S. Government] and the private sector and is consistent with the goals and objectives of the US National Space Policy." Thus, the report recommended, "the USG should encourage and facilitate the commercialization of US ELVs." The report noted that "international and national legal obligations and concerns (including those related to public safety) require the USG to authorize, supervise and/or regulate US private sector operations."

There was, however, one area of ambiguity in this endorsement—the conflict between the government role in nurturing a commercial ELV industry and the reality that NASA was actively seeking fee-paying users of the space shuttle.

A NASA priority at this point was ensuring that there was enough demand for shuttle launches from non-U.S. government users to eventually support up to 24 launches per year, thereby allowing NASA to spread the fixed costs of shuttle operations across those launches and allowing the price per launch to be kept as low as possible. There was also political pressure to keep the post-1988 shuttle price for commercial and foreign users low enough to allow the shuttle to compete with the European *Ariane* booster for commercial payloads.

Ariane's success in attracting customers, including U.S. companies, that otherwise might have contracted for a shuttle launch was a significant irritant to the Reagan administration. A low shuttle price subsidized by the U.S. government, although aimed at competing with the subsidized *Ariane*, would make it difficult, perhaps impossible, for U.S. operators of ELVs to develop a profitable business, since they received no subsidies and thus had to bear the actual costs of launching their boosters. The working group report recommended that "the USG should continue to make the STS [Space Transportation System]" available to nongovernment users, taking into account "the effects that STS pricing for commercial and foreign flights could have on commercial launch operations." Rather than set a shuttle price that reflected actual costs, "the price for commercial and foreign flights on the STS must be determined based on the best strategy to satisfy the economic, foreign policy, and national security interests of the United States."

The working group report recognized that commercializing ELVs "effectively preserves the U.S. ELV launch capability at little or no cost to the government," thereby providing a backup capability if the shuttle were to be unavailable for an extended period, and that "commercialization of US ELVs is totally in consonance with existing national policies and offers a net benefit to the nation." The group thus concluded that "commercialization of ELVs is in the national interest."[18]

One analyst, noting the incompatibility between a policy intended to foster a commercial ELV industry and pricing the space shuttle so that it could compete with *Ariane*, called the report's recommendations a "paradoxical policy that would stymie the creation of a domestic launch industry" and a "fatal flaw."[19] Shuttle pricing was to become during 1984–1985 a major area of contention between NASA and the advocates of ELV commercialization.

The primary focus of the report of the SIG (Space) Space Launch Policy Working Group was commercializing the existing launch vehicles used by the government prior to the availability of the space shuttle. The manufacturers of these vehicles had "to ask themselves whether it made more business sense to close down their production lines or convert them to commercial use at the very time when the largest market, that of U.S. government launches, was slated to be reserved for the shuttle." The companies had to craft a strategy "that would not only maximize their launch vehicle profits, but would also protect their position as NASA contractors." Each manufacturer pursued a different approach. Martin Marietta and its subcontractors, manufacturers of the *Titan*, decided to join with Federal Express to explore commercial opportunities for the vehicle, which previously had been used only for U.S. government launches. General Dynamics Convair, which built the *Atlas*, forecast a viable commercial market for its booster, and decided to enter the competition for commercial launch contracts itself. McDonnell Douglas, the manufacturer of the workhorse *Delta* booster, after internal debate about using the *Delta* to compete with the shuttle for launch contracts, decided not to try to commercialize the vehicle. The company's space division was headed by NASA's former space shuttle manager John Yardley, who "had faith in the shuttle and thought that any commercial expendable launch vehicles that were marketed in competition were doomed to failure." The company did indicate its willingness to continue to manufacture *Delta*s for others, and there were indications of interest from an entrepreneurial firm, Transpace Carriers, founded by the former NASA manager of the *Delta* program, David Grimes, in taking over the marketing and operation of the *Delta*.

Martin Marietta and McDonnell Douglas had to walk a fine line, since they had lucrative contracts with NASA's space shuttle program, and had to balance their advocacy of ELV commercialization with the possibility of alienating a major customer. There were even reports that NASA shuttle program managers had threatened to cancel NASA's multibillion dollar contract with Martin Marietta for the shuttle's external fuel tank if the company persisted in commercializing its *Titan* ELV. McDonnell Douglas was very involved in the shuttle program, developing a family of upper stages to carry communication satellites from the shuttle's payload bay to their final orbit; that may have been another reason that it did not itself seek to commercialize the *Delta*. General Dynamics did not have major space shuttle-related contracts.[20]

The working group as a secondary focus did also address issues associated with the convoluted process of approving the September 1982 launch of the *Conestoga* booster. Its report observed that while the process used to provide approval for the *Conestoga* launch could "unquestionably be streamlined," it was more important to develop "a more efficient long-range regulatory plan," because "requirements and procedures to license, supervise and/or regulate US commercial ELV operations should be developed before the advent of routine commercial ELV operations."

The working group suggested that the Department of State, which had coordinated the *Conestoga* approval process, "should be designated the lead agency for coordinating all U.S. commercial ELV requests within the USG." This was because State was "well suited to coordinate the political approval for commercial launches which involve issues such as international agreements, national security concerns, and technology transfer issues." While the State Department would have the lead role, "operational concerns and supervision would be best handled by the cognizant technical agencies, such as DOD, NASA, FAA, and FCC."

The issue of what government agency should have the lead role in promoting and regulating commercial space launches had been contentious within the working group. Several of those who met with the group argued strongly against giving the FAA the lead role, fearing the impact on an emergent space launch industry of what they saw as the heavy handed regulatory approach the FAA used in overseeing the aviation industry. NASA early on suggested that it, rather than the State Department, should have the lead role. NASA's lead policy person, Norm Terrell, later suggested that "NASA could have had the lead role for the asking," but that NASA Administrator James Beggs was skeptical of the viability of ELV commercialization and did not want to put the agency "in the position of encouraging a concept of questionable validity." The issue of which agency should have the lead role continued to be controversial in subsequent months, and ultimately would be resolved only by a presidential decision.[21]

Attached to the working group report was a proposed National Security Decision Directive (NSDD) on "Commercialization of Expendable Launch Vehicles." National Security Adviser Judge Clark in May 1983 sent the directive to President Reagan for his signature. Clark told the president that "all agencies unanimously support an affirmative policy" with respect to ELV commercialization. Such unanimous agreement meant that there was no need for a SIG (Space) meeting with Reagan to discuss the NSDD, and Reagan soon signed it. On May 16, 1993, Clark distributed to government agencies the approved directive, designated NSDD 94."[22]

The directive stated that "the U.S. Government fully endorses and will facilitate the commercialization of U.S. Expendable Launch Vehicles (ELVs)" and that "the U.S. Government will license, supervise, and/or regulate U.S. commercial ELV operations only to the extent required to meet its national and international obligations and to ensure public safety." It retained the qualification that the space shuttle would continue to be "available for all qualified users," including potential customers for commercial ELVs, and that the current shuttle price would be maintained through 1988 "to provide market stability and ensure fair competition." The directive noted that after 1988 "it is the U.S. Government's intent to establish a full cost recovery policy." These statements represented a victory for NASA and those concerned with competing with *Ariane*, and a defeat for the ELV commercialization advocates. They had hoped for a more rapid switch to a higher shuttle price in order to provide a more level-playing field.

The directive established a SIG (Space) Working Group on Commercial Launch Operations. NASA was named co-chair, together with the Department of State, of the new working group. Members of the working group were to represent SIG (Space) members and observers, plus "other affected agencies," including the FAA and the FCC.[23]

With the approval of NSDD 94, the first step in creating a new commercial ELV industry was taken. But this was a relatively easy move; there had been little disagreement that a policy facilitating a successful commercial business using already-developed and new ELVs was desirable. Translating that agreement into an organizational and regulatory framework and, in particular, selecting the government agency to implement that framework would prove rather more difficult.

Crafting an Approach to ELV Commercialization

There soon was a rather surprising challenge to proceeding with ELV commercialization. The Soviet Union raised objections in a July 7, 1983, *demarche* to the Department of State, complaining that ELV commercialization would provide "access to technology systems constituting a potential source of danger" and that "involvement of non-governmental legal entities in space activities increases the risk of various types of incidents which may have unpredictable consequences." The Soviet Union wanted to hear from the U.S. government "how the granting of the right to private companies to launch ballistic missiles ... is consistent with the obligations assumed by the US side in the area of strategic arms limitations." The State Department responded several weeks later, arguing that "Soviet arms control concerns were not relevant to the commercialization issue."[24]

Working Group Report

In the months following issuance of NSDD-94, the SIG (Space) Working Group on ELV Commercialization prepared a report on "U.S. Government Organization and Process for Handling U.S. Commercial Launch Operations." The report proposed that a "Lead Agency" for ELV commercialization be designated, but noted that the working group had not reached agreement on which government organization should have that role. The two candidates were the Department of Commerce and the Department of Transportation, which housed the FAA. The report recommended that "Lead Agency responsibilities must transcend those of a coordinating role over regulation/licensing of ELV launches to include an expanded and important role in encouraging/facilitating ELV commercialization." NASA alone among the participating agencies opposed a promotional role for the lead agency, arguing that NASA "was already engaged in the promotion of space commercialization" and that "its authorizing 1958 statute gives it an effective monopoly in the field." Thus, NASA suggested, it was inappropriate to give a promotional role vis-à-vis any area of space activity to another government agency. This proprietary attitude on NASA's part was resented by those advocating private sector entry into not only space launch but also commercial activities in general, and became a continuing source of interagency hostility in following years.[25]

The working group report included the detailed justifications prepared by the Departments of Commerce and Transportation on why each was best qualified to become the lead agency for ELV commercialization. The leadership of the Commerce Department saw such a role as a natural extension of the department's historic function in overseeing and promoting new areas of U.S. commerce, ranging from the aviation to the maritime sector. There would be a need for addressing level-playing field issues related to international competition with Europe and potentially with the Soviet Union and China, and the department's leaders believed that it was well organized to address those international trade issues.

The Department of Transportation leaders saw the lead agency role as a way of demonstrating the department's fealty to the free market principles of Ronald Reagan and his administration. They argued that space launch was just another mode of transportation, and thus logically should be overseen by the Transportation Department. Both departments, of course, were also subject to the normal bureaucratic ambition to expand an organization's scope of operations. Whichever department was selected, the report recommended that the successful agency "identify or establish an appropriately visible organizational entity, at the Department [Secretary] level," to lead the ELV commercialization effort.

In the Reagan administration, the policy resulting from the deliberations of a SIG (Space) working group would normally have been embodied in an NSDD. However, the working group noted in its report that "some of the Working Group members, including the candidates for the Lead Agency role, prefer that an Executive Order be issued to designate the Lead Agency and its responsibilities." This preference reflected the fact that "some of the agencies which will be involved in the approval process are not members of the NSC [National Security Council] and do not normally operate with NSDD's. In addition, Executive Orders would get wider circulation within the private sector."[26]

Choosing the Lead Agency

Before an Executive Order with respect to ELV commercialization could be issued, the government department that would perform the lead agency role had to be selected. The working group report did not make a recommendation in this respect, saying only that either Commerce or Transportation "would be suitable." On September 21, NSC's Gil Rye had sent a note to Craig Fuller suggesting that "you or Ed Meese try to informally reach agreement between either Secretary [of Commerce] Baldridge or Secretary [of Transportation] Dole as to who should be Lead Agency. If agreement cannot be reached at your level, I see no recourse but to ask the President to decide." Agreement proved elusive; it thus fell to Ronald Reagan to select the lead agency.[27]

As the date for the presidential decision approached, Baldridge added a personal argument not only for his department taking the lead role in ELV commercialization but also for space commercialization overall, telling National Security Adviser Bud McFarlane that "the Department of Commerce can lead the effort to develop and implement a civil Industry in Space program" by being both "an advocate within Government for industrial interests by removing barriers" and "an entrepreneur to promote development through the support of incentives." Baldridge had become critical of NASA's monopoly perspective, suggesting that "government domination of space development [has] created inertia; now is the time to encourage private industrial development in space." Baldridge argued that as part of an Industry in Space program "the DOC should be the lead agency for ELVs and this decision should be made promptly. DOC has the technical and business expertise to deal with all the issues."[28]

A meeting with the president to hear arguments from both Commerce and Transportation on why they should be designated the lead agency was scheduled for the morning of November 16. That meeting would be held under the auspices of the CCCT, even though the process up to this point had been managed by the NSC's SIG (Space). Rye and Fuller had agreed in late 1983 to use the interagency mechanism best suited to bring a particular issue before the president for decision, irrespective of which forum, SIG (Space) or CCCT, had been to that point dealing with the issue. The lead agency choice for commercializing ELVs was deemed an issue best suited for CCCT presentation. (As was discussed in Chap. 9, a similar approach was used in bringing the space station issue before the president.)

A forcing influence on the timing of the presidential meeting was the plan of a House of Representatives committee to hold hearings on legislation dealing with ELV commercialization. In 1983, in parallel with deliberations within the executive branch on the issue, Representative Dan Akaka (D-HI) had introduced a bill dealing with this issue. The bill was designated H.R. 3942 and titled the "Expendable Launch Vehicle Commercialization Act." Akaka staff member Diana Hoyt had drafted the bill at the urging of, and with the assistance of, SSI and Starstruck and the law firms they had engaged to help with their dealings in Washington. The draft bill designated the Department of Commerce as the lead agency. Both Commerce and Transportation had been invited to testify at a November 10 hearing on the bill. The White House was able to persuade the committee to postpone the hearing for a week to allow time for the president to make the lead agency choice.[29]

The CCCT met in the White House Cabinet Room at 10:00 a.m. on the morning of November 16. Like the meeting two weeks later on the space station decision, this was the kind of gathering Reagan valued as important to his decision-making, with advocates of differing positions making their best case to the president. Secretary Baldridge led the argument for designating Commerce as the lead agency. The case for Transportation was made by Secretary Elizabeth Dole. As was his usual practice, the president did not announce his decision at the conclusion of the 45-minute meeting.

Following the meeting, Craig Fuller prepared a memorandum giving the president three options:

1. Take no position at this time on the lead agency for commercialization of ELVs and await further discussion of broader space commercialization issues. Congress would be told that the Administration has no position at the present time on this matter.

2. Designate the Department of Commerce the lead agency for commercialization of ELVs. An executive order would be prepared and testimony would be drafted in accordance with this decision.
3. Designate the Department of Transportation as the lead agency for the commercialization of ELVs. An executive order would be prepared and testimony would be drafted in accordance with this decision.

Fuller told President Reagan "once you make a decision, we will take the appropriate steps to implement it. If you wish to make a decision today, we will advise the agencies this afternoon in order to allow them an opportunity to testify at tomorrow's hearing." Reagan was ready to decide. On the afternoon of November 16, he initialed "RR" opposite the option "designate the Department of Transportation as the lead agency."[30]

There are various explanations offered for the president's choice. Some suggest that Secretary Dole charmed Reagan into giving her the role she sought. The fact that she was married to an influential Republican senator, Bob Dole (R-KS), may also have played a role. Courtney Stadd from Starstruck, who had been one of those lobbying for Commerce as the lead agency, suggests that Dole told the President that this was a classic case of deregulation. She argued that she could oversee this industry with a small team of regulatory experts and that it was "very appropriate and logical" for the Department of Transportation to take on the job. Stadd was told that during the November 16 CCCT meeting "one of the President's aides passed a note to Secretary Baldridge asking whether he'd be open to dividing the authority with the Department of Transportation so that Commerce would have the promotional role and DOT would have the licensing and regulatory role." Stadd "was informed that the Secretary decided at that instant, no, that he felt a divided authority was a weakened authority, and that if, in their wisdom, the White House—the President—opted to have DOT take over licensing, he was not going to stand in its way." Ronald Reagan himself did not provide a sense of the reasoning behind his decision. In his November 16 diary entry, he noted only "Cab. Council meeting on commercial entry into the Space effort. Question should the govt. department in charge of this be the Dept. of Transportation or Commerce. Both wanted it. I came down on the side of Transportation."[31]

Whatever the reason, a new government actor on space policy issues had been created. Reflecting this reality, Secretary Dole was soon invited to become a member of SIG (Space), positioning the Department of Transportation as an active participant in future space policy debates. While Reagan's decision was a blow to Baldridge's ambition to place the Department of Commerce in a central role with respect to space commercialization, he and his associates continued to try to carve out such a role within the administration. With both Commerce and Transportation now involved in space policy discussions, NASA's clear leadership on civilian space issues would be under continuing challenge.

It took several months before an executive order recording the president's decision was ready for Ronald Reagan's signature. However, the decision did not go unremarked. On January 25, 1984, Ronald Reagan in his State of the Union address declared: "The market for space transportation could surpass our capacity to develop it. Companies interested in putting payloads into space must have ready access to private sector launch services. The Department of Transportation will help an expendable launch services industry to get off the ground."[32]

Two days later, Craig Fuller proposed that the signing ceremony for the executive order be a high-profile event, since the order was a first tangible step in fulfilling Reagan's commitment to stimulate space commercialization. He noted that "both the Congress and industry are looking for visible signs" that the president was serious about this commitment. He noted that "both groups look toward the Administration for leadership in reducing regulatory constraints" and that the order would represent "an important step in building close working relationships between the Government and the private sector."[33]

On February 24, 1984, as he and Vice President George Bush stood before models of several launch vehicles and a large picture of the Earth as seen from space (Fig. 12.1), Reagan noted, "We're doing all we can to encourage space work by American industry. Private enterprise made America great. And if our

Fig. 12.1 President Reagan, flanked by Secretary of Transportation Elizabeth Dole and Vice President George H.W. Bush, discussing commercialization of expendable launch vehicles, February 24, 1984. (Photograph courtesy of Reagan Presidential Library)

efforts in space are to show the same energy, imagination, and daring as those in our country, we must involve private enterprise to the full." He added:

> until today, private industries interested in ELV's have had to deal with 17 Government agencies. From now on, they'll only have to get in touch with the Department of Transportation, and the Department will clear away what Secretary Dole has called "the thicket of clearances, licenses, and regulations that keep industrial space vehicles tethered to their pads." With Elizabeth and her team in charge, private enterprises interested in space won't see red tape; they'll see blue sky.[34]

To carry out its new responsibilities, the Department of Transportation soon established an Office of Commercial Space Transportation. The new office was located in the Office of the Secretary, giving it a top-level position within the department's hierarchy; it was headed by Secretary Dole's top assistant, Jenna Dorn.

Conclusion

The 1983–1984 effort to create a combined promotional and regulatory regime for commercial space launch was premature. The Office of Commercial Space Transportation had little to promote or regulate in its early years, since the commercial launch industry using existing ELVs was quite slow to develop as it tried to compete with both *Ariane* and the space shuttle, and the new entrepreneurial launch companies also attracted neither significant investment nor customers. It took the *Challenger* accident in 1986 to create a window of opportunity for commercial launch, and the first licensed commercial launch did not occur until 1989.

But in a broader sense, the steps taken in these two years created the foundation for the new privately based space launch industry that has emerged in recent years. They also legitimized as actors in the space policy process new government agencies as well as representatives of new private sector actors in addition to the established aerospace firms. The steps toward creating the initial policy and regulatory framework for commercial launch during the Reagan administration thus have had a lasting impact on the U.S. space program.

Notes

1. Ronald Reagan: "Radio Address to the Nation on the Space Program," January 28, 1984. Online by Gerhard Peters and John T. Woolley, APP, http://www.presidency.ucsb.edu/ws/?pid=40349.
2. A copy of NSDD-42, portions of which remain classified, can be found at www.reaganlibrary.gov/sites/default/files/archives/reference/scanned-nsdds/nsdd42.pdf.

3. For an overview of the origin and evolution of the Landsat program, see Pamela Mack and Ray Williamson, "Observing the Earth from Space," in John M. Logsdon et al., eds., *Exploring the Unknown: Selected Documents in the History of the U.S. Civilian Space Program*, Volume III: Using Space, NASA SP-4407 (Washington, DC: Government Printing Office, 1998), 155–177.
4. Memorandum from Allen Lenz to Robert Carlstrom, "Land Remote Sensing Legislation," April 29, 1981, Box 1, Outer Space Files, RRL; memorandum from Ed Harper, Office of Management and Budget, to Craig Fuller and Martin Anderson, "Resolution of Issues Related to Private Sector Transfer of Civil Land Observing Satellite Activities," July 13, 1981, reprinted in Logsdon et al., eds., *Exploring the Unknown*, Volume III, 306; Philip Boffey, "Administration Proposes Selling U.S. Weather Satellites to Industry," *NYT*, March 9, 1983, 1.
5. Cabinet Council on Commerce and Trade, "Decision Memorandum," December 14, 1981, CIA-RDP84B00049R001700060009-4, CREST; memorandum from Chief, Interdepartmental Affairs Staff, OPP to Director of Central Intelligence, "CCCT Meeting – LANDSAT Issue," December 15, 1981, CIA-RDP84B000449R001700060018-4, CREST.
6. Ibid.
7. Memorandum from Director, Intelligence Community Staff, to Director of Central Intelligence, "Civil Satellite Transfers to the Private Sector," August 31, 1982, CIA-RDP84M00396R000300020028-6, CREST. "National Security Concerns Regarding Meteorological Satellite Commercialization," paper attached to Cabinet Council on Commerce and Trade, "Agenda," April 20, 1982, CIA-RDP84M00395R000600220030-9, CREST.
8. Memorandum from Director, Intelligence Community Staff, to Director of Central Intelligence, "Civil Satellite Transfers to the Private Sector," August 31, 1982, CIA-RDP84M00396R000300020028-6, CREST.
9. Transmittal from the Department of Commerce to *Commerce Business Daily*, "Civil Operation Remote Sensing from Space," September 3, 1982 (published September 10, 1982), CIA-RDP84M00127R000200030028-4, CREST.
10. Government Technical Review Panel, Report on Industry Responses on Commercialization of Civilian Remote Sensing Systems," November 10, 1982, reprinted in Logsdon et al., eds., *Exploring the Unknown*, Volume III, 310–311. "Advisory Unit Urges that U.S. Fund Remote Sensing to 1995," *AWST*, November 29, 1982, 26.
11. Memorandum from the Cabinet Council on Commerce and Trade to the President, "Transfer of the Civil Remote Systems to the Private Sector," February 28, 1983, Folder 12772, NHRC.
12. Philip Boffey, "Administration Proposes Selling U.S. Weather Satellites to Industry," *NYT*, March 9, 1983, 1; Philip Hilts, "Reagan Took Solo Action on Weather Satellite Sale," *WP*, March 26, 1983, A1; Hobart Rowan, "Reagan's Satellite Sell-Off Would Rip Off Taxpayers," *WP*, March 13, 1983, F1.
13. House Concurrent Resolution 168, "Transfer of Civil Meteorological Satellites," November 14, 1983, reprinted in Logsdon et al., eds., *Exploring the Unknown*, Volume III, 321–329.
14. For a summary of this period in the evolution in U.S. remote sensing activities, see Mack and Williamson, "Observing the Earth from Space," in Logsdon, *Exploring the Unknown*, Volume III, 172–175 and Brian Jarout, "Lessons of

Landsat: Experimental Program to Commercial Land Imaging, 1969–1989," in Roger D. Launius and Howard E. McCurdy, eds., *NASA Spaceflight: A History of Innovation* (New York: Palgrave Macmillan, 2018), 155–184.

15. Much of this discussion of David Hannah and SSI comes from Michael A.G. Michaud, *Reaching for the High Frontier: The American Pro-Space Movement, 1972–1984* (New York: Praeger, 1986), Chap. 12, and an interview with Charles Chafer, March 6, 2017.
16. Memorandum from David Hannah to George Keyworth, "Government Regulation of Private Enterprise in Outer Space," October 22, 1981, RAC 14, Papers of George Keyworth, RRL.
17. Michaud, *Reaching for the High Frontier*, 260.
18. "Interagency Space Launch Policy Working Group Report," April 1983, Document CIA-RDP92B00181R00170160018-7, CREST, 3–4, 14–16, 20–22.
19. Andrew Butrica, *Single Stage to Orbit: Politics, Space Technology, and the Quest for Reusable Rocketry* (Baltimore, MD: Johns Hopkins University Press, 2003), 25.
20. Joan Lisa Bromberg, *NASA and the Space Industry* (Baltimore, MD: Johns Hopkins University Press, 1999), 115, 129.
21. "Interagency Space Launch Policy Working Group Report," April 1983, 32–33. Memorandum from Norm Terrell to Dr. (James) Fletcher, May 2, 1986, Box 10, Papers of James Fletcher, NARA.
22. Memorandum from William P. Clark for the president, "NSDD on Commercialization of Expendable Launch Vehicles (ELVs)," undated.
23. A copy of NSDD 94 can be found at www.reaganlibrary.gov/sites/default/files/archives/reference/scanned-nsdds/nsdd94.pdf.
24. Memorandum from Jim Chamberlin, Arms Control and Disarmament Agency to SIG (Space) ELV Working Group, "Soviet Interest in ELV Commercialization, with attached diplomatic cables, September 7, 1983, CIA-RDP92B00181R001701600008-8, CREST.
25. Charles Horner, "Personal Note for Crag Fuller," August 19, 1983, Box 7, Outer Space Files, RRL.
26. SIG (Space) Working Group on the Commercialization of ELVs, "Report on U.S. Government Organization and Process for Handling U.S. Commercial Space Launch Operations," September 15, 1983, CIA-RDP92B00181R001701600007-9, CREST.
27. Handwritten note from Gil Rye to Craig (Fuller), September 21, 1983, Box 8, Outer Space Files, RRL.
28. Letter from Malcolm Baldrige to Robert McFarlane, November 4, 1983, National Security Council, Executive Secretariat, 13–82 File, RRL.
29. Memorandum from Constance Horner to Craig Fuller, "Talking Points on H.R. 3942, Expendable Launch Vehicle Commercialization Act," November 8, 1983, Folder 12772, NHRC.
30. Memorandum from Craig Fuller to the president, "Determining the Lead Agency for Commercializing Expendable Launch Vehicles," November 16, 1983, Folder 12772, NHRC.

31. Interview of Courtney Stadd by Rebecca Wright, January 7, 2003, NHRC; Douglas Brinkley, ed., *The Reagan Diaries* (New York: HarperCollins, 2007), 198.
32. Ronald Reagan: "Address Before a Joint Session of the Congress on the State of the Union," January 25, 1984. Online by Gerhard Peters and John T. Woolley, APP, http://www.presidency.ucsb.edu/ws/?pid=40205.
33. Memorandum from Craig Fuller to Fred Ryan, "Schedule Proposal," January 27, 1984, Box 8, Outer Space Files, RRL.
34. Ronald Reagan: "Remarks on Signing an Executive Order on Commercial Expendable Launch Vehicle Activities," February 24, 1984. Online by Peters and Woolley, APP, http://www.presidency.ucsb.edu/ws/?pid=39560.

CHAPTER 13

Commercializing Earth Orbit

In addition to applying to the space sector Ronald Reagan's conviction that the free market was the preferred path to future progress, there were three other influences supporting Reagan administration initiatives toward space commercialization during the 1983–1984 period. One was the emergence, just as the Reagan administration arrived in Washington, of a variety of private sector actors interested in space commercialization. Some were traditional aerospace companies interested in expanding the scope of their activities to include more commercially oriented activities. They were helped along by increasing emphasis in the late 1970s by the National Aeronautics and Space Administration (NASA) on future economic payoffs from space, described as "space industrialization," as a rationale for continuing a government space research and development effort and for developing the in-space infrastructure, including the space shuttle and eventually a space station, needed for the private sector to try out its ideas.

Also influential was the growing recognition that "space need not be a government monopoly and that the private sector could be an independent actor, not just a contractor." This recognition "attracted bright young people of a new generation, impatient with the old, politically dependent ways of getting into space. There was a sudden flowering of new, entrepreneurial commercial space ventures." This flowering, in turn, "reflected a new wave of entrepreneurship emerging throughout the American economy … The emergence of new centers of venture capital in Houston and in Northern California's 'Silicon Valley' was a factor." In addition, "many of these new entrepreneurs had connections with pro-space activism and were extensively intertwined with the new pro-space movement" that had gained steam in the late 1970s.[1]

One analyst of the emergence of these new space entrepreneurs commented:

J. M. Logsdon, *Ronald Reagan and the Space Frontier*,
Palgrave Studies in the History of Science and Technology,
https://doi.org/10.1007/978-3-319-98962-4_13

> A missionary zeal for bringing capitalism to space infused many of the leaders of these companies. Often it took the form of championing private, as against government, enterprise; sometimes it exalted small, entrepreneurial firms as against large aerospace companies. Free enterprise, claimed these men and women, could create technology that was better, cheaper, and more suited to market demands. A nationalistic element entered. They argued that free enterprise was a part of U.S. tradition, so that to promote it in space was to defend the American way.

In many cases, underpinning this zeal was a "new right" ideology that "had nothing to do with the bottom line."[2]

A third influential factor leading to the increased emphasis on commercialization was high enthusiasm regarding the potential economic payoffs from various commercial space activities. While the revenue from privatizing remote sensing and commercializing expendable launch vehicles was expected to be relatively modest, payoffs from new commercial activities carried out in Earth orbit were forecast to be in the multiple billions of dollars. Former Secretary of Commerce Philip Klutznick, writing in *The New York Times*, suggested that private ventures in space "deserve to be taken seriously. They are not castles in the air to distract attention from idle plants and double-digit unemployment on earth; they involve working factories whose commercial potential has been tested. Properly exploited they will help transform the nation's $100 billion investment in the space program into sustained world leadership in an area of technology with as yet unimaginable applications for commerce and manufacture." One extremely optimistic forecast anticipated over $65 billion in annual revenues from commercial space activities in the year 2000. Those activities included:

- $27 billion in pharmaceutical production
- $3.1 billion in gallium-arsenide semiconductor processing
- $11.5 billion in space processed glasses
- $15 billion in advanced space communications revenues
- $2 billion in remote sensing revenues
- $2 billion from in-orbit servicing
- $1 billion from commercial launches
- $3.7 billion from aerospace support activities.[3]

These projections caught the attention of investors willing to accept the high risks, long time spans before returns, and uncertain policy environment of the space sector. Writing in mid-1984, the editor of the trade weekly *Aviation Week & Space Technology* suggested that "once an exotic idea in the realm of science fiction, space as an area for a broad spectrum of commercialization is coming into reality." He noted as "surprising … the ability of new companies, some with relatively young and untested principals and managers, to raise private funding." That such financing was available, he suggested, "is a tribute to the instincts of the investors about the potential of space." Another surprise was "the willingness of industry outside the aerospace mainstream to take seriously the potential for space commercialization."[4]

Some members of the Congress were also intrigued by the economic promise of space. In particular, two young Republican congressmen, Newt Gingrich from Georgia and Robert Walker from Pennsylvania, became outspoken advocates of space commercialization. Sensing that NASA "did not want to engage in the kind of thing where commercial enterprise would bring ideas to them that had not been developed within NASA," Walker sought changes in the 1958 Space Act to urge NASA to "address the broader context rather than the rather narrow way" it was approaching commercial space activities. He was successful in that effort; the Space Act was amended in 1984 to state that a basic goal of NASA was "to seek and encourage, to the maximum extent possible, the fullest commercial use of space."[5]

First Steps Toward a Commercialization Policy

NASA had a vested interest in promoting space commercialization, since almost all ideas for achieving that objective would have to depend, at least initially, on the use of the space shuttle (and eventually a space station) as the platform for investigating the profit-making potential of new orbital activities. Since soon after the space shuttle had been approved by Richard Nixon in 1972, NASA began planning ways to generate new commercial users for the space shuttle, and had developed innovative mechanisms for partnering with the private sector in commercially oriented activities aboard the vehicle. In 1982, as part of its campaign to gain approval for developing the space station, NASA had commissioned a series of studies by the aerospace industry of innovative commercial uses of the capabilities that a permanent orbital platform could provide.[6]

NASA Administrator James Beggs in 1982 hired a bright young lawyer named Llewellyn Evans, Jr., universally known as "Bud," to work with him in promoting space commercialization initiatives. Evans's father had been president of Grumman Aircraft during the Apollo program, when the company built the lunar module that carried crew from lunar orbit to the Moon's surface and back to lunar orbit. Bud Evans was a true believer in the potential of private sector space activities, and over the next several years would play an activist role in putting into place a framework within which those activities might thrive.

Beggs also asked the National Academy of Public Administration in early 1983 to study "potential private industrial and commercial development of space" and to "identify institutional barriers which may inhibit private sector investment; propose policies and programs to foster such investment; and assess alternative ways for government and the private sector to work together." The academy created an 11-person panel to carry out the study. That panel issued its report in May 1983; it concluded that the United States "will fall behind in the contest for leadership in space and the economic rewards associated with that position" if it did not remove stumbling blocks to greater private involvement in space. The report was critical of NASA, noting that commercialization activities were "badly fragmented" among various NASA offices. The panel noted that NASA's "R&D mind set" made "considerations central to commercialization, like cost, market demand, and financial risk, foreign to it."[7] Its report made clear that developing a framework for space commercialization would involve Reagan administration attention that transcended NASA. A government-wide perspective was needed. It was such a perspective that Klutznick, the panel's chair, had advocated in his *New York Times* essay.

The White House Gets Involved

That perspective first emerged in the context of the efforts, beginning in December 1982, to develop a national space strategy (see Chap. 5). Other areas of the strategy were to be developed by interagency working groups, but space commercialization would be the focus of a private sector group working under the guidance of Craig Fuller, director of the White House Office of Cabinet Affairs. Fuller had come to the White House from a public relations background associated in part with the aerospace industry, and he had developed a personal interest in space policy issues in addition to his formal White House portfolio. So his role as an advocate for space commercialization was not all that surprising.

Bud Evans became the link between NASA and Fuller, working quite closely with the White House as the commercialization effort moved forward. Although Fuller's study was formally part of the Senior Interagency Group for Space [SIG (Space)] strategy effort, and Gil Rye, the National Security Council (NSC) staff person in charge of that effort, had hoped that Fuller would provide input on a schedule that would allow it to be integrated in the draft strategy by mid-1983, Fuller and Evans moved forward at their own pace.

Their first step was assembling a private sector group to identify needed government actions to spur space commercialization. Given the leverage of his White House position, Fuller was able to target the top people in the private sector firms he wanted to involve. He and Evans began planning a high-level space commercialization meeting. At that meeting, 15 to 20 industry leaders would meet with senior White House staff to examine space commercialization issues and opportunities. As the meeting was being organized, the White House decided that President Reagan, as a way of demonstrating his personal interest in the space commercialization issue, would attend a portion of the session. Reagan enjoyed opportunities to hear the views of those outside government on issues of interest to him, and such an encounter would provide him a chance to do that. In suggesting that Reagan should join the industry group for lunch, Fuller promised the president "a lively discussion of what commercial opportunities lie ahead for the U.S. in space."[8] The best day on Reagan's calendar for the get-together was Wednesday, August 3, 1983, and so the meeting was set for that date.

Fuller and Evans had been in informal contact with potential invitees to the meeting as it was being planned, but the formal invitation to attend came only on July 26, just over a week in advance of the gathering. Fuller sent an invitation letter to 13 people, saying, "You are invited to join the President, several senior members of the Administration and several business colleagues for a meeting at the White House on commercial space activity." The letter noted that "a group of senior Administration officials has been considering issues related to commercial activity in space and we are at the stage where a discussion with individuals who actually invest private resources in commercial space ventures would be most beneficial." Attached to the letter was a two-page summary of "space commercialization issues." They ranged from the basic question "Should private sector space endeavors be encouraged by the Government?" to "What are the most efficient and effective means for providing encouragement for private sector investment in space commercialization ventures?" and "What types of support are necessary ... What should the U.S. government consider?"[9]

Fuller and Evans, together with several others from the White House staff, on July 28 held a preliminary meeting with representatives of the firms planning to attend the August 3 session. Out of that meeting came a listing of the commercial space activities that those firms were currently undertaking, plus an identification of issues of concern to those firms. For example, the investment firm E.F. Hutton pointed out that "government must ... allow companies to earn returns commensurate with the risks just as government policies offered incentives to induce settlement of the West in the 1800s (Morrill Act, Railroad Right of Ways). Incentives may assist in opening the commercial frontier in space." Grumman Aerospace commented that "NASA must be able to respond rapidly to the needs of the commercial developer or else it will be viewed as a poor and impractical partner." The entrepreneurial Orbital Systems Corporation (later Orbital Sciences Corporation, and now Northrup Grumman Innovation Systems) suggested that "the infrastructure of a healthy commercial space sector would be composed of technical, financial, and institutional components. The Federal Government can and should take concrete steps to bolster the supporting infrastructure of commercial space enterprise in each of these areas."[10]

Ronald Reagan and Space Commercialization

The agenda for the August 3 meeting called for a 10:00 a.m.–noon session during which Fuller, who would chair the meeting, would introduce the day, followed by welcomes from Ed Meese and James Beggs. This would be followed by a general discussion of space commercialization issues. At noon the meeting would break for a 90-minute working lunch with the president. After lunch the meeting participants would have a brief wrap-up session.

Fuller provided a briefing memorandum for Reagan in advance of the event, telling him that "this working luncheon will allow you an opportunity to review space commercialization opportunities with a select group of industry representatives." Fuller reminded the president that "each flight of the space shuttle has carried both scientific and commercial packages into orbit. The success of the shuttle and the interest in new commercial space activities suggests that the Federal government must carefully consider policy options that will affect the nature, scope and direction of U.S. commercial space activity." Fuller's memo listed the policy issues under consideration:

- Reducing institutional risks: need to establish consistent space commercialization policy; need for reduced regulatory burden.
- Reducing technical risks: enhancing space facilities; increasing availability of technical information and expertise; performing basic research in areas with commercialization potential.
- Reducing financial risks: lessening high front-end costs of doing business in space; considering capping insurance risks; reducing antitrust barriers; enhancing protection of patent and other intellectual property rights for space ventures.
- Creating incentives for space commercialization: guarantee a portion of the market by flying government payloads on industry-financed portions of the space infrastructure (space platforms, shuttle orbiters, etc.); reduce shuttle payload costs/price.

Fuller listed those who would lunch with the president. There was a mix of people from established aerospace firms and new entrepreneurial ventures. Attendees included the president or chief executive officer of 11 companies: John Yardley from McDonnell Douglas; Maxime Faget from Space Industries; Robert Hanson from John Deere & Company; Fred Smith from Federal Express; George Jeffs from Rockwell International Space Operations; George Skurla from Grumman Aerospace; David Thompson from Orbital Systems; David Hannah from Space Services; Oliver Boileau from General Dynamics; John Latshaw from E.F. Hutton; and John Townsend from Fairchild Space. Entrepreneur Klaus Heiss would also be present. From the White House, Ed Meese, James Baker's deputy Richard Darman, Jay Keyworth of the Office of Science and Technology Policy, Gil Rye of the NSC, and Craig Fuller would attend. From NASA, Jim Beggs and Bud Evans would join the lunch, as would Clarence "Bud" Brown, newly nominated as deputy secretary of commerce (Fig. 13.1).[11]

At the lunch, Reagan heard about the importance of the space station to full exploitation of the potential for private sector activity in orbit, and told the attendees, "I want a space station, too. I have wanted one for a long time." In addition to his positive words about the space station, the president's "enthusiasm about the outlook for commercial space ventures helping maintain U.S. leadership" was welcomed by the corporate executives present. One was quoted as saying: "It looked like the light bulb really went on when we raised to him the benefits of space pharmaceutical manufacturing, the serious European competition with *Ariane* and the multiple arguments underlying station justification." Reagan was also told that "several hundred million dollars in private investment is on the verge of being poured into commercial

Fig. 13.1 Joining President Reagan for an August 3, 1983, lunch to discuss space commercialization were space industry leaders and senior White House and NASA officials. They included (clockwise from the president) John Yardley, James Beggs, George Jeffs, Fred Smith, unidentified, David Thompson (back of head), David Hannah, unidentified, John Townsend, Craig Fuller, unidentified, Bud Evans, Gil Rye, Max Faget, Klaus Heiss, two unidentifieds, Edwin Meese, and George Skurla. (Photograph courtesy of Reagan Presidential Library)

space activities if the right mix of government policy and space program stability can be obtained." The president assured the group that he wanted to make sure that government policy would not "muck up" the prospects for such investment; writing in his diary that evening, Reagan called the concept of space commercialization "very exciting."[12]

Following Up

The industry participants in the August 3 space commercialization meeting agreed to work together to prepare a series of "white papers" expanding on their views regarding the needed policy changes and initiatives. They formed a 15-person Commercial Space Group composed of representatives from the companies that had been present at the August 3 meeting plus several other firms interested in commercial space. Bud Evans coordinated this effort for Fuller; he saw the white papers as a way of crystalizing industry thinking and getting desired government actions explicitly stated. Evans identified six topics for the white papers:

1. How does the administration insure a consistent Federal commercial space policy?
2. What economic incentives should be considered to promote commercial space activity? Options included tax credits, depreciation allowances, low-cost capital, free R&D flight time on the shuttle, among others.
3. What techniques should be used to expand the market for commercial space activities?
4. What is the appropriate role for NASA? Options included NASA being a "non-regulatory facilitator" or a "regulator."
5. What legal and regulatory barriers exist?
6. What national security issues, such as classification or restrictions on technology transfer, affect commercial space ventures?[13]

As these white papers were being prepared, President Reagan reiterated his support for space commercialization in an October 19, 1983, speech at the National Air and Space Museum on the occasion of NASA's 25th anniversary, suggesting that "private companies are already beginning to look to space … And when profit motive starts into play, hold onto your hats. The world is going to see what entrepreneurial genius is all about and what it means to see America get going."[14]

Twenty industry white papers dealing with the six topics were ready by mid-December, just after Reagan had decided to approve development of the space station that had been a private sector priority. The papers totaled almost 50 pages and were titled "Private Enterprise in Space—An Industry View." The preamble to the papers noted that "the success of recent Shuttle flights has aroused interest in the possibilities of profitable, free-enterprise businesses in orbit" and that "to examine that outlook, several private firms jointly established a Commercial Space Group" to appraise "opportunities for and impediments to the commercial use of space." The conclusions of the group were "straightforward":

> Commercial activities in space by private enterprise need to begin now if our nation is to retain and improve its leadership in science and technology, its high living standards, and its advantages in international trade.
>
> Natural and bureaucratic barriers inhibit the commercialization of space. These handicaps need to be diminished or removed with Government help.
>
> With the Government as partner, private enterprise can turn space into an arena of immense benefits for our nation.

The group, "to assure that the potential benefits are realized," had developed "options for resolving twenty of the most critical issues connected with space commercialization." These options reflected the interests of those companies most likely to benefit from their pursuit; they were not an unbiased assessment of the pros and cons of particular courses of action. The bullish view taken by the Commercial Space Group was evident in the introduction to the issue papers, which suggested that "outer space is perhaps the 21st-century equivalent of a new continent waiting to share its wealth … Space commercialization is perhaps as much our nation's manifest destiny as was the taming of lands earlier in our history."[15]

Fuller forwarded the Commercial Space Group report to those in the White House involved with space issues, saying that "this effort has been extremely beneficial and will greatly advance our efforts to stimulate commercial activity in space." He recommended that "we immediately institute a thorough administration-wide review" of the material. National Security Adviser Bud McFarlane called the report a "superb effort" and suggested that President Reagan "highlight this significant effort during his State of the Union Address" as one of "his main priorities for his National Space Strategy." That suggestion was accepted as Reagan told a joint session of Congress on January 25, 1984, "Just as the oceans opened up a new world for clipper ships and Yankee traders, space holds enormous potential for commerce today … We'll soon implement a number of executive initiatives, develop proposals to ease regulatory constraints, and, with NASA's help, promote private sector investment in space."[16]

A Commercial Space Policy

Despite President Reagan's promise that actions to promote space commercialization would come "soon," it took almost six months to issue the policy incorporating those actions. In the interim, the venue for developing the commercialization policy switched from SIG (Space) to the Cabinet Council on Commerce and Trade (CCCT). This was part of a general shift in jurisdiction between the two bodies over the preceding several months, so that issues with limited national security salience could reach the president through the CCCT route. It was not until April 10, 1984, that Craig Fuller forwarded to the members of the CCCT the December 1983 report of the private sector Commercial Space Group, saying that "with the help of the Government as a partner, private sector enterprise can help turn space into an arena of immense benefits for our Nation." He asked the members of the Council to "appoint a representative to serve on a Cabinet Council for Commerce and Trade Working Group that will be responsible for assuring appropriate coverage of critical agency concerns" in formulating a policy responsive to the recommendations of the group. NASA, in the person of Bud Evans, would chair the working group. Fuller set out a schedule that called for a final working group report in mid-June.[17]

Comments were quick in arriving. William Niskanen of the Council of Economic Advisers characterized the proposals as a "grab bag … While some of these proposals are eminently sensible and obviously consistent with Administration policies, other pose serious questions about the meaning of our commitment to the private commercial use of space." Secretary of Commerce Malcolm Baldridge suggested that "we have considerable trouble with some of the options identified." He added that with respect to the appropriate roles of government agencies, the options were "oversimplified." The Treasury

Department observed that "the draft issue papers are a useful catalog of industry concerns, but do not provide much useful guidance on the major question of when and how the government should move to shift responsibility for commercial space activity to the private sector ... The proposals are one-sided in their concentration on the government side of the equation."[18]

Working Group Deliberations

The working group met through May and June. As its work progressed, one member of the group, Lehman Li from the White House Office of Policy Development, raised some skeptical concerns. He suggested that the working group "has been operating in ways that are unusual and may place the Administration in a difficult position later." Li was worried that "the President will be making decisions, in many cases affecting taxpayers, on the basis of decision memoranda prepared largely by the industry that would benefit most by the decision." He added that "this process bears a great deal of resemblance to Democratic proposals for an industrial policy. Government would be making policy targeted toward a specific industry. Industry would be proposing the government policy changes needed to improve the industry's competitiveness." Li suggested that he was not alone in his concerns, saying that "many people view this ... as a push by NASA and the industry group to circumvent the normal budget and policy process."[19]

Indeed, NASA's Evans was pushing the administration to take actions to stimulate new private sector investment in space; some of those actions would imply financial actions (budget expenditures, tax relief, etc.) on the part of the government. But to Evans, this was not industrial policy. In his notes as he prepared for a meeting with Office of Management and Budget Director David Stockman, Evans wrote: "potential for stimulating private investment good—but need specific action by the Administration—so far only words." He added that "industry proposed a number of specific initiatives that would not cost much." He stressed that any government support was intended "to stimulate private investment in R&D which could result in commercial products—a traditional gov't role—the gov't would not spend money to develop the commercial products ... This is analogous to what NACA and NASA did in the aeronautics field, which has paid off in tremendous positive balance of payments."[20]

Reagan Approves a Commercial Space Policy

By mid-July, recommendations from the working group were ready for presidential approval. In a July 17, 1984, presidential decision memorandum, Fuller noted that "a Cabinet Council on Commerce and Trade Working Group has reviewed an assortment of initiatives designed to encourage commercial activities in space." The memorandum was accompanied by the industry white papers.

There were 13 actions proposed for presidential approval: five economic initiatives, four legal and regulatory initiatives, one research and development initiative, and three initiatives to establish and implement a commercial space policy. Three additional initiatives, the president was told, would require more study and thus were not ready for his approval; these were modifying the research tax credit for the space industry, making sure free government services for space R&D were not subject to taxes, and assuring "reasonably priced" access to the space shuttle. Reagan was advised to reject one industry-proposed initiative, reducing investment risk through government loan guarantees and other financial actions.[21]

Ronald Reagan approved the recommendations as set out in the Fuller memorandum, and on July 20, 1984, at a ceremony honoring the Apollo 11 astronauts on the 15th anniversary of the first lunar landing, reminded the audience that he had said in his January State of the Union address, "we would soon develop initiatives to help promote private sector investment in space." He added: "We're now embarking on that course."

> We'll do all we can to ensure that industry has a routine access to space and a suitable, reliable place to work there. And we'll do this without needless regulatory constraints ... The benefits our people can receive from the commercial use of space literally dazzle the imagination. Together we can produce rare medicines with the potential of saving thousands of lives and hundreds of millions of dollars. We can manufacture superchips that improve our competitive position in the world computer market. We can build space observatories enabling scientists to see out to the edge of the universe. And we can produce special alloys and biological materials that benefit greatly from a zero-gravity environment.

Linking space achievements to his concept of American exceptionalism, Reagan suggested that "by accepting the challenge of space we'll carry forward the same courage and indomitable spirit that made us a great nation and that carried our Apollo astronauts to the Moon."[22]

As Reagan spoke, the White House released the text of a new National Policy on the Commercial Use of Space. The policy statement noted that "tax laws and regulations which discriminate against commercial space ventures will be changed or eliminated." It added that "laws and regulations predating space operations will be updated to accommodate the commercial use of space" and that "in partnership with industry and academia, government will expand basic research and development which may have implications for investors aiming at developing commercial space products and services." The policy statement promised that "entrepreneurs will receive assurances of consistent government actions and policies over long periods," recognizing that "commercial developments in space often require many years to reach the production phase." The policy specified that the Reagan administration would establish "a high-level national focus for commercial space issues by creating a Cabinet Council on Commerce and Trade (CCCT) Working Group on the Commercial Use of Space."[23]

As it was released, Bud Evans briefed reporters on the new policy. *The New York Times* noted that "the Reagan administration plans to remove immediately several regulatory and tax obstacles to industrial use of space" and would "propose legislation to remedy 'discrimination' in the tax code against space ventures." For example, one of the changes proposed in the policy would make the existing 10 percent investment tax credit for research and development projects newly applicable to space efforts. Under the then-current Internal Revenue Service interpretation of tax policy, that credit had been denied experimental space projects because they were considered exports, on the basis that they had left the United States as they were being carried into orbit by the shuttle, even though they subsequently were returned to the country. Of the 13 initiatives proposed in the policy, 7 would require legislative changes. *Aviation Week & Space Technology* noted that "administration officials voiced optimism that the proposed changes will be accepted by Congress, which has already directed NASA to encourage space commerce."[24]

Implementing the Policy

It took three months to form the Working Group on the Commercial Use of Space that had been called out in the July 20 policy, and even then the group did not get immediately to work. The working group would be chaired by Richard Shay, deputy general counsel at the Department of Commerce. Secretary of Commerce Baldridge used his general counsel's office not only to deal with legal issues but also as a policy development unit. Even though Commerce had lost the lead role with respect to the commercialization of expendable launch vehicles, Baldridge had maintained his interest in having Commerce, rather than NASA, take the lead role in this new business sector. He wanted someone who worked closely with his office to be the lead person on the working group; this led to the choice of Shay as the working group chair. Baldridge also hired young space activist Courtney Stadd to support Shay in his dealings with the working group. Vice chair of the working group was NASA's deputy general counsel John O'Brien.

On October 23, Shay sent a memo to working group members calling a first meeting of the working group for "the week of November 5th or November 12th." He attached to his memo the executive summary of the report of the earlier Cabinet Council working group chaired by Bud Evans that had been the basis of the July 20 policy statement. He asked his working group members not only to "review and comment" on that summary but also to "suggest additional initiatives." As it turned out, the working group first met only on November 29 and took up as a first order of business, at Craig Fuller's suggestion, a review of NASA's proposed pricing plan for space shuttle launches of commercial and foreign payloads. (That activity will be discussed in Chap. 17.)[25]

Recognizing that the working group was off to a slow start and was in danger of losing focus, Fuller in December 1984 wrote to Michael Driggs, who, as special assistant to the president for policy development and assistant director for commerce and trade, provided White House oversight of the group. Fuller suggested that Driggs direct the working group to "concentrate its energies on developing action plans and legislative proposals for implementing the president's initiatives" that had been spelled out in the July 20 commercial space policy. He asked Driggs to report back to the CCCT "when these objectives have been met." Once again, there were even more lengthy delays; the working group adopted an "interim report" only a year later, on December 20, 1985, and that report only "identified policy issues which need to be addressed." After almost a year and a half, little momentum had developed with respect to Reagan administration commercial initiatives other than those being pushed by NASA.[26]

NASA and Space Commercialization

In addition to his work with Fuller and others in the White House, Bud Evans since mid-1983 had chaired an internal NASA task force on space commercialization. That task force had finished its work in early 1984 and submitted its report to Administrator Beggs. It recommended the creation of a NASA Commercial Space Policy and of a new office to manage its implementation. Action on these recommendations was delayed as the national space commercialization policy was finalized. NASA in early September 1984 announced that it was forming an Office of Commercial Programs, with veteran NASA manager Ike Gillam as its director and Evans as his deputy. On October 29, Beggs announced the NASA Commercial Space Policy and Implementation Plan. While the national-level commercialization policy had been only a few pages long, the NASA plan ran to some 150 pages. In announcing the plan, Beggs tied it to the stated national policy goal of expanding "opportunities for U.S. private sector investment and involvement in civil space and space-related activities."

The preamble to the NASA policy stated that "the new chapter in the U.S. space program that opened early in this decade with the first flights of the Shuttle is now reaching a new phase: space technology is ripe for its transition from exploration to major exploitation, from experimentation to expanded profitable commercial uses." Implementing the NASA policy would "expedite the expansion of self-sustaining, profit-earning, tax-paying, jobs-providing commercial space activities." The new policy would support commercial space activities by "reducing the risks of doing business in space to levels competitive with conventional investments" and by "reaching out and establishing new links with the private sector to stimulate the development of private business in space." It provided for a small amount of NASA seed money to support commercially oriented research in industry and initiated NASA and industry-funded Centers for the Commercial Development of Space located at U.S. universities and research organizations. NASA would also commit to providing an initial market for space venture products and services that fit the agency's needs and had been developed through significant private investments.[27]

The NASA effort in space commercialization had little success. The January 1986 *Challenger* accident meant that the space shuttle was not available as a platform for commercially oriented research for over two years. By the time the shuttle started flying again, there were limited opportunities for commercial research on board, and industry lost much of the enthusiasm that had been building before the accident. The space station was still years in the future.

In addition, the Office of Commercial Programs had a hard time gaining traction. Bromberg provides a succinct summary of the problems the new office faced in implementing the provisions of the NASA commercialization policy. She notes that "many of the functions it had been awarded had been pulled from other programs, whose staff was not always keen to relinquish them." It ran into "territorial conflicts" over control of research into materials processing in space with NASA's Office of Space Science and Applications, where that research had been lodged since the 1970s. It found itself "in quarrels with the Space Shuttle Office" over its taking over the assignment of marketing shuttle payload bay space. NASA engineers "were comfortable taking one, two, or three years to integrate a new device into the shuttle. Business, especially small business, did not have the money it took to keep a research project alive over such long time periods." The Office of Commercial Space Programs "was a small office with … a low budget. It had to contend with indifference in the business community and a firmly rooted R&D culture within its own agency … In the end it was, at best, a new procommerce voice within the cacophony with which NASA spoke to the private sector."

Reflecting on NASA's mid-1980s commercialization effort, Beggs in 2000 suggested that such an effort "comes hard for the great, good bureaucrats of NASA. They want to do the right things … But this is an area they don't understand. It's an area [that] gets in the way of doing things they are charged to do … We got to go to Mars next year! What do you mean I'm supposed to spend time commercializing space?"[28]

Conclusion

At the 1984 peak of enthusiasm for the promise of new commercial activities in space, *Aviation Week* observed that "space commercialization has captured a share of the public imagination that is reminiscent of the pioneering days of manned spaceflight." But that enthusiasm turned out to be both transient and misplaced. During the 1980s, none of the commercialization initiatives put forward by the Reagan administration resulted in significantly increased economic activity in space. It turned out that in almost every case, those suggesting that Earth orbit could quickly become an arena for widespread and rewarding commercial businesses were at best premature in their enthusiasm and at worst mistaken. Writing at the peak of the enthusiasm for space commercialization, this author suggested that "large, new, commercial payoffs are still 15–20 years away; before then, the cupboard will be almost bare." Even that comment turned out to be optimistic. Still, it was Ronald Reagan and his administration that laid the policy and institutional framework for the advent of a commercial space sector in areas other than the already-established communications satellite business. It would take several decades and a number of false starts for that sector to emerge, but the foundation for commercial space activities was laid while Reagan was president.[29]

Notes

1. Michael A. G. Michaud, *Reaching for the High Frontier: The American Pro-Space Movement, 1972–1984* (New York: Praeger, 1986), 247.
2. Joan Lisa Bromberg, *NASA and the Space Industry* (Baltimore, MD: Johns Hopkins University Press, 1999), 120.
3. Philip Klutznick, "Outer Space Profits," *NYT*, September 10, 1983, http://www.nytimes.com/1983/09/10/opinion/outer-space-profits.html. Also, Craig Covault, "Unique Products, New Technologies Spawn Space Business," *AWST*, June 25, 1984, 40–41.
4. William H. Gregory, "Space as a Business," *AWST*, June 25, 1984, 11.
5. The Walker quote comes from Neil Dahlstrom, ed., "Commercial Space Policy in the 1980s: Proceedings of a Roundtable Discussion," May 2000, 9. National Aeronautics and Space Act of 1958, as amended, https://history.nasa.gov/spaceact-legishistory.pdf.
6. See John M. Logsdon, "Selling the Space Shuttle: Early Developments," in Roger Launius and Howard McCurdy, eds., *Seeds of Discovery* (New York: Palgrave Macmillan, 2018) for an account of early steps in promoting the space shuttle as a platform for commercial activities, and Howard E. McCurdy, *The Space Station Decision: Incremental Politics and Technological Choice* (Baltimore, MD: The Johns Hopkins University Press, 1990) for an account of attempts to demonstrate the commercial potentials of a space station.
7. NASA, "Panel to Study Commercial Enterprise in Space," NASA News Release 83-12, February 7, 1983. Bromberg, *NASA and the Space Industry*, 121–122.
8. Memorandum from Craig Fuller to Fred Ryan, "Schedule Proposal," July 20, 1983, Box 7, Outer Space Files, RRL.
9. Letter from Craig Fuller to John Yardley, McDonnell Douglas Astronautics, July 26, 1983, Files of Valerie Neal.
10. Memorandum from Craig Fuller to various White House and Executive Office staff, "Space Commercialization," August 2, 1983, with attached listing of private sector activities and concerns. Folder 12772, NHRC.
11. Craig Fuller, "Briefing Paper for the President – Meeting on Space Commercialization," August 2, 1983, Box 12, Subject Files, Papers of Craig Fuller, RRL.
12. Douglas Brinkley, ed., *The Reagan Diaries* (New York: Harper Collins, 2007), 172. Craig Covault, "Reagan Briefed on Space Station," *AWST*, August 8, 1983, 16.
13. Memorandum from Bud Evans to Craig Fuller, August 15, 1983, Folder 12772, NHRC.

14. Ronald Reagan: "Remarks at the 25th Anniversary Celebration of the National Aeronautics and Space Administration," October 19, 1983. Online by Gerhard Peters and John T. Woolley, APP, http://www.presidency.ucsb.edu/ws/?pid=40662.
15. Commercial Space Group, "Private Enterprise in Space – An Industry View," draft dated December 13, 1983, Box 11, Subject Files, Papers of Craig Fuller, RRL.
16. Memorandum from Craig Fuller to Ed Meese and Robert McFarlane, "Report and Proposal from the Commercial Space Group," December 15, 1983, Box 8, Outer Space Files, RRL. Memorandum from Robert McFarlane to Craig Fuller, "Report and Proposals from the Commercial Space Group," December 23, 1983, Folder 12772, NHRC. Ronald Reagan: "Address Before a Joint Session of the Congress on the State of the Union," January 25, 1984. Online by Peters and Woolley, APP, http://www.presidency.ucsb.edu/ws/?pid=40205.
17. Memorandum from Craig Fuller to the Cabinet Council on Commerce and Trade, "Commercial Space Initiatives," April 10, 1984, Box 2, Outer Space Files, RRL.
18. Memorandum from William Niskanen to Craig Fuller, "Commercial Space Initiatives," April 16, 1984; memorandum from Malcolm Baldridge to Craig Fuller, "Commercial Space Initiatives," April 16, 1984; memorandum from David Chew to Craig Fuller, "Commercial Space Initiatives," April 16, 1984, Box 2, Outer Space Files, RRL.
19. Memorandum from Lehman Li to Roger Porter, "Space Commercialization," May 30, 1984, Box 9, Outer Space Files, RRL.
20. Unsigned but likely Bud Evans, "Overview Thoughts for Stockman Meeting," June 5, 1984, Box 9, Outer Space Files, RRL.
21. Memorandum from Craig Fuller to the President, "Commercial Use of Space," July 17, 1984, Box 3, Outer Space Files, RRL.
22. Carole Shifrin, "Reagan Backs Space Commerce," *AWST*, July 30, 1983, 17. Ronald Reagan: "Remarks at a White House Ceremony Marking the 15th Anniversary of the Apollo 11 Lunar Landing," July 20, 1984. Online by Peters and Woolley, APP, http://www.presidency.ucsb.edu/ws/?pid=40180.
23. Office of the Press Secretary, the White House, "National Policy on the Commercial Use of Space – Fact Sheet," July 20, 1984.
24. Stephen Engelberg, "Reagan Set to Remove Tax Obstacles to Commercial Use of Space," *NYT*, July 21, 1984, 12.
25. Memorandum from Richard Shay to members of the Cabinet Council Working Group on Space Commercialization, October 23, 1984, CFOA 1318, Papers of Michael Driggs, RRL.
26. Memorandum from Craig Fuller to Michael Driggs, "CCCT Working Group on Space Commercialization," December 14, 1984, Folder 12772, NHRC. Memorandum from Michael Driggs to Economic Policy Council Members, "Commercial Space Working Group," undated but December 1985, with attached memorandum from Robert Brumley to Economic Policy Council, "Interim Report," undated but December 20, 1985, CFA 19828, Papers of Michael Driggs, RRL.
27. Memorandum from administrator to officials-in-charge of Headquarters Offices and Field Installations, "NASA Commercial Use of Space Policy," with attached "NASA Commercial Space Policy," October 29, 1984, NHRC. The quotes are

from pp. v–vi. Craig Covault, "NASA Formulates Policy to Spur Private Investment," *AWST*, November 26, 1984, 18.

28. Bromberg, *NASA and the Space Industry*, 122–124. Beggs's remark is from Neil Dahlstrom, ed., "Commercial Space Policy in the 1980s: Proceedings of a Roundtable Discussion," May 2000, 17.
29. William Gregory, "Policy for Commercialization," *AWST*, December 3, 1984, 13. John M. Logsdon, "Space Commercialization: How Soon the Payoffs?" *Futures*, February 1984, 71–78.

CHAPTER 14

Space Shuttle Issues: Round Two

During 1983 and 1984, debates regarding the space station and space commercialization were the primary focus of White House space policy attention. In parallel, as the space shuttle completed its four test flights and entered operational service, there were also diverse shuttle-related issues that required administration attention. How best to take advantage of the shuttle program's successes while also dealing with its problems and with the shuttle's appropriate role in the nation's space efforts became and would continue to be a continuing theme in Reagan administration space policy.

The first operational mission of the shuttle took place in November 1982; the crew deployed two commercial satellites. During 1983, there were four flights; five more missions were flown in 1984. Shuttle orbiter *Columbia* was used for all 1982–1983 missions, and then taken out of service for major modifications. Orbiters *Challenger* and *Discovery* were first launched during 1984. In its first ten operational flights, the shuttle, in addition to deploying commercial and National Aeronautics and Space Administration (NASA) payloads, demonstrated its ability to rendezvous with and retrieve orbiting satellites and to repair and redeploy them or to return them to Earth. A late 1983 mission carried the European-provided Spacelab pressurized laboratory module, providing additional space for in-orbit experiments. There were several extravehicular activities (spacewalks), including some using the Manned Maneuvering Unit, a device that allowed an astronaut to operate in space without being tethered to the shuttle. During its early missions, then, the space shuttle demonstrated most of the attributes that had been touted as reasons for developing a highly capable spaceplane. The shuttle was increasingly seen as an iconic symbol of U.S. technological leadership; on almost every mission, the shuttle achieved a space "first" and reinforced its image as being on the cutting edge of space achievement. During 1983–1984, the space shuttle, at least in public perception, was an overwhelming success.[1]

J. M. Logsdon, *Ronald Reagan and the Space Frontier*,
Palgrave Studies in the History of Science and Technology,
https://doi.org/10.1007/978-3-319-98962-4_14

However, there were some troubling realities accompanying this string of accomplishments. Early experience with operating the shuttle indicated that the original goal of routine, low-cost access to space would most likely be impossible to achieve. Launch delays, frequent refurbishment of shuttle components between missions, lack of spare parts, and persistent problems with the orbiter's thermal protection system suggested that the optimistic early expectations regarding shuttle performance, cost of operation, and turnaround time had been quite unrealistic. Far from becoming "fully operational," the shuttle remained an experiment in reusability and routine access to space, requiring a large workforce with accompanying high costs and long turnaround times; the trade publication *Aviation Week & Space Technology* suggested: "The officially declared operational space shuttle still has a sizeable element of research and development to digest."[2] On a few missions, there were close calls with respect to potential catastrophic accidents. Recognition of the shuttle's limitations was of increasing concern to those in the national security community worried about depending on the shuttle to launch the most critical intelligence payloads. For advocates of commercializing expendable launch vehicles (ELVs), the shuttle, with its artificially low price for commercial customers backed up by NASA's marketing efforts, was seen as a barrier to their potential business success.

The NASA leadership, even though aware of the vehicle's problems, was in the uncomfortable position of having to continue, both inside of and outside of the government, to defend the shuttle's original promises, both in order to maintain the vehicle's role as the primary system for launching U.S. government missions and to market the shuttle to paying customers as an affordable and reliable means of launching commercial payloads. As it struggled to meet its early launch schedule, NASA was still publicly projecting that by the late 1980s the shuttle would be flying 24 times a year. The anticipation was that payments from commercial and foreign users of the shuttle would return up to $1 billion each year to the U.S. Treasury.

The gap between this objective and the reality of the shuttle's early problems at times led NASA's managers to discount warnings that the shuttle was a risky system, as they pressed, even in the face of constrained budgets, to increase the system's flight rate. Looking back after the 2003 breakup of the orbiter *Columbia* at the shuttle's early history, the *Columbia* Accident Investigation Board observed that as NASA in the 1970s had gained approval to develop the space shuttle, "it had accepted the bargain to operate and maintain the vehicle in the safest possible way. The Board is not convinced that NASA has completely lived up to the bargain."[3]

Marketing the Space Shuttle

As NASA had adjusted shuttle pricing policy in 1982, making the shuttle competitive with the European *Ariane* launch vehicle had been a significant consideration. *Ariane*'s threat to the shuttle's claim to be the world's premier launch vehicle led NASA to propose a series of actions to actively market the shuttle to potential commercial and foreign users. Another consideration in this effort was to make sure that after 1988, when "full cost recovery" would be the basis for shuttle pricing, there would be enough shuttle customers to spread the fixed cost of shuttle operations across frequent flights, thereby keeping the cost per flight as low as possible. Adding a marketing orientation to a research and development agency such as NASA was not a natural fit, but the space agency took a number of steps in that direction.

"We Deliver": Astronauts as Salesmen

One of those steps was to dramatize the shuttle's role as a "space truck" carrying commercial payloads into orbit by using the shuttle crews as promoters of that capability. The first operational launch of the space shuttle took place in November 1982; the primary payloads were two commercial communication satellites. The satellites were successfully deployed, and the four-astronaut crew, who had advertised themselves as the "We Deliver" team, posed in the shuttle middeck holding a placard with that motto and declaring themselves and their vehicle the "Ace Moving Company."

In addition to delivering satellites into an initial orbit, the shuttle in 1984 demonstrated the capability to retrieve them. Two communication satellites were successfully deployed from the shuttle payload bay during a February 1984 mission, but on both deployments there was a failure of the rocket stage intended to transfer the satellites to their final position. The failures left the satellites stranded uselessly in low Earth orbit. The satellites were insured, and the insurance companies paid compensation to both satellite owners, thereby becoming owners of the satellites. During a November 1984 mission, the insurance companies paid NASA to retrieve the satellites and return them to Earth for resale and relaunch. This retrieval mission was successful, and astronauts Dale Garner and Joe Allen proudly displayed a "For Sale" sign as they returned from their rescue spacewalks.

A Marketing Brochure

The "We Deliver" motto also served as the title of a 12-page glossy brochure that NASA in 1983 prepared to tout the shuttle's advantages. The document was printed in several languages, reflecting the worldwide character of NASA's marketing effort; one target audience for the brochure were the attendees at the June 1983 Paris Air Show. The brochure proclaimed that the space shuttle was "the most useful and versatile space transporter ever built. It has also demonstrated a remarkable suitability for delivering communication satellites to earth orbit." It claimed that "in all the world, you won't find the Shuttle's equal" and that "you can't get a better price." NASA was venturing far away from its heritage as an organization dedicated to space science and exploration.[4]

The Space Shuttle on Tour

In spring 1983, NASA decided, after an assessment of the technical and security risks involved, to fly the shuttle test orbiter *Enterprise* to Europe in order to make it the centerpiece of the U.S. exhibit at the Paris Air Show and to showcase the vehicle in other European countries. *Enterprise* had been used for approach and landing tests in 1977 and for a variety of other test purposes; it was too heavy to be modified so that it could be launched into orbit. Ferrying the orbiter to Europe would require the use of the Shuttle Carrier Aircraft, a modified Boeing 747. NASA had only one of these aircraft, and a major risk of the European trip was the possibility of an accident or loss involving the carrier aircraft; such an incident could disrupt the overall shuttle program. Beggs in notifying the White House of NASA's plans for the "symbolic goodwill flight" had indicated that "the loss of *Enterprise* would not be serious but the loss of the 747 would have a very severe impact." There was also a risk of a terrorist attack; the orbiter would be a very visible and tempting target. That the White House was willing to accept these risks suggests the priority given to both the political and commercial objectives of the shuttle program.

There was discussion between the White House and NASA on where the orbiter would land for public viewing. NASA narrowed down 15 possible destinations to three—Paris, Bonn, and London. The White House then pressed NASA to add a fourth stop in Rome, since "omission of a stop in Rome will be seen by the Italians as one more in a series of slights." National Security Adviser Judge Clark wrote to Beggs on May 23, 1983, saying that "it is important that we add a stop in Italy … to avoid the perception that we value our relations and aerospace cooperation in particular with Italy less than with the three other large countries of Europe." Clark added: "I understand that the additional stop will increase the exposure of the shuttle and the Boeing 747 carrying it to the danger of terrorist attack," but "foreign policy considerations" outweighed that risk. A stop in Rome was added to the itinerary. NASA estimated that some two million people saw *Enterprise* during its time in Europe, either when it was on ground display or when it flew over European cities at a low altitude. That visibility helped cement the idea that the shuttle was both a potent symbol of U.S. leadership and a vehicle ready to open up space to a wide variety of users.[5]

A Marketing Plan

These 1983 efforts to attract paying customers to the space shuttle were intended as only first steps. NASA's Office of Space Flight in 1984 produced a shuttle "marketing plan" that noted that "shuttle marketing activities have met with much success, but they are still in a developmental stage … The NASA marketing team needs to be stronger and more sophisticated in order to successfully compete with other marketing efforts." The marketing plan identified the need for

> an aggressive promotional effort, tailored to the needs of the marketplace … emphasizing NASA's extensive experience and how the Shuttle can be used to help accomplish our customers' scientific and business objectives. Sales efforts will be directed toward those identified as potential STS customers to obtain launch commitments. This effort will take place through presentations and regular contact with individual customers as well as targeted audiences. Other government agencies who have influence with customers will also be targets of this activity. Promotional programs will also be directed towards others who influence customer decisions such as spacecraft manufacturers, trade associations, payload operators, and customers of payload products.[6]

Discussions of "sales teams," "potential customers," and "targeted audiences" were certainly departures from NASA's Apollo-era heritage. They reflected a NASA struggling to fulfill the promises made when the space shuttle was approved and to adapt to the Reagan administration emphasis on commercializing space activities. Some of those from other government agencies involved in the commercialization push, in particular those promoting commercial ELVs, were not pleased by NASA's aggressive pursuit of commercial launch contracts, seeing NASA's marketing efforts and artificially low shuttle prices as direct competition posing almost insurmountable obstacles to ELV business success. Conflict between NASA and those agencies, with shuttle pricing policy as its focal point, was by 1984 inevitable, presenting the Reagan administration with a bitter policy debate that was to extend over the next two years.

What Price for a Shuttle Launch?

In his May 1983 National Security Decision Directive 94 on "Commercialization of Expendable Launch Vehicles," Ronald Reagan had indicated that beginning after October 1, 1988, "it is the U.S. Government's intent to establish a full cost recovery policy for commercial and foreign STS flight operations."[7] How to implement that "intent" and how to define "full cost recovery" became points of contention in 1984, and continued to be controversial as the second Reagan term began in 1985.

In November 1983, Reagan had decided that the Department of Transportation (DOT) would have the lead role in both promoting and regulating ELV commercialization. Based on this decision, Secretary of Transportation Elizabeth Dole was added to the membership of the Senior Interagency Group for Space [SIG (Space)], and her top assistant Jenna Dorn, who had been given responsibility for overseeing the department's new Office of Commercial Space Transportation, was made a member of the lower-level Interagency Group for Space [IG (Space)]. The Department of Commerce (DOC) had competed with DOT for the lead commercialization role; even though it lost that competition, Commerce Secretary Malcolm Baldridge remained determined that his agency would have significant influence over space commercialization overall. He assigned the lead space policy role within the department to Deputy Secretary Clarence "Bud" Brown. Baldridge also moved staff support for the department's engagement with space policy decisions from the National Oceanic and Atmospheric Administration to his Office of the General Counsel, thereby moving space issues much closer to the department's top leadership.

By early 1984, Dole and Baldridge, and particularly their staff, were prepared to oppose NASA as the space agency attempted to maintain NASA's near-monopoly position as the U.S. supplier of launch services, not only for the U.S. government but for commercial and foreign customers. During the 1983 ELV commercialization debate, both Commerce and Transportation pushed for immediate adjustments in shuttle pricing policy. They had not succeeded in that effort, and by 1984 were seeking a second chance at a pricing policy adjustment.

The National Security Council (NSC) was striving to get interagency agreement on a statement of National Space Strategy, an effort that dated back to December 1982. In January 1984, SIG (Space) circulated for review a new draft of such a statement. Pricing of space shuttle launches and associated services was not an issue called out in that draft. The DOT reaction suggested "serious concerns that the National Space Strategy, as currently drafted, fails to give proper recognition to the President's ... policy to encourage and facilitate the commercialization of expendable launch vehicles." Dole in early April told Fuller that it was "fundamentally important" that the issue of full cost recovery shuttle pricing be included in any space strategy directive.[8]

Dole was successful in pushing her suggestion. The controversy over shuttle pricing became a sticking point in finalizing the national space strategy, with a compromise solution finally adopted. That process will be described in the following chapter.

Shuttle Policy "A Serious Mistake"

The conflict over shuttle pricing was just the first foray in an increasingly antagonistic relationship between NASA and advocates of a greater private sector role in U.S. space launch activities. That conflict was not the only challenge to the future role of the space shuttle in U.S. space affairs. Another key to the viability of the shuttle program was the continued commitment of the national security community to use the shuttle as its exclusive means of gaining access to space. That commitment was becoming increasingly a matter of dispute in the 1983–1984 period.

The first formal space policy decision of the Reagan administration had been National Security Decision Directive 8, "Space Transportation System." The directive was issued on November 13, 1981, and stated that "the STS will be the primary space launch system for both United States military and civil government missions. The transition to the Shuttle should occur as expeditiously as practical." The directive in essence reiterated the 1978 Carter administration decision that the space shuttle would become the sole launch system for all government payloads and that NASA and Department of Defense (DOD) would no longer purchase the ELVs previously used to launch government missions.[9]

Questioning the Commitment to the Space Shuttle

As the space shuttle began its operational flights, the commitment of the national security community to using it as its sole means of access to space increasingly came into question. Taking the lead in this questioning was Undersecretary of the Air Force Edward "Pete" Aldridge, Jr. Aldridge was not only in charge of the space activities of the Air Force, which included responsibility for launching all DOD satellites; he was also, although publicly unacknowledged, the director of the National Reconnaissance Office (NRO), the very existence of which was classified. In that role, he was responsible for the development, launch, and operation of U.S. intelligence satellites. By mid-1983, confidence in the shuttle among Aldridge and his NRO associates had begun "to seriously erode"; Aldridge had become worried that the decision to end ELV procurement and to place sole reliance on the shuttle was ill-conceived. In response, during 1983 and 1984, Aldridge "waged a mostly secret and very difficult, but eventually successful, campaign" aimed at getting approval to develop an alternative to the space shuttle for launching the nation's largest intelligence satellites.[10]

There were three aspects of Aldridge's concern. One was the shuttle's flight rate. Early experience with the shuttle had indicated that NASA would have trouble meeting the 24-flights-per-year rate that NASA had adopted after the shuttle began flying. Aldridge thought that 18 flights per year, or even perhaps 12, was a more realistic rate. The problem was that the national security community had a "hard requirement" for 12 launches per year, and it would be very difficult to meet that requirement using only the shuttle, and still allow NASA to have launch capability available for its own and commercial and foreign missions. This "could result in DOD being close to the exclusive user of the Shuttle, with the total cost being born by DOD"; the alternative was "DOD payloads would be delayed in order to launch important scientific payloads."

A second concern was the ability of the shuttle to launch the heaviest national security payloads. NASA had promised that the shuttle could launch 65,000 pounds of payload from Kennedy Space Center and 32,000 pounds to polar orbit from a new west coast launch site. The NRO in the late 1970s began to design the next generation of its various satellites to take advantage of this capability, which would exceed the lifting power of the Titan III ELV then in use.

The problem was that shuttle orbiters *Columbia* and *Challenger* were themselves too heavy to be able to carry the heaviest NRO payloads to the desired orbit. That meant that there were only two orbiters, *Discovery* and *Atlantis* (which would enter service in 1985), capable of carrying out all national security missions, and they would have to operate at more than normal power during launch to do so. There would be a serious national security problem if one or both of those orbiters were not available when needed.

Finally, the costs of each shuttle mission were turning out to be much higher than anticipated, and it appeared to Aldridge that the national security community could end up paying an unanticipated greater share of the shuttle's operating expenses. For the first six years of shuttle operations, the price that NASA had committed to charge DOD for a flight was much lower than actual costs. The DOD was concerned that as shuttle prices were adjusted after 1988, the price charged DOD to use the shuttle would be as high, if not higher, as the cost of using ELVs, rather than the significantly reduced price that had been promised and that DOD had used in its long-term financial planning.

Faced with this situation, Aldridge communicated his concerns to Secretary of Defense Caspar Weinberger. He told Weinberger: "We should not terminate the production of expendable launch vehicles until the Shuttle can prove itself, that it can fly at least 24 flights per year, and it can meet the performance demands of the Department of Defense." Aldridge's warning was one of the reasons that Weinberger during the 1983 debate over approving the space station had argued that NASA should be giving priority to making the shuttle fully operational rather than shifting its focus to the space station as a major new project. Weinberger's concern, as expressed to NASA's Beggs, was that "a major new start of this magnitude would inevitably divert NASA managerial talent and resources from the priority task of making the Space Transportation System fully operational and cost effective. With all our national security space programs committed to the Shuttle and dependent on it for their sole access to space, I am sure you can appreciate my concern."[11]

Aldridge's worries had continued to increase; on December 27, 1983, he briefed Weinberger about his concerns with respect to the "shuttle only" policy and proposed a plan to address his concerns. That plan involved DOD procuring at least two heavy-lift ELVs each year to "complement" the shuttle as a way of providing access to orbit for the most critical intelligence payloads.

Agreeing with Aldridge, Weinberger wrote President Reagan on January 23, 1984, saying that the country had made a "serious mistake" in depending on the shuttle for sole access to space. He proposed that DOD be given permission to procure a limited number of ELVs to complement the Shuttle, but said that "DOD remains committed to the STS. Our limited use of ELVs will not undermine the validity of the STS program."

On the same day, Weinberger approved a new "Defense Space Launch Strategy" based on Aldridge's plan. The strategy, issued on February 7, 1984, stated that "the DOD has a validated requirement for an assured launch capability under peace, crisis and conflict conditions. Assured launch capability is a function of satisfying two specific requirements: the need for complementary launch systems to hedge against unforeseen technical and operational problems, and the need for a launch system suited for operations in crisis and conflict situations." The strategy stated that in the near term

> existing Defense space launch planning specifies that DOD will rely on four unique, manned orbiters for sole access to space for all national security systems. DOD studies and other independent evaluations have concluded that this does not represent an assured, flexible and responsive access to space. While the DOD is fully committed to the STS, total reliance on the STS for sole access to space in view of the technical and operational uncertainties, represents an unacceptable national security risk. A complementary system is necessary to provide high confidence of access to space ... In addition, the limited number of unique, manned Shuttle vehicles renders them ill-suited and inappropriate for use in a high risk environment.

The strategy directed the Air Force to "take immediate action to acquire a commercial, unmanned, expendable launch vehicle capability to complement the STS ... These vehicles must provide a launch capability essentially equal to the original STS weight and volume specifications." They would be known as complementary expendable launch vehicles (CELV).[12]

NASA Is "Furious"

According to Aldridge, when it learned of his plans to use a few ELVs in addition to the shuttle, "NASA, especially its Administrator, was furious." Beggs told Aldridge that they were only "a ploy of the Air Force to abandon the Shuttle." Beggs was still angry at DOD's refusal to support NASA's space station plans, and that anger carried over into the DOD launch strategy. Protesting the strategy, Beggs wrote to Weinberger in May 1984, insisting that the Air Force flight requirements could and would be met by the shuttle, and thus a back-up or complement to the shuttle was not needed. Beggs suggested that if the Air Force still felt compelled to have a complement to the shuttle, it should be based on elements of the shuttle system such as the external propellant tank, shuttle main engines, and its solid rocket boosters, rather than a totally separate launch system. Beggs's suggestion that the Air Force consider a shuttle-derived vehicle that NASA would manage under contract with the DOD created an awkward situation. Pursuing this path would put NASA into direct competition with potential industrial suppliers of the CELV. It was not even clear whether DOD was legally able to contract with another government agency when a similar product or service was available from a commercial supplier.

The Air Force decided to split the CELV competition into two parts. First, a preferred industrial supplier would be chosen. Then the winning proposal would be compared with the NASA offer. The competition took place during summer 1984, and it soon became widely known that NASA did not fare well. Aldridge says the NASA-Air Force relationship at this point was "bitter"; he suggested that "NASA officials placed a great deal of pressure on the contractors [General Dynamics and Martin Marietta] not to bid the Air Force competition, since, they were told, their relationship with NASA would be jeopardized." As the summer ended, "NASA continued to fight the Air Force's plan, openly and underground." This NASA-Air Force conflict would not be resolved until early 1985, and its impacts would linger for the remainder of the Reagan administration's time in the White House.

Who Gets to Fly?

Beginning in 1983, NASA began flying the new astronauts for the shuttle program it had selected in 1978. The first post-Apollo class of 35 astronauts included 6 women, 3 African-Americans, and 1 Asian-American. It was thus inevitable that other than the white males who had flown all preshuttle U.S. spaceflights and the four shuttle test missions would go into orbit aboard the shuttle. That happened during 1983, as Sally Ride became the first U.S. woman to go into orbit on a June 1983 flight and, on the next mission in August, Guion "Guy" Bluford became the first African-American space flyer. In addition, NASA early on took the position that the shuttle was safe enough to host a diverse mix of individuals, not just carefully selected and extensively trained professional astronauts. In 1982, the space agency had broadened its policy on shuttle crews to allow the sponsors of major space shuttle payloads to nominate a person to fly along with that payload. Several countries that had contracted for a shuttle launch of a communication satellite planned to avail themselves of that opportunity, and as part of its increasing focus on space commercialization, one company, McDonnell Douglas, in 1984 and subsequently flew a company employee, Charles Walker, on a shuttle flight carrying the company's microgravity experiment. The White House, for foreign policy reasons, was issuing to key U.S. allies an invitation to fly one of their countrymen aboard the shuttle.

Citizens in Space

Beggs in June 1982 had asked the NASA Advisory Council to examine the issues associated with selecting private citizens to fly aboard the space shuttle. The Council had formed an ad hoc task force to carry out such a study. That task force finished its work in May 1983; in transmitting its report to the chair of the NASA Advisory Council, Dan Fink, task force lead John Naugle said that a "flight of private citizens is both feasible and desirable, provided that their flight be for purposes included within the scope of the Space Act." That 1958 legislation had directed NASA to provide for "the widest practicable and appropriate dissemination of its [NASA's] activities." The task force concluded that flying a person who could share his or her experience with a broad audience fit the Space Act mandate, and thus recommended "a modest program to fly private citizens as observers who would then communicate their experiences to a more general public." In making this recommendation, the task force had considered but rejected the possibility of NASA conducting a national lottery in which any U.S. citizen could participate.[13]

The task force report was sent to Beggs; it took more than a year for him to accept the report's recommendations. Beggs's approval came only on June 21,

1984; the reason for the delay is not clear from available records. NASA identified two possible categories of individuals, teachers and journalists, to be offered the initial opportunity for a shuttle flight. Beggs in a 2002 interview commented that "we had been thinking about the idea of taking a journalist up for a long time, because we thought if we could get a journalist up there, at least somebody would be able … to give us, if not good publicity, at least a lot of publicity." He recalled that NASA's top lawyer, Neil Hosenball, said "Why not a teacher?" Beggs "thought about that for a little bit" and concluded, "Well, why not? Because the biggest receptive audience we have in this country are the kids. Kids love space. A teacher could give you an introduction to those kids that no one else could."

By July 1984, NASA decided that the first flight would be offered to a full-time teacher in a U.S. elementary or secondary school. What NASA named the Spaceflight Participant Program was intended to be a continuing effort, with subsequent flight opportunities made available to additional categories in addition to a journalist, who would be second in line. NASA informed Craig Fuller in the White House of its decision; up to this point, the effort had gone forward without White House involvement. On July 14, Fuller sent a message to his assistant saying, "Beggs has been trying to get a sign-off on the first category of non-astronauts to fly in the shuttle. He wants to choose teachers. That is ok with us."[14]

The White House soon embraced the NASA initiative, deciding that it was the sort of idea that would provide political benefits to the president during an election year and that Ronald Reagan should personally announce the flight opportunity. In an August 27, 1984, speech to a group of secondary school students, Reagan said:

> It's long been a goal of our space shuttle, the program, to some day carry citizen passengers into space. Until now, we hadn't decided who the first citizen passenger would be. But today I'm directing NASA to begin a search in all of our elementary and secondary schools and to choose as the first citizen passenger in the history of our space program one of America's finest—a teacher.
>
> When that shuttle lifts off, all of America will be reminded of the crucial role that teachers and education play in the life of our nation. I can't think of a better lesson for our children and our country.

Discussing the announcement, Beggs reported that "NASA sent Reagan a list of options in which the agency sought a schoolteacher as its first choice," and that there were "no political considerations were involved in citing a profession critical of the president for budget cuts involving education." He also said that he expected "journalists, artists, writers, entertainers, lawyers and blue-collar workers" to be offered future flight opportunities, with "two, three or even four passengers a year" on shuttle flights later in the decade. The Spaceflight Participant Program was an immediate success in terms of public attention; by the end of 1984, NASA had received over 21,000 requests for the application to be considered for the teacher's flight opportunity.[15]

Politicians in Space

On November 7, 1984, Senator Jake Garn (R-UT) announced that he had received a letter from Beggs inviting him to make "an inspection tour and flight aboard the shuttle." Garn had been asking NASA when he might fly aboard the shuttle even before his first interaction with Beggs in mid-1981. Garn was an experienced pilot, with more flying hours than all but one NASA astronaut. But that was not the basis for Beggs's invitation; Garn chaired the Senate appropriations subcommittee that controlled NASA's funding, and in his letter Beggs noted that it was "appropriate for those with Congressional oversight to have flight opportunities to gain a personal awareness and familiarity" with NASA's activities. NASA indicated that "other Congressional leaders 'directly responsible for NASA activities would be given consideration if they are interested' in making space trips."[16]

Even Young Astronauts

In his July 6, 1984, column in *The Washington Post*, investigative journalist Jack Anderson wrote that he had "called on Reagan last October," leaving with him a proposal to endorse a privately created Young Astronauts Program intended to "attract youths, stir their imaginations and stimulate their spirit of adventure." The program, suggested Anderson, could include scholarships for educational and internship opportunities, prizes for "special achievements," trips to space-related facilities in the United States and overseas, and eventually might even include "America's outstanding Young Astronaut on a future space flight."[17]

The Young Astronaut proposal, like the idea of sending a teacher to space, was the kind of initiative that Ronald Reagan's White House was likely to welcome, since it resonated with the president's linking of the space program to a positive American future. On June 19, 1984, as he helped dedicate the new headquarters building of the National Geographic Society, Reagan told the audience that he was "particularly pleased to learn of the efforts … to develop and support a new national Young Astronauts Program to involve young Americans more directly in our space program. This new organization will expand their appreciation of space as a place in which people can live and work and learn."[18]

An October 17 ceremony on the south lawn of the White House allowed President Reagan to kick off the program. In his remarks at the event, the president suggested that "just as our past achievements in space reassure us of our greatness, the Young Astronauts Program reassures us that we will keep dreaming new dreams and keep moving forward." Reagan made no mention in his remarks of a short-term opportunity for a Young Astronaut to fly on the shuttle; rather, he told the young people at the ceremony "maybe one day, one of you young astronauts can follow" in the footsteps of the 17 new NASA astronauts in attendance. He added "but you must be ready. And that means mastering science, math, and computers—the wonderful world of high tech."[19]

Conclusion

The policy decisions to develop the space station, to invite international participation in that development, and to place added emphasis on space commercialization, especially with respect to commercializing ELVs, were significant space accomplishments as President Ronald Reagan stood for re-election in November 1984. But perhaps it was the apparent success of the space shuttle program that most publicly was associated with the first Reagan term. The 13 shuttle flights between April 1981 and Election Day on November 4, 1984, reminded the American public that the Reagan administration had gotten U.S. astronauts back into orbit on a frequent basis. Ronald Reagan had interacted with the crew on most of those shuttle flights, in the process identifying himself with their accomplishments. While there were problems and conflicts associated with the space shuttle program in the 1983–1984 period, to the American public and indeed the world's population they served as evidence supporting Ronald Reagan's claim in his 1984 State of the Union speech that "America is back, standing tall."

Notes

1. For an encyclopedic account of shuttle missions, and indeed of the space shuttle itself, see the three-volume work by Dennis Jenkins, *Space Shuttle: Developing an Icon* (Forest Lake, MN: Specialty Press, 2017).
2. William Gregory, "Growing Pains for Shuttle," *AWST*, February 20, 1984, 11.
3. *Columbia* Accident Investigation Board, *Report*, Volume 1 (Washington, DC: Government Printing Office, August 2003), 97.
4. The text of the brochure is reprinted in John M. Logsdon et al., eds., *Exploring the Unknown: Selected Documents in the Evolution of the U.S. Civil Space Program*, Volume IV, Accessing Space, NASA SP-4407 (Washington, DC: Government Printing Office, 1999), 423–426.
5. See Hans Mark, *The Space Station: A Personal Journey* (Durham, NC: Duke University Press, 1987), 155–161 for an account of the *Enterprise* trip. Letter from James Beggs to Gil Rye, March 16, 1983, Box 7, Outer Space Files; letter from William Clark to James Beggs, May 23, 1983, Box 30, Subject Files, Executive Secretariat, National Security Council, both in RRL.
6. Office of Space Flight, NASA, "The Space Transportation Marketing Plan," June 1984, 5–6, NHRC.
7. Ronald Reagan, National Security Decision Directive 94, "Commercialization of Expendable Launch Vehicles," https://www.reaganlibrary.gov/sites/default/files/archives/reference/scanned-nsdds/nsdd94.pdf.
8. Memorandum from Jenna Dorn to executive secretary, National Security Council, "Comments on Draft National Space Strategy," February 2, 1984, https://fas.org/irp/offdocs/nsdd/nsdd-144.htm. Memorandum from Elizabeth Dole to Craig Fuller, "Shuttle Pricing Policy," April 4, 1984, Box 9, Outer Space Files, RRL.
9. National Security Decision Directive 8, "Space Transportation System," November 13, 1981, https://www.reaganlibrary.gov/sites/default/files/archives/reference/scanned-nsdds/nsdd8.pdf.

10. Unless otherwise cited, material in this section is drawn from an oral history of Edward "Pete" Aldridge, Jr. conducted by Rebecca Wright as part of the NASA Headquarters Oral History Project, May 29, 2009, https://www.jsc.nasa.gov/history/oral_histories/NASA_HQ/Administrators/AldridgeEC/AldridgeEC_5-29-09.htm, and E.C. Pete Aldridge, Jr., "Assured Access: 'The Bureaucratic Space War," Robert Goddard Historical Essay, no date, ocw.mit.edu/courses/aeronautics-and-astronautics/16-885j-aircraft-systems-engineering-fall-2005/readings/aldrdg_space_war.pdf. This essay was Aldridge's entry in a space history competition organized by the National Space Club. These accounts naturally reflect Aldridge's perspective on the conflict over the DOD initiative to complement dependence on the shuttle with a few expendable launch vehicles. The quote in the final sentence of the paragraph is from Peter Hays, "NASA and the Department of Defense: Enduring Themes in Three Key Areas," in Steven Dick and Roger Launius, eds., *Critical Issues in the History of Spaceflight*, NASA SP2006-4702 (Washington: Government Printing Office, 2006), 234.
11. Letter from Caspar Weinberger to James Beggs, January 16, 1984, File 12905, NHRC.
12. Caspar Weinberger, Memorandum for the Secretaries of the Military Departments, Chairman of the Joint Chiefs of Staff, Under Secretaries of Defense, Assistant Secretaries of Defense, General Counsel, "Defense Space Launch Strategy," with attached "Defense Space Launch Strategy," February 7, 1984, reprinted in John M. Logsdon with Dwayne Day and Roger Launius, eds., *Exploring the Unknown: Selected Documents in the History of the U.S. Civil Space Program*, Volume II: External Relationships, NASA SP-4407 (Washington, DC: Government Printing Office, 1996), 401–402.
13. Letter from John Naugle to Dan Fink, with attached "Report of the Informal Task Force for the Study of Issues in Selecting Private Citizens for Space Shuttle Flight," June 16, 1983, Folder 4261, NHRC.
14. Interview of James Beggs by Kevin Rusnak, March 7, 2002, NHRC. Interview with Alan Ladwig, May 6, 2016. Ladwig was selected in 1983 as manager of the Space Flight Participant Program. Memorandum from CLF (Craig Fuller) to L. Herbolsheimer, July 14, 1983, Outer Space-1 File, RRL.
15. Ronald Reagan: "Remarks at a Ceremony Honoring the 1983–1984 Winners in the Secondary School Recognition Program," August 27, 1984. Online by Gerhard Peters and John T. Woolley, APP, http://www.presidency.ucsb.edu/ws/?pid=40300. Ann Bradley, Associate Deputy Administrator, NASA, Special Announcement, "NASA's New Space Flight Participant Program," August 27, 1984, Folder 12782, NHRC. Thomas O'Toole, "Reagan Tells NASA to Choose Schoolteacher for Shuttle Flight," *WP*, August 28, 1984, A1. "Garn, Teacher Orbiter Flights Expected in Next Realignment," *AWST*, December 3, 1984, 21.
16. John Noble Wilford, "Garn, Head of Senate Space Panel, is Chosen to Fly aboard Shuttle," *NYT*, August 8, 1984, B13.
17. Jack Anderson, "President Hopes to Spur Youth into Space Age," *WP*, July 6, 1984, Box 3, Outer Space Files; memorandum from Jack Anderson to the President, "Young Astronauts," October 14, 1983, Box 8, Outer Space files, RRL.

18. Ronald Reagan: "Remarks at Dedication Ceremonies for the New Building of the National Geographic Society," June 19, 1984. Online by Peters and Woolley, APP, http://www.presidency.ucsb.edu/ws/?pid=40063.
19. Memorandum from Tom Gibson for Michael Deaver, "Status of Young Astronauts Program Development; Components for a Presidential Event for a Young Astronauts Kick-off," September 28, 1984; memorandum from James Coyne to Michael Deaver, "Young Astronauts Program," September 28, 1984, Box 9, Outer Space Files, RRL. Ronald Reagan: "Remarks at a White House Ceremony Launching the Young Astronaut Program," October 17, 1984. Online by Peters and Woolley, APP, http://www.presidency.ucsb.edu/ws/?pid=39259.

CHAPTER 15

Finishing the First Term

During its first term in the White House, the Reagan administration took seriously the advice of its 1980 National Aeronautics and Space Administration (NASA) transition team that "NASA and its civil space program represent an opportunity for positive accomplishments by the Reagan administration." Between 1981 and 1984, the White House made a series of space policy and program decisions that created the foundation for achieving the objective that the transition team had advocated: "NASA can be many things in the future—the best in American accomplishment and inspiration for all citizens."[1]

There was one area of unfinished business as the 1984 presidential election approached. In December 1982, the National Security Council (NSC) had begun a study intended to lead to a national space strategy (see Chap. 5). The intent of such a study had been to create "a broad action-oriented plan for a more vigorous and focused U.S. Space Program." That plan would include "three alternative space programs: current, enhanced and significantly enhanced."[2] Given Ronald Reagan's top priority domestic goal of reducing government spending, the notion that he would be willing to approve an expanded space effort requiring significant budget increases was from the start likely unrealistic. Instead, by the time in August 1984 that the space strategy reached Ronald Reagan's desk for his approval, it was primarily an accounting of the space decisions made during first Reagan term. These decisions, including approval of going ahead with a multiyear and multibillion dollar space station program, expanded international cooperation in the space station and other NASA projects, and greatly increased emphasis on facilitating private sector involvement in commercially motivated space activities, did add up to a more ambitious space program than had been in place when Reagan had come to the White House in January 1981. But that program would be conducted employing the same share of the Federal budget as had been the case when Reagan took office. Most Reagan space initiatives did not require, at least in

J. M. Logsdon, *Ronald Reagan and the Space Frontier*,
Palgrave Studies in the History of Science and Technology,
https://doi.org/10.1007/978-3-319-98962-4_15

the short term, an increase in the space budget, and there had been some budget reductions to offset new programs. Ronald Reagan "liked" civilian space, but he was not willing to increase the NASA budget by anything more than a token amount to finance major new space initiatives.

Finishing the National Space Strategy

It took a lengthy struggle among contending interests to reach agreement on the final statement of National Space Strategy. As the strategy formulation effort got underway in 1983, there were deep divisions between NSC ambitions for the strategy and the views of the national security community; those views questioned the viability of the concept of a single overarching strategy for U.S. civil, national security, and commercial space activities. The community's position was that such a single strategy was artificial and not needed. One concern was that a unified approach to space might facilitate trade-offs among the civilian, military, and intelligence space efforts that would work to the detriment of national security programs. That resistance continued throughout late 1983 and into the following year.

Steps Toward a Strategy

It fell to Gil Rye, the NSC lead for space policy, to attempt to find a path that could overcome that resistance. On January 25, 1984, the same day that President Reagan announced his approval of the space station in his State of the Union Address, Rye circulated for comment a draft National Security Decision Directive (NSDD) on the national space strategy. That document set out "four major areas for priority emphasis":

- "Establish a Permanent Manned Presence in Space"
- "Increase International Cooperation in Space"
- "Stimulate Private Sector Investment in Space"
- "Respond to the Changing National Security Environment in Space."

The directive elaborated on each of these priorities. With respect to the civil and commercial priorities, the elaboration was basically a restatement of the conclusions reached in the ongoing discussions of those issues. The final part of the statement set out steps to implement the strategy. New here was a provision, reflecting science adviser Keyworth's position that the space station was just a means to broader space goals, that "NASA and OSTP [Office of Science and Technology Policy] will conduct a study to assess the feasibility of establishing more ambitious, long range goals for the civil space program such as a manned lunar station or manned expedition to Mars." Also new was a requirement, reflecting the national security community's questioning of sole dependence on the space shuttle for access to space, that "DOD and NASA will jointly conduct a study to assess the nation's launch capability and organization."[3]

Rye had seized the initiative by drafting the strategy statement, but it could not move forward until the various Senior Interagency Group for Space [SIG (Space)] member agencies had a chance to comment on that draft. Those comments were not long in coming, and most were critical. Rye's reaction to the criticisms was "We got their attention!" by circulating the draft strategy. He added that "judging from the comments on the draft NSDD, we now apparently have the agencies clamoring to provide inputs to the document." To discuss the criticisms, Rye scheduled a February 3, 1984, meeting of the Interagency Group for Space [IG (Space)]. That meeting would be chaired by Rear Admiral John Poindexter, a senior NSC staff member since 1981 who had moved into the deputy national security adviser position as Bud McFarlane had been promoted to the top job in October 1983. The goal of the February 3 meeting was to get IG (Space) members engaged in quickly creating a second draft of the NSDD; Rye still hoped to get the document to the president before the end of February.[4]

Moving forward on that fast schedule once again proved impossible; the February 3 meeting was contentious. New to the IG (Space) was the Department of Transportation (DOT). Its representative, Jenna Dorn, suggested that issues related to the future commercial use of the space shuttle, particularly with respect to shuttle pricing policy, needed to be part of the strategy discussion; adding this issue to the strategy-crafting process resulted in slowing down the effort. On February 23, Poindexter told the IG (Space) members that "one NSDD and one Strategy will be produced," that the NSDD "will be approximately two to three pages in length" and would be unclassified, and would "build upon the three civil/commercial priorities for space announced by the President in his State of the Union address."

The Intelligence Community interpreted Poindexter's message as suggesting, as they had urged, that national security issues would not be a focus of the NSDD. The director of the Intelligence Community staff, Admiral E.A. Burkhalter, replied to Poindexter that he agreed to working on a civil-oriented strategy, but that the Intelligence Community would object if the NSDD were to "expand beyond the civil approach." This response provoked a sharp reply from Poindexter, saying that his intent "was not to focus on a short unclassified civil space strategy only." Poindexter found it "discouraging that elements of the bureaucracy have continuously resisted" the space strategy effort and that he did not understand "your substantive concern with spelling out in one document what your strategy is for the use of space in carrying out your intelligence missions." He argued that "to properly focus the U.S. involvement in space we need a national strategy that spells out what the President wants done … Presidential guidance in the classified mission areas will … be useful and appropriate. One might conclude from your memorandum that the Intelligence Community does not want any Presidential guidance."[5]

Finalizing the Strategy

Whether or not Poindexter's strong language had any effect, tensions between the NSC and the Intelligence Community and the Department of Defense (DOD) lessened in spring 1984, and the process of agreeing on the content of the national space strategy gained momentum. Even so, other areas of disagreement remained. In particular, the debate over shuttle pricing policy replaced national security community concerns as a sticking point. The IG (Space) met again on April 4 to discuss a second draft of the strategy. Participants in that meeting agreed to insert shuttle pricing policy as an issue in the space strategy statement, and a second draft of the statement was circulated on April 7. It posed the question, "Should NASA commit to a firm policy on full cost recovery for commercial and foreign STS flight operations beyond FY 1988?" It soon became evident that all executive agencies except NASA answered the question in the positive, but that NASA was not yet prepared to make such a commitment.

After the April IG (Space) meeting, there were three remaining areas of disagreement with respect to the content of the NSDD. They were:

- How to monitor the technology transfer implications of international participation in the U.S. space station?
- Should a joint DOD/NASA study be conducted to assess capabilities to satisfy the need for U.S. government launch services?
- Should NASA commit to a firm policy on full cost recovery for commercial and foreign space shuttle flight operations after October 1, 1988?

NASA in the Minority

On all three issues, NASA was advocating one position; all other agencies supported an alternative. There were a variety of reasons for this situation. With respect to the technology transfer issue, NASA had been arguing ever since international participation in the space station had been approved that the space agency was fully capable of ensuring that there would be no unwanted technology transfer as a result of that participation. If there were any questions about NASA's stewardship in this respect, NASA suggested that the SIG (Space), with NASA as a member, would be the appropriate forum for resolving disputes. The technology transfer possibility associated with space station cooperation was a serious concern of highly placed individuals in the Department of State, the DOD, and the Central Intelligence Agency (CIA), and they were insistent that that the SIG on the Transfer of Strategic Technology, of which NASA was not a member, was the appropriate forum for overseeing NASA's performance in making sure that station cooperation did not lead to the transfer of sensitive technology.

It was NASA that had proposed that the space strategy call for a NASA/DOD study of how best to meet government space transportation needs in both the short and the longer term. As noted in the preceding chapter, the DOD and the National Reconnaissance Office (NRO) had become increasingly concerned about sole dependence on NASA's space transportation system, and in early 1984 were beginning to push for developing an expendable launch vehicle (ELV) as a complement to shuttle use. The joint study was a tactic by NASA both to sustain as much as possible of the policy that all government payloads would be launched on the shuttle and to try to get the national security community committed to using any follow-on, shuttle-derived launch system that NASA might develop. In commenting on the second draft of the national space strategy, Brown from the Department of Commerce (DOC) suggested that "DOD has a mission need for ELV's. They are in the best position to judge what advice they need from NASA. No Presidential direction is needed to make this happen." He added that "a Presidentially directed study of launch vehicle technology by the Government will intimidate the private sector and deter investment in commercial ELV's because it will create uncertainty as to the U.S. Government's ultimate intention with respect to ELVs. Industry is not prepared to compete with NASA on ELVs."[6]

With respect to the shuttle pricing issue, NASA found itself facing opposition led by the DOT and DOC. Having two agencies interested in commercial space policy as members of the interagency process had changed the dynamics of that process. NASA was used to being regarded as *the* civilian space agency, with almost a monopoly position in deciding space issues not involving national security concerns. It did not welcome having other civilian agencies involved in what it considered its "turf." With the space shuttle serving national security users, and the growing dissatisfaction among those users with dependence on the shuttle, there was the possibility of bureaucratic alliances between the national security community and the Transportation and Commerce Departments in opposition to NASA views within the interagency process. NASA perceived itself as a bit beleaguered.

The Home Stretch

The SIG (Space) met again on April 23, 1984, with the intent of carrying out a final review of the space strategy directive before forwarding it to the president for his approval. That meeting did not reach consensus on the directive's wording, particularly with respect to shuttle pricing and technology transfer. The directive was once again revised to reflect discussions at the meeting and sent out for review on May 12. There were still significant disagreements regarding the directive's language. In providing comments to McFarlane on the May 12 draft, the CIA's McMahon noted that "cooperative international space programs and activities may provide opportunities for technology transfer that are not present in other US programs," and thus that, because "the Soviets place a high priority on the acquisition of US space technology," there was a need to "examine our current approaches for controlling the adverse transfer of US space technology and develop a long-term plan that is adequate to protect the government's legitimate interests." With respect to shuttle pricing, McMahon noted that the directive contained "language which could liberally be interpreted effectively to reverse the President's full cost recovery decision." Although this critique came from the CIA, it also reflected the Commerce, Transportation, and Office of Management and Budget (OMB) positions on shuttle pricing. Surprisingly, in view of the long-running Intelligence Community resistance to the concept of a national space strategy, McMahon told McFarlane: "We are particularly pleased with the elements of the space strategy that apply directly to the foreign intelligence space program. In this regard, we fully support the guidance to complete the approved modernization; improve capabilities in support of the operational military forces, continuity of government, and the National Command Authority in crisis and conflict; and maintain a vigorous national security space technology program."[7]

The shuttle pricing controversy dragged on for the next two months. At one point, McFarlane told Dole: "We have evidence [a classified CIA report] to suggest that the French *Ariane* ELV would be the primary beneficiary of an increase in Shuttle price ... If NASA is arbitrarily forced to raise its Shuttle prices, it appears that *Ariane*, and not U.S. ELVs, will benefit through increased demand from payload customers, Such a result would obviously undercut the President's primary goal of maintain U.S. space leadership." He suggested that "removing the Shuttle as a viable contender would also have other serious implications." These included diminishing the ability of the shuttle to serve as "a significant and highly visible instrument of our foreign policy"; it could also "be counterproductive to our other space commercialization goals," including increasing the commercial use of the shuttle "to spawn new industries." Finally, "a reduction in foreign and domestic commercial Shuttle launches resulting from increased prices could possibly result in increased prices charged for U.S. government launches." Based on such considerations, said McFarlane, "the bottom line is that we must proceed prudently and cautiously in resolving this issue."[8]

The White House was trying to develop a shuttle pricing statement that would satisfy all parties. NSC deputy Poindexter and Alton Keel of OMB met on July 5 and came up with revised language that deleted the "consistent with the need to maintain international competitiveness" phrase that had been in earlier drafts; they also added a requirement that the Department of DOT and the DOC should carry out a study to assess the role of ELVs in maintaining international competiveness. The NSC's Rye speculated that "NASA will strongly object to the deletion of the key phrase, 'consistent with the need.'" He suggested that the space agency would argue that the deletion "will send a signal that the President is backing off on his commitment to the Shuttle and that the fear of price increases will drive away Shuttle customers."[9]

Rye was correct. McFarlane in mid-July sent a revised statement on shuttle pricing to NASA's Beggs, Dole at DOT, and Baldridge at DOC. This statement produced a strong defensive reaction from Beggs, who noted that the shuttle pricing issue "has momentous implications for virtually every aspect of the President's space policy." Beggs had become "increasingly concerned over the past months that this complex policy issue was being relegated to treatment as a more narrow budget issue. The latest language you have circulated heightens my concerns." Beggs told McFarlane that he wanted "an opportunity to discuss this issue with you and the President before he makes his decision" on shuttle pricing. He pointed out that "we are in a very fluid situation" with respect to the viability of a commercial ELV industry; that "there is much private capital tied up in the Shuttle's capability to launch payloads into space," since several companies had collectively invested over $150 million to develop upper stages to use with the shuttle; and that "it is the unique capability of the Shuttle as a workplace that has opened up new possibilities for the commercial uses of space."[10]

The suggested modifications in the language of the draft NSDD also did not satisfy the advocates of ELV commercialization. Dole of DOT was joined by Commerce Secretary Baldridge in a letter to McFarlane. They insisted on the need for "an unambiguous commitment ... to implement a policy that will allow the [ELV] industry to begin making the investment and planning decisions critical to its future growth and prosperity." Responding to the competition posed by *Ariane*, they suggested that "it is imperative that a full range of U.S. space transportation capabilities be available to meet the needs and preferences of all potential customers. Domestic ELVs provide technology comparable to that of *Ariane* but with highly proven reliability, making them a natural competitor to *Ariane*." Alluding to the problems that the shuttle was having in meeting its launch schedule, Dole and Baldridge suggested that "even with the current heavily subsidized Shuttle prices, loss of payloads to *Ariane* has been the result of customer insistence on launch date assurance and vehicle availability," and that "experience has demonstrated that U.S. government subsidies are simply not an effective mechanism to maintain U.S. leadership in world markets."[11]

McFarlane was getting impatient; he was eager to get the overall space strategy statement before the president. On July 25, he suggested that "the existing language is as close as we can come to a final statement of policy until the studies come in." He felt the White House had been "bending over backwards to maintain peace in the family," but it was urgent to move ahead with completing the space strategy since "there are several unrelated issues which are hanging fire while we discuss this [the shuttle pricing] issue ... We simply need more facts to determine how both the STS and ELV programs can both be sustained." McFarlane on August 7 wrote Dole and Baldridge, telling them that he was preparing to send the strategy statement to the president with the pricing issue unresolved and reminding the two cabinet members "this is a complex issue for which there are no easy answers."[12]

There were continuing negotiations and compromises among concerned agencies with respect to the language of the NSDD. These negotiations, by deferring decisions on several contentious issues, were able to produce a directive that gained the concurrence of all involved agencies, and the SIG (Space) did not have to meet again. By mid-August there was agreement that the directive was ready to go to Ronald Reagan for his approval.[13]

Finally, a National Space Strategy

Summarizing the situation with respect to shuttle pricing as he forwarded the space strategy statement to the president for approval, McFarlane told Reagan:

> DOC, DOT and OMB argue that the NSDD 94 language is not sufficiently definitive in that it allows NASA to establish full-cost recovery sometime beyond FY1988 and does not provide a date certain for the new pricing policy to be implemented. They also argue that without a date certain, the Government will continue to subsidize the price of Shuttle launches, NASA will not be motivated to make the Shuttle truly cost-effective and commercial ELVs will be priced out of the marketplace. NASA argues that establishing a date certain does not provide sufficient pricing flexibility to assure the Shuttle's competitive advantage for international launches and, as a result, European and Soviet launch vehicles … could capture a greater share of the market.

McFarlane suggested to the president that "in the interest of 'desubsidizing' the Shuttle and driving down operating costs, we want to ensure that the international competitiveness of the Shuttle is not adversely affected. The SIG (Space) will monitor this situation closely. In the interim, the policy will send a positive signal that hopefully will enhance the competitive position of commercial expendable launch vehicles."[14]

McFarlane also told the president that "the National Space Strategy identifies selected, high priority efforts and responsibilities and provides implementation plans for major space policy objectives." The directive set out 17 separate priorities. One was related to the space shuttle; four, to the civilian space program; two, to commercial space efforts; and ten, to national security space programs. McFarlane noted that all agencies now supported the NSDD and recommended that the president sign the directive; Reagan agreed.[15]

The strategy statement dealt with shuttle pricing with carefully worded compromise language:

> On October 1, 1988, prices for STS services and capabilities provided to commercial and foreign users will reflect the full cost of such services and capabilities. NASA will develop a time-phased plan for implementing full cost recovery for commercial and foreign STS flight operations. At a minimum, this plan will include an option for full cost recovery for commercial and foreign flights which occur after October 1, 1988. OMB, in consultation with DOC, DOT, DOD, NASA and other agencies will prepare a joint assessment of the ability of the U.S. private sector and the STS to maintain international competitiveness in the provision of launch services … Both the time-phased plan and the OMB analysis will be submitted for review and comment by the SIG (Space) and the Cabinet Council on Commerce and Trade.[16]

With this outcome, the issue of shuttle pricing was far from resolved, since NASA still had to put forward its plan for full cost recovery and the OMB-led study had to provide a perspective on the relationship between the shuttle and commercialized U.S. ELVs in maintaining a U.S. competitive advantage in the global competition for launch contracts.

Because it dealt with military and intelligence space efforts, the NSDD was classified "Top Secret/Codeword," a level of classification applied only to the most sensitive U.S. government documents. News that the NSDD had been approved soon spread, and Rye reported to McFarlane that "there have been numerous requests for copies." In addition, some of the contents of the directive had been leaked to *The New York Times* and the trade journal *Aviation Week & Space Technology*. In response, Rye, working with DOD and Intelligence Community staff, prepared an unclassified version of the directive for press and public release.

Highlights of the unclassified "fact sheet" included:

- Restatement of the policy that the space shuttle would remain "the primary space launch system for both national security and civil government missions," but acknowledgment that "in order to satisfy the requirement for assured launch, the national security sector will pursue the use of a limited number of ELVs to complement the STS." This provision ratified the Defense Space Launch Strategy that had been developed by the Air Force and NRO's Pete Aldridge and approved by Secretary of Defense Weinberger earlier in 1984.
- Restatement of the decision to "establish a permanently manned presence in space" and to seek as a "centerpiece" of increased international cooperation overall "agreements with friends and allies to participate in the development and utilization of the Space Station," making "every effort to obtain maximum mutually beneficial foreign participation." SIG (Space) would "review all major policy issues" involved in space station agreements, including technology transfer issues, thereby giving the national security community a voice in the character of those agreements but keeping SIG (Space), with NASA as a member, as the primary forum for overseeing those issues.
- Earlier drafts of the policy had indicated that NASA and OSTP would study long-term goals in space. Reflecting the fact that in July the Congress in passing the Fiscal Year 1985 NASA Authorization Act had included a provision calling on the president to appoint a National Commission on Space, the policy stated that the president would appoint such a commission to "identify goals, opportunities, and policy options for the next twenty years."
- Restatement of the November 1983 decision to assign to the DOT the responsibility of encouraging "commercial Expendable Launch Vehicle activities" and of the intent expressed in the July 1984 commercial space policy to take various initiatives to "stimulate private sector commercial space activities."
- With respect to national security space activities, there were statements of intent to "maintain assured access to space," pursue a high priority "long-

term survivability enhancement program," make government-wide efforts to "stem the flow of advanced western space technology to the Soviet Union," continue "to study space arms control options," and insure that "DOD space and space-related programs will support the Strategic Defense Initiative."[17]

National Commission on Space

One missing element in the National Space Strategy was a statement of an overarching long-term goal for the civilian space program. One of the controversies as President Reagan in December 1983 had approved developing a space station was whether that decision should be deferred until the administration set out such a long-range goal in space, such as establishing a base on the Moon or sending astronauts to Mars, on the grounds that the station should be designed to serve those goals. Science adviser Jay Keyworth had been the leading advocate of that position within the White House, and he had gotten some support from other senior members of the administration, skeptical for their own reasons about going ahead with the space station.

The need for setting out long-range goals in space was also perceived by some members of Congress interested in future space activities. In particular, Senator Ernest "Fritz" Hollings (D-SC) almost from the time the space shuttle started flying had proposed a study to identify such goals. By early 1984, Hollings had introduced the "National Commission on Space Act," setting up a presidentially appointed commission to conduct such a study. NASA communicated to Congress Reagan administration opposition to such a commission, saying that "the proposed Commission and its proposed investigations are unnecessary" and that "sufficient mechanisms already exist for the formulation and implementation of policy and the resolution of major space issues."[18]

Administration opposition to a space commission was not successful. By June 1984, both the Senate and the House of Representatives had incorporated setting up a such a commission into their versions of the NASA authorization bill. Title II of the NASA FY 1985 Authorization Act, signed into law on July 16, 1984, thus established a 15-person National Commission on Space "to define the long-range needs of the Nation that may be fulfilled through the peaceful uses of outer space" and "to articulate goals and develop options for the future direction of the Nation's civilian space program."[19]

The White House had hoped to name former NASA official George Low to chair the commission, but Low died on cancer on July 17. The White House then turned to former NASA Administrator Tom Paine as the commission chair. This appointment foreshadowed the commission's conclusions; Paine in the 1969–1970 period when he was NASA administrator had become an advocate of a visionary program of space exploration focused on human journeys to Mars. In the succeeding 15 years, his commitment to such a bold vision had strengthened. Negotiations over commission membership were contentious, and the other members of the commission were not named until early 1985.

The executive order establishing the commission, issued only on October 12, 1984, had stipulated that no more than three of the members of the commission could be government employees. It turned out that only one government employee, NASA astronaut Kathryn Sullivan, was a commission member; the other 14 commissioners came from a mix of private sector backgrounds. Scientist Laurel Wilkening from the University of Arizona was commission vice-chair. Well-known personalities such as Neil Armstrong, Chuck Yeager, former United Nations ambassador Jean Kirkpatrick, space entrepreneur Gerard O'Neill, and "father" of the U.S. military space program Bernard Schriever were among other commissioners.[20]

The National Commission on Space worked throughout 1985 to develop its recommendations and prepare a final report. By the time the report was completed in early 1986 and formally presented to Ronald Reagan on July 22 of that year, the idea of setting bold new goals had been replaced by debates over how best to recover from a tragic space shuttle accident.

The First-Term Reagan Record on Space

Ronald Reagan had announced in January 1984 that he would stand in November for reelection to a second term as president, and, uncontested, had won the Republican nomination as the party's presidential candidate in August. Speaking on August 30, 1984, at NASA's Goddard Space Flight Center near Washington, a few hours after shuttle orbiter *Discovery* had been launched for its first flight, Reagan "hitched his reelection campaign to the launch." He said: "The space age is barely a quarter of a century old, but already we have taken giant steps for all mankind. And our progress is a tribute to American teamwork and excellence. We can be proud that we're first, we're the best; and we are so because we're free. There's nothing that the United States of America cannot accomplish, if the doubting Thomases would just stand aside and get out of our way." Reagan did not have to say that one of those "doubting Thomases" was Walter Mondale, his Democratic opponent for the presidency, who had led the fight against the space shuttle as a member of the Senate during the 1970s before becoming Jimmy Carter's vice president in 1977. Reagan added, again referring to Mondale, "permit me to suggest that the fraternity of pessimists, who today insist strong growth will ignite high inflation, are looking at abstract statistics, theories, and models, not the reality of a changing world. They do not see that as we acquire more and more knowledge from new technologies, we no longer move forward in inches or feet; we begin to leap forward ... Only by challenging the limits of growth will we have the strength and knowledge to make America a rocket of hope shooting to the stars."[21]

To emphasize Reagan's first-term space achievements, on September 7 the White House organized a high-profile White House press briefing on "America's Future in Space." The event was attended by over 200 media representatives. Science adviser Keyworth, NASA's Beggs, and Secretary of Transportation Dole provided their perspectives on the nation's space efforts. Gil Rye, representing the White House, was the final speaker. Rye reminded the assembled media representatives that President Reagan "is a tremendous supporter and leader of the U.S. space program. He takes great pride in our past accomplishments and looks forward to even greater ones." Rye suggested that Reagan "believes that a vigorous and forward-looking space program is one of the most highly visible and tangible demonstrations of world leadership," with the potential to "advance the state-of-the-art in high technology, elevate the human spirit, capture our imaginations, demonstrate our pioneering initiative and hold out a progressive future for all mankind." After Rye spoke, the president made an unscheduled drop by, described as "a clear laying-on of hands on space as a Reagan Administration pillar."[22]

As Reagan was reelected on November 6, 1984, winning the popular vote in 49 of the 50 states, there was optimism that the positive momentum with respect to the U.S. space program that had been built up between 1981 and 1984 would be sustained and even increased. Despite lingering interagency disagreements between NASA and other space-involved agencies, the combination of flying a series of seemingly successful space shuttle missions, getting started on designing a space station, engaging international partners in the station and in other space programs, and encouraging increased private sector involvement in space activities seemed to foreshadow a vibrant U.S. space effort during the second Reagan term.

In a 2016 interview, President Reagan's first-term top policy adviser Ed Meese remembered that in 1984 he had formed a task force on issues that would face Reagan should he be reelected. According to Meese, "We had spent the first term fixing problems with the space program; the second term would focus on laying out a vision for the future." Elements of that vision included:

- Foster and promote the mastery of advanced technologies essential to progress and prosperity;
- Expand human knowledge and imagination so as to bolster innovation, and continually renew the inquisitive spirit;
- Strengthen Free World capabilities by providing experience and jointly managing large and complex programs;
- Demonstrate the ability to work cooperatively together in productive and peaceful pursuits as a symbol of Free World unity;
- Expand the application of space-devised techniques and products to the objective of economic and social development;
- Increase the role of private industry in the development and utilization of space systems.[23]

These goals were a good indication of the Reagan administration's high aspirations as it began a second term in power for the future of the U.S. space program.

For a variety of reasons, most of all a tragic 1986 space shuttle accident that was a major setback for the U.S. space program, this ambitious agenda was not actively pursued. The team of White House and NASA officials who had shaped space decisions during the first Reagan term underwent an almost total turnover in 1985, and their replacements were not nearly as effective as had been their predecessors in dealing with conflicts and still moving the space program forward. The accomplishments during Ronald Reagan's first term would end up being the high point in establishing his space legacy.

Notes

1. Letter from George Low to Richard Fairbanks, Director, Transition Resources and Development Group, December 19, 1980, Papers of George M. Low, Folsom Library, Rensselaer Polytechnic Institute, Troy, NY.
2. A copy of NSSD 13-82 can be found at https://www.reaganlibrary.gov/sites/default/files/archives/reference/scanned-nssds/nssd13-82.pdf.
3. A copy of the draft National Space Strategy, dated January 25, 1984, with OMB comments included, is attached to a memorandum from David Stockman to Robert McFarlane, "OMB Comments on Draft NSDD on National Space Strategy," February 2, 1984, CIA-RDP92B001181R001501530088-0, CREST.
4. Memorandum from Gil Rye to John Poindexter, "IG (Space) Meeting – February 3, 1984 – 4:00–5:00 P.M. – Rm. 208, OEOB," February 2, 1984, https://fas.org/irp/offdocs/nsdd/nsdd-144.htm.
5. Memorandum from John Poindexter to members of the Interagency Group for Space, "Follow-on to IG (Space) Meeting," with attached Terms of Reference, February 23, 1984, CIA-RDP86M00017R001501130037-8, CREST; memorandum from Director, Planning and Policy Staff to Director, Intelligence Community Staff, "Follow-on to IG (Space) Meeting," March 1, 1984, CIA-RDP92B00181R00150130075-4, CREST; memorandum from John Poindexter to director, Intelligence Community Staff, "National Space Strategy," March 1, 1984, CIA-RDP92B00181R00150130077-2, CREST.
6. Memorandum from Clarence Brown to John Poindexter, "Draft NSDD on National Space Policy," April 12, 1984, https://fas.org/irp/offdocs/nsdd/nsdd-144.htm.
7. Memorandum from John McMahon to Robert McFarlane, "Final Draft of National Space Strategy," May 25, 1984, CIA-RDP86M00886R001000070010-6, CREST.
8. Memorandum from Bud McFarlane to Elizabeth Dole, "STS Pricing Issue," June 21, 1984, Folder 12772, NHRC.
9. Memorandum from Al Keel to John Poindexter, July 9, 1984, and memorandum from Gil Rye to Robert McFarlane, "STS Pricing Issue," July 10, 1984, both at https://fas.org/irp/offdocs/nsdd/nsdd-144.htm.
10. Memorandum from Robert McFarlane to Malcolm Baldridge, Elizabeth Dole, and James Beggs, "STS Pricing Issue," July 12, 1984, https://fas.org/irp/offdocs/nsdd/nsdd-144.htm; letter from James Beggs to Robert McFarlane, July 17, 1984, Folder 12772, NHRC.
11. Letter from Elizabeth Dole and Malcolm Baldridge to Robert McFarlane, July 19, 1984, https://fas.org/irp/offdocs/nsdd/nsdd-144.htm.

12. Note from Robert McFarlane to John Poindexter, July 25, 1984, and memorandum from Robert McFarlane (signed by John Poindexter), "National Space Strategy," August 7, 1984, https://fas.org/irp/offdocs/nsdd/nsdd-144.htm.
13. Memorandum from Gil Rye to Robert McFarlane, "NSDD on Space Strategy," August 7, 1984, https://fas.org/irp/offdocs/nsdd/nsdd-144.htm.
14. Memorandum from Robert McFarlane to the President, "NSDD on National Space Strategy," August 15, 1984, with attached "STS Pricing Issue," https://fas.org/irp/offdocs/nsdd/nsdd-144.htm.
15. Memorandum from Robert McFarlane to the President, "NSDD on National Space Strategy," August 15, 1984, https://fas.org/irp/offdocs/nsdd/nsdd-144.htm.
16. Memorandum from Gil Rye to Robert McFarlane, "NSDD on National Space Strategy," August 7, 1984; letter to Robert McFarlane from Elizabeth Dole, August 14, 1984; letter from "Mac" (Malcolm Baldridge) to Robert McFarlane, August 13, 1984, https://fas.org/irp/offdocs/nsdd/nsdd-144.htm. The White House, "Fact Sheet – National Space Strategy," released September 7, 1984, Folder 12905, NHRC.
17. Ibid.
18. Interview with Marcia Smith, who was executive director of the National Commission on Space, May 2, 2016; letter from John Murphy, NASA, to Senator Pete Domenici, January 25, 1984, Folder 12776, NHRC.
19. Memorandum from John Murphy to Administrator, "National Space Commission," June 1, 1984, Folder 12776, NHRC. Title II, National Commission on Space, Public Law 98-961, July 16, 1984.
20. Executive Order 12490, "National Commission on Space," October 12, 1984.
21. David Hoffman, "President Uses Launch to Attack 'Pessimists,'" *WP*, August 31, 1984, A3. Ronald Reagan: "Remarks During a Visit to the Goddard Space Flight Center in Greenbelt, Maryland," August 30, 1984. Online by Gerhard Peters and John T. Woolley, APP, http://www.presidency.ucsb.edu/ws/?pid=40309.
22. Remarks by Gil Rye, White House Space Briefing, September 7, 1984, provided to author by Rye. William Gregory, "More Decisive than It Looks," *AWST*, October 29, 1984, 11.
23. Interview with Edwin Meese, III, October 12, 2016. "Long-Range Space Goals," attached to a memorandum from Robert McFarlane to George Shultz, "International Space Initiatives," December 24, 1984, CIA-RDP86M00886R001000070002-5, CREST.

CHAPTER 16

Changing of the Guard

In the months following the beginning of Ronald Reagan's second term as president in January 1985, there were many significant changes in Reagan's White House staff, including those individuals dealing with the space program. In addition, by the end of 1985 the two top officials at the National Aeronautics and Space Administration (NASA) had left their positions. As a result, few of the people who had shaped space policy and program decisions during the first Reagan term remained in place.

The U.S. space program between 1985 and 1988 suffered, not only in the aftermath of the January 1986 *Challenger* accident that grounded the shuttle for over two years and forced a conflict-filled reevaluation of its role as the government's primary launch vehicle, but also overall. The space station program and commercialization efforts ran into problems and delays. Attempts to craft a post-*Challenger* national space policy set no long-term direction for the American space effort other than a continued quest for U.S. space leadership. No individual was able to take the lead in reestablishing in the post-*Challenger* period the forward momentum that had characterized space policy decisions during the first Reagan term, and Ronald Reagan himself both had to struggle to counter the aftermath of a major White House scandal and gave policy priority to dealing with broad issues in the U.S.-Soviet strategic relationship. Space issues commanded far less of Reagan's attention in the final years of his presidency than had been the case in his earlier years in the White House.

Reagan speechwriter Peggy Noonan, who had come to the White House in 1984, summarized the White House transition in her book *What I Saw at the Revolution:* "It was a presidency cut in half, with two very different administrations. The first term was an inefficient, yeasty, fractious administration of high quality people with a passion for policy and ideas … who produced for the president on a continual basis ideas and initiatives and who provided him with real options." Of the three men closest to Reagan, "Deaver covered the

J. M. Logsdon, *Ronald Reagan and the Space Frontier*,
Palgrave Studies in the History of Science and Technology,
https://doi.org/10.1007/978-3-319-98962-4_16

institution of the presidency ... Meese covered the president's ideological base ... Baker was running things ... That was the mix. It was a mess and it really worked and it served the president well."

Noonan added: "The first administration was rocky and successful, the second rocky and less so." Central to the lack of success in the early years of second Reagan term was Donald Regan, the man who replaced James Baker as White House chief of staff. Noonan suggests that Regan was "temperamentally unsuited" to that position and also "unsuited in terms of experience." In addition, "he had second-rate people around him." The result, in the space realm as well as in other policy areas, was a second Reagan term of limited accomplishments. What success there was came in the foreign policy realm, as international participation in the space station was finalized and Reagan and new Soviet leader Mikhail Gorbachev made meaningful steps toward forging a less-hostile U.S.-Soviet relationship, including resumption of space cooperation between the two nations.[1]

The End of the "Troika"

At the start of the first Reagan term in 1981, the president's inner circle was composed of the three men he met with at the beginning of almost every work day—Chief of Staff James Baker, Deputy Chief of Staff Michael Deaver, and Presidential Counselor Edwin Meese. These three, dubbed the "troika" by the press, were the links between a president who wanted to see his fundamental beliefs reflected in policy choices and the various elements of the administration and the Congress that could make that desire become reality. When William "Judge" Clark came to the White House as national security adviser in January 1982, he had become the fourth member of the inner group.

Clark had left the inner circle when he became Secretary of Interior in November 1983; Meese was the next to attempt to leave. Attorney General William French Smith submitted his letter of resignation in January 1984, and President Reagan quickly nominated Meese to be his successor. However, a variety of questions were soon raised about whether Meese had violated ethics rules after coming to the White House, and it took over a year, with an investigation by a special prosecutor and two lengthy confirmation hearings, before he was finally confirmed for the attorney general position by the Senate. French stayed on at the Department of Justice during that period, and Meese stayed in the White House. He still on most days met with the president together with Baker and Deaver, but as a lame duck, his policy influence was diminished. Meese left the White House in February 1985 after being sworn in as attorney general by the president.[2] Reagan did not name a replacement for Meese as presidential counselor, leaving a void in policy advice to the president that Don Regan intended to fill.

As Ronald Reagan was preparing to be sworn in for his second term, Mike Deaver was "overworked and exhausted." He also later admitted fighting "abuse of alcohol." Deaver "realized it was time to get out," and on January 3, 1985, the White House announced that he would resign in the coming months.

Deaver left the White House in May 1985 to start his own public relations firm. While he had not been visibly involved in space policy debates during the first Reagan term, he was very conscious of how Reagan's support of the space program had burnished the president's image as a forward-looking, optimistic leader, and thus had supported those who wanted to identify Reagan with U.S. space achievements.[3]

While the departures of Meese and Deaver meant that Ronald Reagan no longer had the advice of two of his long-time and trusted California associates, it was the loss of James Baker as his chief of staff that would have the most impact on the conduct of the presidency in the 1985–1986 period. Baker later in life admitted that he had "hated" trying to run a White House "riven by different philosophies and ideological outlooks," one with "a lot of tension … between the so-called true believers and my side of the White House, the get-the-job-done, pragmatic type." He added: "I had to deal with Meese and the ideologues shootin' at me all the time." He was "bone tired" and "didn't think he could endure the internecine warfare much longer." Baker was also thinking about some day running for president, and believed that he needed experience in running a cabinet department, preferably one involved in foreign affairs, as part of his preparation for that attempt.

Baker was thus open to an idea suggested in late 1984 by Secretary of Treasury Donald Regan—that Regan and Baker switch jobs as the second Reagan term began. Baker and his deputy Dick Darman thought through the idea, found it attractive, and Baker and Regan, with the assent of Deaver, by mid-December agreed it was a "done deal," if only the president would approve. On January 7, Baker and Regan were taken into the Oval Office by Deaver, who told Reagan that "I've finally brought you someone your age to play with (Regan was 66 years old)" and "Don has something he wants to discuss with you that he's talked to me and Jim about … We'd like to know what you think about it." Regan explained what he and Baker had in mind, and noted that "Reagan listened without any sign of surprise. He seemed equable, relaxed—almost incurious. This seemed odd." In fact, Reagan had been alerted to the switch by his wife Nancy, who had learned about it from Deaver, a close confidant. At the end of the meeting, Reagan said, "Yes, I'll go for it." The next day, the president went to the White House press room and announced the switch to a surprised press corps.[4]

It would turn out that Don Regan was a poor fit to the role of White House chief of staff. Regan was "imperious" and "accustomed to doing things his way." He saw himself as the White House chief executive officer, with Reagan as the chairman of the board. Regan intended to oversee all aspects of White House operations, a responsibility that had been shared among Baker, Meese, Deaver, and Clark and his successor Bud McFarlane. He was quoted as saying, "Not a sparrow falls on the White House lawn without me knowing it." Regan's style and the requirements for success in his new position were a serious mismatch.[5]

Regan soon made one change that would prove particularly relevant to subsequent space policy debates. He decided to replace Meese's seven cabinet councils, including the Cabinet Council on Commerce and Trade that had become deeply involved in space policy issues during the first term. Regan as secretary of the treasury had found the cabinet council system "cumbersome and redundant." He reduced the interagency mechanisms dealing with domestic policy to two—a Domestic Policy Council and an Economic Policy Council. Ronald Reagan announced the changes on April 11, saying, "Today I am announcing the creation of two Cabinet-level bodies—the Economic Policy Council and the Domestic Policy Council—to assist me in the formulation and execution of domestic and economic policy … The new entities will streamline policy development and decision making. Together with the National Security Council, they will serve as the primary channels for advising me on policy matters." The Economic Policy Council, with the now Secretary of the Treasury James Baker as its chair, would become a major forum for discussing space policy and program issues in the second Reagan term.[6]

One reporter would later write: "The day Reagan agreed to the Baker-Regan switch was the day his White House lost its magic." With Regan in charge, Ronald Reagan and his passive approach to the presidency did not have "the benefit … of a White House staff driven by inner competition. The first-term feuding and jockeying for power between the pragmatist and conservative factions … had been a blessing in disguise to Reagan. By challenging one another and exposing Reagan to conflicting options, Baker, Deaver, Meese, and Clark had also challenged their uncurious president." Regan was determined to end conflicts among the different White House offices. He saw himself as "a corporate chief executive who would make the tough decisions and permit Reagan to enjoy the harmony he craved." This approach would not produce the policy successes of the first Reagan term.[7]

Other White House Staff Changes

There were at the next level of White House staff a number of 1985 departures with implications for the space program. Baker's deputy Dick Darman, who had played a behind-the-scenes role in advocating a vigorous space effort, joined Baker in moving to Treasury. Director of Cabinet Affairs Craig Fuller, who had been active in pushing the space station and space commercialization issues to a positive outcome, in early 1985 became Vice President George H.W. Bush's chief of staff, a position that he hoped would lead, if Bush was elected as Reagan's successor in 1988, to his becoming White House chief of staff in a Bush administration.

The roles Darman and Fuller had played in presidential policy-making were filled by individuals who Don Regan brought with him from treasury. They were described as "a coterie of Regan sycophants … who spent all their time saluting." They were "clueless in the ways of White House governance"; in Peggy Noonan's view, "they were just second-rate." While Regan "did not know what he did not know," his aides were "oblivious to what they were oblivious to." Noonan soon gave them the nickname "the mice," because "they were always running across the floor" following Regan.[8]

The "mouse" who replaced Craig Fuller as Cabinet Secretary was Alfred Kingon, who had been assistant secretary of treasury for policy planning and communications. Kingon had a stock market background, serving as a stock analyst and then for ten years editor of the business magazine *Financial World* before joining the Reagan administration in 1983. In 1985, he was 64 years old, almost twice Fuller's age. Kingon would be the White House link to NASA and other executive agencies on space issues, and would play an active role in the 1986 space policy debates regarding recovery from the *Challenger* accident.

David Stockman left his position as director of the Office of Management and Budget (OMB) on August 1, 1985, to take a Wall Street job. While both President Reagan and Don Regan were strong opponents of a tax increase to deal with the rapidly rising government deficit, Stockman was suspected as being open to that possibility. He was replaced by James Miller, "a more reliable conservative." Miller became known as "the anti-Stockman," because he publicly opposed tax increases, and "the Abominable No-Man," for his persistent opposition to budget increases. Under Miller, OMB would be a determined opponent of additional spending to recover from the *Challenger* accident and the leader in bringing dramatic increases in space station cost estimates to the president's attention.[9]

Finally, science adviser George "Jay" Keyworth left his position at the end of 1985 to found a Washington-based corporate intelligence firm. Keyworth from the time he came to the White House in 1981 had espoused ideas for the civilian space program that were often at variance with the path actually pursued by the Reagan administration, in particular arguing that a long-range goal for the space program should be set before the president approved space station development. He had had only marginal influence on civilian space policy decisions, and from the time Reagan in March 1983 had announced his plan for what became the Strategic Defense Initiative, Keyworth had devoted almost all his efforts to advancing that proposal. Keyworth was not immediately replaced as science adviser; John McTague, Keyworth's deputy, was named to the position on an acting basis.[10]

McFarlane and Rye Leave National Security Council

The shifts in the White House domestic policy staff were paralleled by changes at the National Security Council (NSC). Air Force Colonel Gil Rye, the NSC director for space programs since early 1982, had been an insistent advocate for an enhanced national space effort, and had incurred the enmity of the national security community for his suggestion of a possible redistribution of roles and budgets among the civilian, defense, and intelligence space efforts and then for his advocacy of NASA's space station proposal. This advocacy, when combined with the untimely death in an early 1983 airplane accident of Rye's champion within the Air Force, four-star General Jerome O'Malley, had effectively eliminated Rye's chance to get promoted up the Air Force ranks. Rye was thus ready to retire from military service, and in September 1985 left the NSC to accept a position as president of a newly created subsidiary of the Communications Satellite Corporation.

Rye's successor as NSC's space person was another Air Force colonel, Gerald "Jerry" May. Prior to joining the NSC staff, May had worked at the Pentagon, handling space policy issues for the office of the Joint Chiefs of Staff. May would not turn out to be the effective bureaucratic operator that Rye had been, and would leave the NSC after less than two years in his position.

While Rye left the White House on good terms, such was not the case with his boss, National Security Adviser Robert "Bud" McFarlane. From the start of Don Regan's tenure as White House chief of staff, Regan and McFarlane had clashed as Regan had tried to extend his control into foreign affairs. In a dispute over McFarlane's failure to wake him when the White House learned of the killing in East Germany of a U.S. military officer, Regan "berated" McFarlane, accusing McFarlane of "insubordination" and telling him that "you work for me." McFarlane's response was, "I work for the president," to which Regan replied, "The hell you do. You work for me and everything you do will come through me or you'll be out of here." Regan, recognizing the important role McFarlane played, soon apologized, but the relationship between the two had been poisoned. According to Secretary of State George Shultz, McFarlane's "relations with Don Regan were tense and strained." McFarlane was a high-strung individual and found the relationship with Regan more than he could handle; on December 4, 1985, "a beaten-down McFarlane called it quits."[11]

McFarlane's deputy, Vice Admiral John Poindexter, was immediately announced as McFarlane's successor. Poindexter had joined the NSC as its military assistant in 1981, and had become deputy national security adviser after Judge Clark left in 1983. Poindexter's appointment was "a somewhat curious choice … His substantive experience in national security and foreign policy was somewhat limited." Poindexter was clearly very intelligent; he had graduated first in his class at the Naval Academy and then received a doctorate in physics from the California Institute of Technology. But he had not gotten "the broad political and public relations experience" needed for success in his new position. Rather, he was seen "as a low key administrator, skillful in handling operational matters (especially those related to military activities or computers)." Poindexter "lacked the stature or influence with Reagan to act as a mediator in resolving policy disputes or tensions."[12]

Poindexter had been previously involved in space issues, often chairing meeting of the Interagency Group for Space in the 1983–1985 period. But he was not well prepared for the challenges he soon faced. Less than two months after taking over the NSC, Poindexter had to assume a lead role in crafting a recovery strategy from the *Challenger* accident. His performance in that role would come under widespread criticism.

The White House point men on space policy issues during Reagan's first term had been Gil Rye and Craig Fuller. Their replacements, Jerry May and Alfred Kingon, were far less able to handle difficult policy issues and bureaucratic conflicts, and there were no others in the White House to compensate for their shortcomings.

Mark and Beggs Leave NASA

NASA Deputy Administrator Hans Mark reports in his book *The Space Station: A Personal Journey* that "on April 14, 1984, I made the decision to leave NASA and government service." Mark's reasons for resigning his position were a mixture of professional and personal. He believed that he had been party to "significant achievements" with respect to the U.S. space program and that "it was time to go and do something else." By the time he made his decision to leave, the space shuttle had been operational since mid-1982 and President Reagan had given initial approval to the space station; these had been Mark's top two priorities when he came to NASA in 1981. Mark had been approached to take the position of chancellor of the University of Texas system. He had accumulated over $40,000 in personal debt, much of it related to sending his children to expensive private schools, and the Texas salary was almost twice his government salary. Mark's wife also "was not happy in Washington."

While this is Mark's explanation for his departure from NASA, journalist Joseph Trento in his book *Prescription for Disaster* provides an alternative view. According to Trento, by 1984 "Mark's relations with the White House were not going well." NASA Administrator Beggs suggested to Trento that "Mark left because he was afraid that people in the White House would either fire him or take other actions that would destroy his career ... They were going to assassinate him sooner or later."[13] It is not clear who in the White House Mark had antagonized; certainly his relationship with the NSC's Gil Rye remained positive. Most likely, it was science adviser Jay Keyworth, who had been a vocal opponent of the space station that Mark had pushed so hard. Whatever the reasons for his departure, it took Mark several months to wrap up his NASA tenure; his resignation took effect only on September 1.

Graham as Mark's Replacement

Mark's deputy administrator position was not filled until November 25, 1985, after being vacant for 15 months. Mark later suggested that one reason for the delay was that James Beggs "through arrogance and hubris, thought he could manage the place without a deputy."[14] Whether or not this was a valid assessment, Beggs also got into an extended conflict with the White House over the individual the administration proposed as deputy administrator, a Ph.D. physicist and engineer named William Graham. Graham was founder of a California-based national security consulting firm and since 1982 had been serving as the Senate-confirmed chairman of the administration's General Advisory Committee on Arms Control and Disarmament (Fig. 16.1).

Fig. 16.1 William Graham in December 1985 was named NASA acting administrator as James Beggs took a leave of absence. In this photograph, Graham chats with President Reagan and Cabinet Secretary Alfred Kingon aboard Air Force One after the January 31, 1986, memorial service in Houston for the seven *Challenger* astronauts. (Photograph courtesy of the Reagan Presidential Library)

In Trento's telling of the conflict, which is at variance with Mark's account, "Beggs wanted to fill Mark's position as quickly as possible." Beggs asked the White House personnel office for a list of possible people to replace Mark; he did not put forward candidates himself. A response came back with several people whom Beggs thought well qualified. Graham's name was also on the list, put there by Keyworth, who was actively promoting Graham for the NASA job. Keyworth knew Graham from his work in the nuclear weapons field. Beggs told the White House "some of these guys are completely acceptable to me and I would be happy to take anyone of them ... but this guy Graham has no qualification ... I have read his resume and he is not qualified for this job." Graham had no space-related experience and he had never managed more than a dozen people. Another of Beggs's concerns was Graham's experience had been totally within the national security field, and his appointment might add to the sense that NASA was being "militarized."

Beggs in February 1985 interviewed all of those the White House had suggested, including Graham, and then told the White House that all candidates were acceptable—except Graham. But the White House was insistent, and Beggs over the next months learned to his surprise that President Reagan had actually already signed the form nominating Graham as NASA deputy administrator, almost certainly without being made aware of Beggs's opposition. Beggs was thus presented with a *de facto* choice of Graham, with no alternative offered.

By this time, Don Regan had taken over as White House chief of staff. Apparently, Keyworth had convinced Regan that Graham was the right person for the NASA job. While Beggs saw Keyworth as "a right wing nut," Keyworth had more influence on this issue than Beggs realized. Keyworth, having crossed swords with Beggs and Mark during the space station debate, "wanted someone I trusted" as Mark's replacement. He added, "so did Don Regan," although Regan soon told Beggs that Graham "was not his choice for the job."

Beggs went to see Regan to explain his concerns about Graham. He soon heard back; Regan told him: "I can't do anything ... It's done ... This guy has support on the West Coast from friends of the president." Keyworth's mentor Edward Teller was supporting Graham, and his views carried weight in the Reagan White House. Other Graham supporters were members of Reagan's California "kitchen cabinet." Beggs still refused to agree to Graham's appointment. The White House called Beggs in August, saying that Regan was demanding that a decision on Graham be made. Beggs agreed to talk to Graham again to explain his reservations. He told Graham: "I have nothing against you personally ... but you are not qualified." This conversation did not resolve the issue; Graham subsequently told Beggs that he still wanted the deputy position.

Once again the White House called Beggs, saying, "You have just to take this guy and do us a favor." In a "weak moment" and after months of resistance, Beggs agreed. The White House sent Graham's nomination as NASA deputy administrator to the Senate on September 12. His "cordial 15-minute" confirmation hearing took place shortly thereafter. Graham was confirmed by the Senate on November 18 and he took up his position as NASA deputy administrator on November 25.[15]

Beggs Indicted, Takes Leave of Absence

On December 2, eight days after Graham arrived, Beggs learned that he had been one of four people named in a grand jury indictment in which the Department of Defense was charging his former company, General Dynamics, with improper expense accounting during the 1970s related to an Army air defense weapon. Beggs had known since early September that he was under criminal investigation, but had been assured by the top lawyer for General Dynamics that the firm was not taking the investigation, which had been going on for several years, seriously. By October 1985, in Trento's version of the affair, "it was clear that Beggs was the one the prosecutors were after." Beggs's first thought was that his opposition to Graham and the investigation had to be linked, although he eventually came to believe that they were not. More likely, he thought, his being targeted in the investigation was an aftermath of his antagonistic relationships with respect to the space station and space shuttle with the Department of Defense and Air Force leadership.

After briefly considering seeking an out-of-court resolution to the dispute, the government had proceeded to seek the indictments. Learning that he was being charged, Beggs called Don Regan. Beggs asked Regan to grant him a leave of absence from NASA, but Regan instead asked Beggs to resign. Beggs refused unless President Reagan personally asked for his resignation. That did not happen, and Beggs's request for a leave was granted. The leave was announced on December 4, which was coincidentally the same day that Bud McFarlane's departure from the NSC was made public. White House spokesperson Larry Speakes made the Beggs announcement, saying that President Reagan "while reluctantly acceding to his request for a leave of absence," had asked Beggs to assist in an "orderly transition ... at this critically important agency." Responding to a reporter's question as he announced McFarlane's resignation, Ronald Reagan said, "With regard to Mr. Beggs, I don't know of anyone who could have done a finer job than he has done and is doing at NASA. And we're talking about something that is supposed to have happened prior to government service. And, also, if you read it correctly, not something in which he in any way was doing anything—if he was doing this at all—that would redound to his benefit personally or enrich him in any way."[16]

It would take 18 months, but on June 19, 1987, Beggs was cleared of all charges. Beggs suggested that an apology from the Department of Justice was in order, but at that point it was not forthcoming. It was not until a year later, on June 29, 1988, that Attorney General Ed Meese sent Beggs an unusual and gracious personal letter of apology, saying, "I have had an opportunity to carefully review the circumstances surrounding the wrongful indictment against you. Accordingly, I wish to offer you a profound apology on behalf of the Federal Government and the Department of Justice ... I understand that there is no way to undo the pain you have suffered ... Your fellow citizens should now be more aware than ever that your character is untarnished and your behavior unblemished."[17]

An Uneasy Transition

With James Beggs granted the leave of absence, the White House quickly designated William Graham as NASA acting administrator. Beggs did not leave the NASA building; rather, he moved out of the administrator's suite to an office on the same floor, but down the hallway. As Beggs told the NASA staff about his situation, at Graham's request he also introduced Graham, saying, "He's going to be acting administrator and I have full confidence he'll do a good job." According to Trento's report, between that moment in early December and January 28, the day of the *Challenger* accident, Beggs and Graham did not speak. Graham told Trento that Beggs was "aggressive and hostile," and that Beggs was undermining him, telling others at NASA that he would soon be back in charge. Trento observes that "the atmosphere at NASA Headquarters was divisive, and morale was reaching its lowest ebb in the entire history of the space agency."[18]

This clearly was not a healthy situation as NASA was preparing for a particularly ambitious year in 1986, with launches of several planetary missions and the Hubble Space Telescope scheduled, not to mention the January launch of teacher-in-space Christa McAuliffe and her six colleagues on space shuttle *Challenger*.

Notes

1. Peggy Noonan, *What I Saw at the Revolution: A Political Life in the Reagan Era* (New York: Random House, 1990), 213–214.
2. Leslie Maitland Werner, "Senate Approves Meese to Become Attorney General," *NYT*, February 24, 1985, https://www.nytimes.com/1985/02/24/us/senate-approves-meese-to-become-attorney-general.html.
3. Michael Deaver, *A Different Drummer: My Thirty Years with Ronald Reagan* (New York: HarperCollins Publishers, 2001), 200–201.
4. Chris Whipple, *The Gatekeepers: How the White House Chiefs of Staff Define Every Presidency* (New York: Crown Publishers, 2017), 105, 113, 132–133. Donald T. Regan, *For the Record: From Wall Street to Washington* (New York: Harcourt Brace Jovanovich, 1988), 227. Douglas Brinkley, *The Reagan Diaries* (New York: Harper Collins, 2007), 292. Jack Nelson, "Regan, Baker to Trade Jobs as Reagan Aides," *The Los Angeles Times*, January 9, 1985, http://articles.latimes.com/1985-01-09/news/mn-11927_1_treasury-secretary.
5. Whipple, *The Gatekeepers*, 134–136.
6. Regan, *For the Record*, 235. Ronald Reagan: "Statement on the Establishment of the Economic Policy Council and the Domestic Policy Council," April 11, 1985. Online by Gerhard Peters and John T. Woolley, APP, http://www.presidency.ucsb.edu/ws/?pid=38462.
7. Whipple, *The Gatekeepers*, 135. Lou Cannon, *President Reagan: The Role of a Lifetime* (New York: Public Affairs Press, 2000), 495.
8. Whipple, *The Gatekeepers*, 135. Noonan, *What I Saw at the Revolution*, 76.
9. Steven F. Hayward, *The Age of Reagan: The Conservative Counterrevolution, 1980–1989* (New York: Three Rivers Press, 2009), 406–407.
10. "Keyworth Quits White House Post," *Science*, December 13, 1985, 1249.
11. John P. Burke, *Honest Broker? The National Security Adviser and Presidential Decision Making* (College Station, TX: Texas A&M Press, 2009), 216–217.
12. Ibid., 218.
13. Joseph J. Trento, *Prescription for Disaster: From the Glory Days of Apollo to the Betrayal of the Shuttle* (New York: Crown Publishers, 1987), 252. Trento is an investigative journalist with a reputation for not always being historically balanced or accurate. However, his book is based on a series of contemporary interviews with those involved, and is the only published account of these events, and thus I have chosen to use it as a source in this study.
14. Interview with Hans Mark, October 16, 1991. A copy of the interview, most likely conducted by T.A. Heppenheimer, is in NHRC.
15. Most of this account is drawn from Trento, *Prescription for Disaster*, 253–263. On Graham's confirmation hearing, see "Washington Roundup," *AWST*, October 7, 1985, 15.

16. Trento, *Prescription for Disaster*, 263–270. Trento reports Beggs as being convinced that William Graham "was part of a setup to get him out of NASA" and knew of the pending indictment as he refused to drop his quest for the NASA job. There is no evidence available to support Trento's report. See also Thomas O'Toole, "NASA Chief Takes Leave to Fight Fraud Charges," *WP*, December 5, 1985, A3 and Michael Dornheim, "Beggs, General Dynamics Named in Federal Fraud Indictment," *AWST*, December 9, 1985, 24. The Speakes announcement is at Office of the Press Secretary, The White House, "Statement by the Principal Deputy Press Secretary," December 4, 1985, Folder 12781, NHRC. Reagan's comment is at Ronald Reagan: "Remarks Announcing the Resignation of Robert C. McFarlane as Assistant to the President for National Security Affairs and the Appointment of John M. Poindexter," December 4, 1985. Online by Peters and Woolley, APP, http://www.presidency.ucsb.edu/ws/?pid=38114.
17. "General Dynamics, Beggs Cleared of Fraud Charges," *AWST*, June 29, 1987, 25–26. Letter from Edwin Meese III to James Beggs, June 29, 1987, Box 10, Papers of James Fletcher, NARA.
18. Trento, *Prescription for Disaster*, 272–273.

CHAPTER 17

Shuttle Wars

The tensions which had emerged in 1983 and 1984 between National Aeronautics and Space Administration (NASA) as operator of the space shuttle and other elements of the space community carried over into the new year. The National Security Council (NSC) had to step in between NASA and the Air Force to mediate their dispute over a backup to the shuttle, and an acceptable compromise was reached early in 1985. In contrast, the antagonism between advocates of commercializing expendable launch vehicles (ELVs) and shuttle supporters increased, focusing both on how best to compete with foreign providers of commercial launch services and on the issue of the post-1988 price for commercial and foreign users of the shuttle. A July 1985 presidential decision on pricing policy did not smooth over the hard feelings.

Meanwhile, the space shuttle program moved ahead. NASA launched nine shuttle missions during 1985, pushing to increase the flight rate toward its publicly stated target of 24 launches annually. Payload specialists from Saudi Arabia and Mexico and Senator Jake Garn were carried into orbit, and NASA invited another congressman, Representative Bill Nelson (D-FL), to make a flight. The teacher to be the first participant in NASA's Spaceflight Participant Program was announced, and preparations got underway for selecting a second participant, this time a journalist. Ronald Reagan was personally lobbied by one of his longtime friends for a chance to ride on the shuttle. Reagan did what he could to accommodate that request even though he told the friend, screenwriter Doug Morrow, that it was "absolutely incomprehensible to me. Why would anyone want to get any higher off the earth than you can get by sitting on the back of a horse?"[1]

J. M. Logsdon, *Ronald Reagan and the Space Frontier*,
Palgrave Studies in the History of Science and Technology,
https://doi.org/10.1007/978-3-319-98962-4_17

NASA and Air Force Argue[2]

The August 15, 1984, National Space Strategy had declared that the space shuttle would remain "the primary space launch system for both national security and civil government missions," but acknowledged that "in order to satisfy the requirement for assured launch, the national security sector will pursue the use of a limited number of ELVs to complement the STS."[3] NASA Administrator James Beggs saw the Air Force plan to employ ELVs for some of its launches as only the first step toward getting all national security payloads off of the shuttle. NASA, according to Undersecretary of the Air Force Pete Aldridge, continued "to fight the Air Force's plan, openly and underground." The space agency even entered the competition to be what the Air Force was calling the complementary expendable launch vehicle (CELV), proposing an expendable heavy-lift vehicle based on space shuttle components and launched from space shuttle facilities; NASA would develop that vehicle under contract with the Air Force.

In early 1984, the Department of Defense (DOD) had asked the National Research Council, the operating arm of the National Academies of Science and Engineering, to carry out a six-month study of the wisdom of having an expendable alternative to shuttle launches for the largest national security payloads. The study report was issued in September 1984, and strongly supported the DOD position, suggesting that "the complementary ELV as a means of more assured access to space has unique attributes of operational flexibility and security not provided by the STS [Space Transportation System] alone." The report noted that the two launch systems being proposed by industry, Martin Marietta's upgraded Titan III and the Atlas Centaur from General Dynamics, would have "an important advantage" over NASA's proposed shuttle-derived launcher "in that they are launched independently of the complex STS launch environment."[4]

Even with this endorsement, NASA continued to resist the DOD plan. NASA's supporters in Congress, before they would agree to reprogram the funds needed to issue a CELV contract during Fiscal Year 1985, levied a requirement for a joint NASA-DOD study "of the requirements, options and costs of the CELV concept"; the study was to be submitted by January 15, 1985. Neither NASA nor DOD could agree on the focus and content of the report, and the deadline was missed. Aldridge suggested that the delay was due to NASA refusal "to coordinate on the CELV study" because "they knew that the Shuttle-derived ELV had lost the Air Force competition and took the position that another year of study was required … This delaying tactic would of course kill the effort since, without a commitment within the next year the ELV production lines would close and the cost to reopen them would be prohibitive." Beggs had a different perspective, telling National Security Adviser Bud McFarlane that NASA "had been working closely with the Defense Department" and had concluded that DOD had "simply not done the work necessary to make a decision that will relate the near-term CELV choice to the long-term needs of the country." Beggs suggested that NASA had "not been successful in persuading DOD to pursue a joint course" in preparing the report to Congress. He added that, if it was not possible to find a way to forge NASA-DOD agreement, "Cap [Secretary of Defense Weinberger] and I will need to discuss our fundamental differences with the President and have his decision on how he wishes to resolve them."[5]

The NSC was "concerned with the 'war' between NASA and the Air Force." McFarlane decided to have the NSC step in as a mediator. After first trying, without success, to find areas of agreement between senior NASA and Air Force staff, the NSC director for space, Gil Rye, called a meeting early in the week of February 11, 1985, between Beggs and Aldridge, without their staff, to try to defuse the conflict. Rye was "astounded by the depth of hostility and almost total lack of previous communications between the two men on this issue and many more. They raised their voices on numerous occasions, with each accusing the other of lobbying on the Hill against Administration policy, hidden agendas, sabotaging previous agreements, etc." Once the two had vented their feelings, Rye discovered that there was in fact a basis for compromise between their two positions. Aldridge reported that Beggs "kept saying 'You guys can't get off the Shuttle,'" and that he replied, "We will sign up that we will buy at least one third of all the mission the Shuttle can fly in any given year, we'll guarantee at least one third." That assurance apparently was what Beggs was seeking. Rye adjourned the session and quickly scheduled a second meeting between the two men for February 14. In the interim, Rye drafted what Aldridge described as a "treaty" reflecting the areas of agreement (Fig. 17.1).

Fig. 17.1 Edward "Pete" Aldridge, Jr., was Undersecretary of the Air Force and director of the National Reconnaissance Office from 1981 to 1986. In this position, he was the leading opponent of the policy of using the space shuttle as the sole launch vehicle for critical national security payloads. (Air Force photo)

By the February 14 meeting, Rye had turned the Aldridge-Beggs compromise into a draft National Security Decision Directive (NSDD). Aldridge and Beggs agreed to the text, and Rye transmitted the document to McFarlane, who took it to President Reagan. The president signed the document on February 25. There was no Senior Interagency Group for Space [SIG (Space)] or other interagency review of the NSDD before it went to the president, and Reagan signed the directive with no questions.[6]

The document was issued as NSDD-164, "National Security Launch Strategy." It provided that

- NASA and the DOD "will work together to insure that the STS is fully operational and cost-effective at a flight rate sufficient to meet justified needs. (The target rate is 24 flights per year.)"
- "The Air Force will buy ten expendable launch vehicles (ELVs) and will launch them at a rate of approximately two per year."
- "DOD will rely on the STS as its primary launch vehicle and will commit to at least one-third of the STS flights available during the next ten years."
- "NASA and DOD will jointly develop a pricing policy for DOD flights that provides a positive incentive for flying on the Shuttle."[7]

As the White House released the NSDD, the Pentagon announced that the CELV contract would go to Martin Marietta for its Titan 34D7 vehicle, later to be known as Titan IV. This announcement ratified what was already widely known, which is that the CELV would not be a shuttle-derived booster developed by NASA.

While the Beggs-Aldridge agreement and subsequent NSDD may have ended the war between NASA and DOD, skirmishes regarding space launch continued over the next several years. As part of their February compromise, Aldridge thought that Beggs had also agreed to Air Force plans to modify some number of decommissioned Titan II intercontinental ballistic missiles to serve as launchers for relatively small satellites, rather than launch them on the shuttle with all the attendant preparations required. Aldridge believed that this action "was not a critical issue for NASA and there was little argument over its implementation." This turned out not to be Beggs's perspective. On February 25, the day that the NSDD was issued, Beggs was quoted in *The Washington Post* as critical of the Titan II plan, saying, "I don't want to suggest that anything dark is going on here, but some people think this whole affair may be a heavy-handed scheme by the Air Force to give the shuttle a black eye … I don't like it." To add to Beggs's unhappiness, the National Oceanic and Atmospheric Administration (NOAA) announced that it also planned to launch three of its future weather satellites on Titan IIs rather than on the shuttle. Learning of this step on NOAA's part, Beggs "became incensed." His comments to the *Post* reflected that anger.

Rye observed that "with Jim's comments to the *Washington Post*, the Air Force now considers there has been a setback in NASA/Air Force relations."

He added that "unless the relations between NASA and the DOD improve, we could have an even more serious, headline-grabbing problem concerning the interface between the shuttle and SDI [Strategic Defense Initiative] … Unless this interface is handled carefully, we will open ourselves to claims of 'weaponizing' the Shuttle." Rye commented that "we should not air dirty laundry in public before we have washed it internally."[8]

Coke Versus Pepsi

Another "war" of a quite different character involved the two leading U.S. soft drink companies. One of the discoveries on early space shuttle flights was that the members of the astronaut crew were not drinking enough liquids to keep adequately hydrated. As one approach to dealing with this issue, in June 1984 mid-level officials at NASA's Johnson Space Center had initiated discussions with the Coca-Cola Company about the potential of flying the company's carbonated beverages on the shuttle. Press reports of these discussions quickly came to the attention of PepsiCo, the parent company of Pepsi-Cola. PepsiCo's president, Donald Kendall, called NASA's Beggs to complain. His call was followed by a letter from PepsiCo Vice President Max Friedersdorf, who had been a White House assistant early in the Reagan administration. The letter communicated a "very strong objection" to the NASA-Coke interactions, and said that if a soft drink were to be selected to fly aboard the shuttle, "competitive bidding should be instituted." Friedersdorf, politicizing the issue, also pointed out that "PepsiCo, Inc. is strongly identified with the Republican Party and the support of President Reagan and his Administration. At the same time, Coca-Cola was a very strong supporter and advocate for President Carter and it is closely identified with the Democratic Party." On August 6, 1984, NASA sent a letter to both Coke and Pepsi saying that discussions over flying a soda were being terminated.[9]

Coca-Cola persisted in trying to convince NASA to fly its beverage. After several months of back and forth on the issue, in April 1985 NASA agreed to fly the drink containers developed by Coca-Cola on a July shuttle flight. What was to be demonstrated, NASA insisted, was only whether the container would allow a carbonated soda to be consumed in the shuttle's weightless environment. Beggs said that agreeing to fly the Coca-Cola container first, presumably filled with Coke, was "being done purely for the purpose of recognizing Coca-Cola's initiative in the development at their expense of the drink container technology. We believe that they are deserving of such recognition." But Coca-Cola would have to agree "not to advertise in any way that NASA selected the Coca-Cola beverage to be flown in space based on crew preference" and not "to imply a NASA endorsement of the Coca-Cola beverage."[10]

This agreement was not acceptable to Pepsi. The company continued to pressure NASA to also fly its beverage on an early flight. By the end of June, NASA acceded to this pressure, and announced that, if PepsiCo could meet the same conditions as had been imposed on Coke, the forthcoming shuttle flight, scheduled for launch on July 12, would carry containers provided by both Coca-Cola and Pepsi, each filled with the company's soda. This agreement disturbed both senators from Georgia, Sam Nunn and Mack Mattingly; their interest in the issue was linked to the fact that Atlanta was the location of Coca-Cola's headquarters. Nunn wrote to Beggs on June 27, saying that he was "surprised to hear from a NASA press release issued this week that the other soft drink technology is now being considered for the space shuttle flight scheduled for July 12." Nunn sent a copy of his letter to new White House chief of staff Don Regan; he had earlier discussed the situation with Regan in a telephone conversation. Mattingly called Regan directly and followed his call with a July 3 letter. He told Regan that "PepsiCo should have the same opportunity as Coke to make the July flight."[11]

Director of Cabinet Affairs Alfred Kingon, Craig Fuller's replacement, wrote Regan on July 3, saying, "This matter is clearly a problem for NASA and in no way should the White House or you get on the record on one side or the other." Kingon recommended a call to Sam Nunn "explaining that this must be Jim Beggs' decision." Regan responded to Kingon that "I agree" that this should stay a NASA issue, and that "I'm not going to call Senators, as it should stay out of the WH [White House]."[12]

The Spacelab-2 shuttle mission was launched on July 29, 1985, over two weeks after its scheduled date. It carried containers containing both Coke and Pepsi in what NASA dubbed a Carbonated Beverage Dispenser Evaluation test. NASA honored the commitment to test Coke "first" by taking a photograph of one of the seven crewmembers drinking from the Coke-supplied container, with a time stamp that showed it preceded any crewmember sampling Pepsi.

After all the controversy, the crew did not like the beverages; they were too fizzy, and because the shuttle did not have an onboard refrigerator, they had to be drunk warm. Beggs reported that the crew thought that the colas tasted "terrible." It would be some years before there was another attempt to fly a carbonated beverage in space. The "cola war" thus came to an anticlimactic end.[13]

How Best to Compete?

Even as Gil Rye in early 1985 was brokering a compromise between NASA and Air Force over the shuttle/CELV issue, the White House was also dealing with another interagency conflict, this one with respect to commercializing U.S. ELVs. During the 1984 dispute over shuttle pricing policy, NASA and advocates of commercializing ELVs had taken strongly opposing positions on the question of how best to ensure a U.S. advantage in competing with the European *Ariane* and other emerging launch service providers. The NASA position was that it was very unlikely that commercialized U.S. ELVs would be successful in competing with *Ariane*, and thus it was essential that the shuttle price for launching a commercial or foreign payload be kept low enough to be competitive with the European launcher. Commercialization advocates argued that U.S. ELVs could succeed in such competition, but only if they did not have to also compete with the subsidized space shuttle. They argued that government subsidies that kept shuttle prices low should be removed, with the shuttle price being set to reflect the "full costs" of a launch. At that price point, a shuttle launch would be more expensive than either *Ariane* or U.S. ELVs.

To examine these differing perspectives, the August 1984 National Space Strategy had called for the Office of Management and Budget (OMB), in cooperation with relevant government agencies, to "prepare a joint assessment of the ability of the U.S. private sector and the STS to maintain international competitiveness in the provision of launch services." That assessment arrived at the White House in early February 1985. The report had been prepared by an interagency working group formed soon after the space strategy was issued. The report of the working group was much more supportive of the advocates of commercializing ELVs than of NASA. It suggested that shuttle competiveness in the commercial competition for launch contracts was not relevant to the U.S. policy goal of "maintaining world leadership in space transportation." Rather, the unique capabilities of the shuttle were "the proper focus" for defining that leadership, and shuttle success "need not be determined by its share of the commercial communications satellite market," since "competition for routine services that could be performed by the private sector is not a threat" to U.S. leadership. The report concluded that the United States "has a better option" than depending on the shuttle to achieve U.S. competitiveness in the commercial launch market. That option was "encouraging the competitiveness of U.S. commercial ELVs … U.S. commercial ELVs provide the best hedge available for dealing with the uncertainties in the international marketplace for commercial launch services." But this conclusion came with a warning: "The option for U.S. commercial ELVs will disappear unless action is taken now to implement fully a pricing policy for commercial and foreign users that recovers all of the costs" of the space shuttle.[14]

The OMB report was thus an almost total endorsement of the position of commercial ELV advocates. Not surprisingly, NASA disagreed with the report's conclusions, and NASA's Peggy Finarelli prepared a rebuttal. She argued that "U.S. ELV prices are not competitive with *Ariane* in any payload weight class," while at NASA's suggested price the shuttle would provide "some competition … particularly for smaller payloads." In taking this position, Finarelli was arguing that the ability of U.S. ELVs to compete with *Ariane* had nothing to do with the price of shuttle launches; rather, it had to do with the inability of U.S. ELVs themselves to be marketed at a competitive price, and that this inability would not change, whatever shuttle price was set. She added "If Shuttle prices are escalated excessively, the Europeans will be in a position to capture as much of the commercial market as they want." Finarelli added: "Every U.S. [shuttle] launch means money and jobs for U.S. industry … Loss of launches to the Europeans denies this business to U.S. companies."[15]

The OMB study and NASA's rebuttal to its conclusions reflected a broader and increasingly bitter conflict that had been festering for several years over what price to set for shuttle launches of commercial and foreign payloads after October 1, 1988. To NASA, the space shuttle's ability to remain in play as a potential launcher of commercial payloads was essential to the agency's future plans and budgets, and also to its self-image as the central actor in U.S. civilian space activities. To those opposing NASA on shuttle pricing, that issue had become a surrogate for their broader goal, moving NASA out of its central position in U.S. space activities by providing an ELV alternative to NASA as the monopoly supplier of access to space. In that way, they thought, the private sector could explore the profit-making potentials of space activities without contending with what they saw as a recalcitrant NASA bureaucracy.

Pricing the Space Shuttle

A 1985 congressional report noted that "the price that the National Aeronautics and Space Administration (NASA) charges foreign and commercial customers to use the shuttle's launch services has important implications for the development of space in general and for the future of the U.S. space program in particular." As far back as NSDD-93 in May 1983, there had been a decision that after October 1, 1988, the price for commercial and foreign users of the space shuttle would be set at a "full-cost recovery" level. But the precise meaning of "full cost recovery" had not been spelled out then or subsequently, and remained a point of contention. The August 1984 National Space Strategy had directed NASA to provide a "time-phased plan for implementing full cost recovery for commercial and foreign STS flight operations." NASA provided its plan in a September 17 letter from James Beggs to President Reagan that was accompanied by a 53-page explanation of the agency's pricing proposal. Beggs told the president that "to take advantage of power of the shuttle" to project world leadership, "we must fly it. And, in particular, we must use it for commercial and foreign missions which present highly visible opportunities for private citizens and foreigners to fly with us."

NASA proposed charging "a list price of $87 million per Shuttle flight, with the flexibility to adjust that list price up or down by as much as 5%." That flexibility was intended to help the shuttle compete with *Ariane*. The proposed price was stated in 1982 dollars; Beggs compared it to the $71 million price that had been set in 1982 for the 1986–1988 period. NASA's $87 million price was characterized as "surprisingly low" by *Science* magazine, which commented that "if NASA were charging full cost recovery today, the price would be roughly $155 million per launch." NASA had assumed in its proposal that the shuttle would be flying 24 missions per year by 1989, and thus divided the fixed costs of shuttle operations across that many flights, lowering the cost per flight.[16]

White House Turf Wars

Although the shuttle pricing issue had previously been part of the portfolio of the NSC's SIG (Space), the first body to review the NASA pricing proposal was the Working Group on Space Commercialization of the Cabinet

Council on Commerce and Trade (CCCT). That working group had been created by the July 20, 1984, Commercial Space Policy (see Chap. 13). It was chaired by the Department of Commerce, in the person of the department's deputy general counsel, Richard Shay.

A first meeting of the Working Group on Space Commercialization took place on November 29, 1984. The focus of the meeting was on the approach to reviewing the shuttle pricing plan. The group agreed to consider the "ranges of pricing of the shuttle services using varied macroeconomic assumptions and using a low, medium, and high interpretation of full-cost recovery" and the "implications for U.S. industry of the likely alternatives for pricing the services of the shuttle." This was a much broader approach than NASA thought appropriate. Beggs wrote McFarlane on December 7, telling him that "my understanding" was that "NASA's Shuttle pricing plan would be brought forward for the President's consideration through the SIG (Space) mechanism" and that the jurisdiction of the CCCT working group would be limited to general commercialization initiatives, not including shuttle pricing. Beggs pointed out that the implications of shuttle pricing policy "extend to a broad range of space policy concerns reaching far beyond the narrow question of commercialization." Beggs was concerned that the CCCT working group, led by an agency advocating ELV commercialization, "will produce for the President a paper that addresses only one facet of the matter and lacks the balance and perspective which the President requires for his decision." He said that "it's very important" that the shuttle pricing review "be done through the SIG (Space) mechanism." Beggs asked McFarlane "helping to get this straightened out."[17]

NASA and the NSC-led SIG (Space) prevailed on this issue, but it took a presidential directive to sort out the jurisdictions. After over a month of negotiations, Ronald Reagan on January 24, 1985, signed National Security Study Directive (NSSD) 3-85 on "Shuttle Pricing," which specified that "because of the impact of the shuttle pricing issue on the overall U.S. Space Program, the Senior Interagency Group (SIG) for Space will assume lead responsibility on this issue." The shuttle pricing review would be carried out by a SIG (Space) working group chaired by the NSC. The CCCT Working Group on Space Commercialization review would be limited to the impact of shuttle pricing "on commercial activity in space."[18]

Shuttle Pricing Reviews Go Forward

The CCCT Working Group review of shuttle pricing policy was completed by April 1985. It concluded that it was in the U.S. national interest "to maintain an environment where our domestic ELV industry has the potential to compete in the launch services market." The group thought that "the spectrum of justifiable full cost recovery [shuttle] prices runs from $116M [million] per flight to $135M per flight (depending on the assumed flight rate)." NASA "nonconcurred" with the report, since those shuttle prices were substantially higher that what NASA was proposing. The working group was dominated by those eager to see a commercial launch industry emerge, and its report reflected that point of view.[19]

The SIG (Space) working group on shuttle pricing began its work only in early April. Rye described this review process as "slow and tortuous" because of the strong views on the part of NASA and its adversaries on the pricing issue. Another participant in the process noted that "the various agencies have inflexible positions from which the Working Group members are not authorized to deviate, and there is very little agreement on even a basic set of facts and assumptions." The working group was thrown into what Rye characterized as "a bit of a frenzy" when on April 24 NASA's Beggs sent a letter to the president proposing that, rather than set the shuttle price at $87 million as he had suggested the previous September, it be set at $71.4 million per flight. Beggs claimed that this figure represented the costs of a shuttle flight for a commercial or foreign user, which were lower than for a government user, and thus was still in accordance with a "full cost recovery" policy. Beggs told the president: "I am taking this action because I have become increasingly convinced since last September that the Shuttle will not be able to compete effectively with the European *Ariane* launch vehicle at a price of $87M per flight." He added that "competition in the marketplace is highly visible and failure in that public forum would relegate us to second class status in the area of space transportation."[20]

As part of its review, SIG (Space) asked the Central Intelligence Agency (CIA) to do a study of the *Ariane* ELV. That study concluded that "if Shuttle prices are increased, Arianespace [the French-owned firm marketing *Ariane*] is expected to reassess its pricing practices. Under these circumstances, CIA believes that Arianespace would have a strong economic incentive to maximize its return by raising prices to the maximum consistent with its objectives for market share and utilization of capacity. Arianespace is expected to target its prices at about 5% below the price of the Shuttle and any U.S. commercial ELVs."[21]

Setting the Price

The pricing issue was so contentious that it took three meetings of the SIG (Space) principals to reach agreement on what pricing proposal to put before the president. The primary antagonists were the Department of Transportation, supported by the Department of Commerce, on one side, and NASA on the other. Because of the economic dimensions of the pricing issue, the White House Council of Economic Advisers (CEA) was added to the usual SIG (Space) member organizations participating in the review, and it was a CEA proposal that ultimately carried the day.

The first of the three meetings was held on May 28, 1985. Meeting participants considered a 15-page issue paper that had been prepared by the SIG (Space) working group. The paper noted that "the Shuttle pricing decision will have impacts in four principal areas of national interest: space commercialization, budget, trade, and national security." The paper analyzed the pros and cons with respect to these areas in terms of three price options: the $71.4 million suggested by NASA, $112 million, and $129 million. Advocates of the lowest cost argued that "any Shuttle price above about $80M would be noncompetitive against *Ariane*" and that loss of launches to *Ariane* "could also mean loss of related U.S. sales such as telecommunications satellites" and could "inhibit private sector investment and research in space manufacturing." Advocates of a higher Shuttle price argued that accepting a continued low price would "eliminate the private U.S. ELV industry," with accompanying "significant trade, budgetary, national security and other policy risks." Participants in the May 28 meeting were unable to reach agreement on what recommendation to make to the president, and asked the working group to revise the issue paper.[22]

A second SIG (Space) meeting was held on June 17. In addition to the three pricing options that had been considered at the May 28 meeting, an OMB-sponsored option that varied the shuttle price between $71 million and $129 million, depending on the number of missions flown per year, was also discussed, as was a "flexible price" option that had been developed by the CEA. In preparing his boss, CEA chairman Beryl Sprinkel, for the meeting, staffer Hayden Boyd suggested that Sprinkel should oppose the high shuttle prices proposed by the Departments of Commerce and Transportation, on the grounds that the CEA "should be in favor of private enterprise, but against industrial policy which fosters a particular industry [in this case, commercial launchers] for its own sake." The June 17 SIG (Space) meeting also failed to reach consensus of which pricing option to recommend to the president.[23]

A breakthrough in the pricing debate appeared in early July. Working together, the staff of the CEA and OMB agreed on a CEA-originated concept of "auctioning" launches on the shuttle, rather than adopting a fixed price. On July 8, OMB deputy director Joe Wright, in what he characterized as "draft number infinity," sent a description of that concept to the principals in the pricing debate. The next day, Beggs was reported to have agreed to the concept. The "auction pricing" approach, thought OMB and CEA, would permit "market forces, not the Government, to determine the price for Shuttle commercial services." In this approach, "Shuttle flight capacity available for foreign and commercial users would be sold at auction. A minimum acceptable bid (auction floor) would be established … However, the NASA Administrator may price up to two Shuttle equivalent flights per year below the auction floor if necessary to match bona fide offers from foreign competitors."[24]

The "auction option" was one of the three choices contained in the revised issue paper discussed at the third SIG (Space) meeting on the pricing issue, held on July 22. The two other options in the issue paper were "additive costs," supported by the Department of Commerce, and "total cost," supported by the Department of Transportation. The "auction pricing" option was supported by OMB, NASA, and the CEA. The "additive cost" option "would establish a price based on the additive costs associated with adding Shuttle flights for commercial/foreign payloads plus depreciation for an Orbiter." That price was estimated to be $65 million, even lower than the price suggested by NASA. Commerce, in a reversal of its earlier position that revealed a disconnect between the department's leaders and those at the working level participating in the review, argued that "maintaining U.S. competiveness in world markets is the primary national interest" in setting a shuttle price and that "the commercialization of ELVs should not be subsidized through inappropriate pricing of Shuttle operations."

The "total cost" option would recover "direct and indirect operating costs, depreciation and interest." The price under this option would be $129 million. Transportation representatives argued that this price would "encourage and facilitate development of a U.S. commercial ELV industry ... a much needed element in the U.S. space transportation [policy] to maintain leadership." The "auction price" option in the issue paper reflected the OMB/NASA/CEA position; the minimum acceptable bid was set at $82.5 million. The sponsors of this option argued that it "will permit continued U.S. competitiveness for launch services, encourage space commercialization, and maximize the revenue to the U.S. Government for selling commercial launches." The issue paper noted that all three options "purport to meet the policy objective of 'full cost recovery.'"[25]

The SIG (Space) decided to recommend some version of auction pricing to the president. The issue paper was revised to include only two options. Option 1 was "Low Auction Pricing," and option 2 was "High Auction Pricing." In a July 27 decision memorandum to President Reagan, McFarlane recommended to President Reagan that he approve Option 1 because it "best satisfies the overall national interest as reflected in the space policies which you have previously promulgated." This was the case even though "the adoption of Option 1 will diminish the prospects for the commercialization of U.S. expendable launch vehicles." But "any significant increase in the Shuttle price will benefit the French-built *Ariane* ELV rather than any prospective U.S. ELV due to the commitment by France to underbid both the Shuttle and U.S. ELVs." McFarlane told Reagan: "All agencies, except for the Department of Transportation, recommend that you approve Option #1."[26]

Ronald Reagan followed that advice and, on July 30, McFarlane issued NSDD-181, "Shuttle Pricing for Foreign and Commercial Users."[27] The directive adopted auction pricing and set the minimum acceptable bid for a shuttle launch at $74 million, even lower than the figure in the mid-July issue paper. Winning the competition with *Ariane* for launch contracts was thus explicitly given higher priority than supporting the commercialization of U.S. ELVs. This was a clear victory for NASA and advocates of the space shuttle and a defeat for the Department of Transportation and other advocates of ELV commercialization. It would take a tragic shuttle accident to reverse that outcome.

Who Gets to Fly?

As NASA picked up the pace of shuttle missions in 1985, another new shuttle orbiter, *Atlantis*, entered service, joining *Challenger* and *Discovery*. *Columbia* also returned from major modifications, so at the end of the year the shuttle fleet was at its authorized four-orbiter complement. Among other things, this meant that there was space available for nonastronaut individuals on several shuttle flights, and during the year Senator Jake Garn, commercial payload specialist Charles Walker, several other U.S. payload specialists, and payload specialists from France, Saudi Arabia, Germany, the Netherlands, and Mexico went into orbit. Although the nine missions during the year almost doubled the flight rate in 1984, the shuttle was still experiencing developmental problems on almost every flight.[28] These problems were not well known outside of the space shuttle community, and with the anticipation of even more shuttle flights in subsequent years as NASA built up its flight rate, there were a number of aspirants for a place on a future flight.

Doug Morrow Wants a Flight

Ronald Reagan liked to write personal letters, both to his friends and to those whom he did not know but whose letters to the White House were selected for him to read. He handwrote more than 10,000 letters in his lifetime to a wide array of friends and family, politicians, private citizens, and children.[29] One of Reagan's correspondents was longtime friend and Hollywood screenwriter Douglas Morrow, and Morrow used his friendship with Reagan to plead for an opportunity to go into orbit.

Morrow, who in 1985 was 72, two years younger than Reagan, had in the early 1980s climbed to the 23,000 foot level of Mount Everest, and Reagan had apparently asked him, "What do you do for an encore?" Morrow had replied, "What about the Space Shuttle?" On February 11, 1985, Morrow wrote to the president, saying, "I wasn't kidding" and "I think it is long overdue that a mission have a flight historian." He added, "What a shot in the psyche it would be for so many men and women if a guy my age made a flight!"

Whether to humor Morrow or because he took the request seriously, the president had his staff check with NASA about the requirements for a person to be qualified for a spaceflight. The response was that "stringent medical standards" had been established for future applicants to the Spaceflight Participant Program and that NASA was "resisting the mounting pressure from numerous special interest groups to select [a spaceflight participant] from their constituencies." But President Reagan was not a normal "special interest group"; on February 28, he sent a copy of Morrow's February 11 letter along with his own letter to NASA Administrator Beggs. He told Beggs that he had found the Morrow letter "interesting" and sketched Morrow's background for Beggs. He added: "I'm making no personal request … I bring it to your attention only because it sounded, to this lowly horse cavalryman, as if it might have merit." Reagan also wrote to Morrow, saying, as noted earlier, "why would anyone want to get any higher off the earth than you can get by sitting on the back of a horse?" but adding "you have made a very impressive case and I intend to put it in the hands of the Director of NASA myself … I'm sure you understand I can't dictate in a matter of this kind but I can be enthusiastic."

Morrow continued to pressure Reagan, saying in a May 31 letter that "the suspense is stupefying." Reagan again responded, saying, "I don't have any answer to your request to go off into the wild blue yonder. The top man at NASA knows this has my approval, but I think you understand why I can't make this an order. I've gone out of my way to let him know of my continued interest and will keep on doing that." Morrow's response was that he was "thrilled that the flight is still a live possibility, and that you are continuing to nudge NASA's top dog ... I am continuing to train for it." Reagan told Morrow in a June 26 letter: "I'll continue to plug but don't overtrain. What with the international program plus such things as the 'first teacher' idea it looks like there is a waiting line building up."

Morrow's request remained active throughout the rest of the year. On December 16, the president asked: "How are we coming on the response to Doug Morrow regarding Jim Beggs?" Six weeks later, *Challenger* was launched on its fatal mission, and any thought of flying private citizens evaporated.[30]

A Teacher in Space

Ronald Reagan had announced on August 27, 1984, that the first private citizen to have the opportunity to fly aboard the space shuttle would be an elementary or high school teacher. NASA received 10,463 applications for that opportunity, and by May 1985 had narrowed the field to 114 semifinalists. Those individuals traveled to Washington in June to meet with the review committee that would narrow the field to ten finalists; the decision that they would fly aboard *Challenger* during a January 1986 mission had already been made. On June 26, they came to the East Room of the White House, where they were addressed by the president. Reagan told them: "When one of you blasts off from Cape Kennedy next January, you will be representing that hope and opportunity and possibility – you'll be the emissary to the next generation of American heroes. And your message will be that our progress, impressive as it is, is only just a beginning; that our achievements, as great as they are, are only a launching pad into the future. Flying up above the atmosphere, you'll be able to truly say that our horizons are not our limits, only new frontiers to be explored." Reagan in a handwritten insert had added to the speech text that had been prepared for him a poem titled "Teacher," composed by journalist Clark Mollenhoff. It read:

> You are the molders of their dreams—the gods who build or crush their young beliefs in right or wrong.
> You are the spark that sets aflame a poet's hand, or lights the flame in some great singer's song.
> You are the gods of the young—the very young.
> You are their idols by profession set apart.
> You are the guardians of a million dreams.
> Your every smile or frown can heal or pierce a heart.

> Yours are 100 lives, 1,000 lives.
> Yours is the pride of loving them, the sorrow too.
> Your patient work, your touch, make you the gods of hope that fills their souls with dreams and make those dreams come true.

The president added, with his characteristic humor, "For the lucky one who does go up in the shuttle, I have only one assignment: Take notes. There will be a quiz after you land."

Reagan had recently traveled to the National Air and Space Museum to see the wide-screen movie *The Dream is Alive*. He told the teachers: "It's just about as close to being in space as I think you can be and still have your feet on the ground. I was really carried away with that. We never had anything like that in the horse cavalry."[31]

NASA's review panel identified 10 of the 114 teachers as finalists, and announced their names on July 1. The ten were flown to Houston for a round of medical evaluations, briefings, and microgravity flights on NASA's "vomit comet" zero-gravity airplane. They then returned to Washington for interviews with senior NASA officials, and on July 19 Beggs selected the winner of the competition and a backup.[32]

Cabinet Secretary Alfred Kingon had in early July suggested, because the teacher-in-space program was "a very popular Presidential initiative," that "the President announce the name of the teacher and the backup teacher … at a White House ceremony in the Rose Garden." But Reagan had undergone a routine colonoscopy on July 12; the procedure had uncovered a large, potentially cancerous polyp. The decision was made to remove the growth, and Reagan was still in the hospital on July 19. So the announcement of the teacher-in-space selection was made by Vice President George H.W. Bush. The White House had wanted the announcement to be a total surprise to the winner and backup, but NASA's Alan Ladwig, the manager of the Spaceflight Participant Program, decided to let the ten finalists know the outcome before the group went to the White House. Still, as Bush announced Christa McAuliffe of New Hampshire as the chosen teacher and Barbara Morgan of Idaho as her backup, their reactions were ones of delight.[33]

Conclusion

Despite recurring technical problems on almost every space shuttle flight, by the last months of 1985 NASA was planning on an even-more ambitious flight schedule for 1986 and subsequent years. Having prevailed in its resistance to setting a higher shuttle price for foreign and commercial users and having gotten DOD agreement to employ one-third of available shuttle flights, NASA's path to regular and frequent shuttle launches seemed clear.

Ronald Reagan had never attended a shuttle launch, although he had been present at one landing and had made plans to attend at least one other. This situation was a result of both his staff being told that the launches often did not happen on schedule, making it difficult to arrange a presidential trip, and that there was a higher chance of an accident on launch than at landing. With launches becoming more regular, at least one staff person in the White House thought that it was time for Reagan to witness a launch in person. Kingon in October told his boss, Chief of Staff Don Regan, that "one of the greatest experiences I have had since coming to Washington is watching the shuttle being launched from Cape Canaveral. No movie or other pictures do it justice." Kingon was "amazed to find out that the President had never seen the shuttle go off." Kingon had been one of those in the White House who had urged Reagan to see the movie "The Dream is Alive"; he noted that the president "had gone and enjoyed it immensely. I am sure he and Mrs. Reagan would be even more thrilled at watching a real launch."

Kingon had asked NASA what month would have "the greatest probability of success" in launching on schedule. NASA responded that "January has the greatest weather predictability." Thus, Kingon suggested that plans be made to have the Reagans attend the January shuttle launch, at that point scheduled for January 22. The shuttle orbiter *Challenger* would be used for that flight, and one of its seven-person crew would be teacher Christa McAuliffe.[34]

Notes

1. Letter from "Ron" (Ronald Reagan) to Doug Morrow, March 4, 1985, Box 11, Outer Space Files, RRL.
2. Unless otherwise referenced, information in this account is derived from an oral history interview of Edward "Pete" Aldridge, Jr. conducted by Rebecca Wright as part of the NASA Headquarters Oral History Project, May 29, 2009, https://www.jsc.nasa.gov/history/oral_histories/NASA_HQ/Administrators/AldridgeEC/AldridgeEC_5-29-09.htm, and E.C. Pete Aldridge, Jr., "Assured Access: 'The Bureaucratic Space War,'" Robert Goddard Historical Essay, no date, ocw.mit.edu/courses/aeronautics-and-astronautics/16-885j-aircraft-systems-engineering-fall-2005/readings/aldrdg_space_war.pdf. This essay was Aldridge's entry in a space history competition organized by the National Space Club. These accounts naturally reflect Aldridge's perspective on the conflict over the DOD initiative to complement dependence on the shuttle with a few expendable launch vehicles.
3. "Fact Sheet, National Space Strategy," September 7, 1984, Folder 12905, NHRC.
4. National Research Council, *Assessment of Candidate Expendable Launch Vehicles for Large Payloads*, September 1984, https://www.nap.edu/catalog/19350/assessment-of-candidate-expendable-launch-vehicles-for-large-payloads.
5. Letter from James Beggs to Robert McFarlane, February 13, 1985, Box 12, Outer Space Files, RRL.
6. Memorandum from Gil Rye to Robert McFarlane, "NASA Administrator's Comments on ELVs," February 26, 1985, provided to author by Gil Rye. Interview with Gil Rye, April 25, 2016.

7. Ronald Reagan, National Security Decision Directive 164, "National Security Launch Strategy," February 25, 1985, https://www.reaganlibrary.gov/sites/default/files/archives/reference/scanned-nsdds/nsdd164.pdf.
8. Thomas O'Toole, "NASA Chief Hits Air Force Plan to Spurn Shuttle," *WP*, February 25, 1985; memorandum from Gil Rye to Robert McFarlane, "NASA Administrator's Comments on ELVs."
9. Letter from Max Friedersdorf to James Beggs, June 20, 1984; memorandum from Lawrence Herbolshiemer to Alfred Kingdon, "The Coke/Pepsi/NASA Debacle," June 27, 1985, both in Box 11, Outer Space Files, RRL.
10. Letter from James Beggs to Sam Nunn, May 29, 1985, Box 12, Outer Space Files, RRL.
11. Letter from Sam Nunn to James Beggs, June 27, 1985, Box 11, Outer Space Files, and letter from Mack Mattingly to Donald Regan, July 3, 1986, Box 12, Outer Space Files, RRL.
12. Memorandum from Alfred Kingdon to Donald Regan, July 3, 1985, with handwritten note initialed "DTR," Box 11, Outer Space Files, RRL.
13. See https://www.csmonitor.com/Science/2011/0708/The-9-weirdest-things-ever-flown-on-the-Space-Shuttle/Cans-of-Coca-Cola-Pepsi and Joseph Trento, *Prescription for Disaster: From the Glory Days of Apollo to the Betrayal of the Shuttle* (New York: Crown Publishers, 1987), 250–251, for discussions of the astronauts' reaction.
14. Office of Management and Budget, "International Competitiveness in Launch Services," February 1985, CIA-RDP92B00181R001701630002-1, CREST, i–iii.
15. Letter from James Beggs to Robert McFarlane, February 11, 1985, Box 12, Outer Space Files, RRL.
16. Congressional Budget Office, "Pricing Options for the Space Shuttle," March 1985, ix. Letter from James Beggs to the President with attached "NASA Pricing Plan," September 17, 1984, CFOA 1318, Papers of Michael Driggs, RRL. M. Mitchell Waldrop, "NASA Suggests a New Shuttle Price," *Science*, November 9, 1984, 674–675.
17. Memorandum from Craig Fuller to Michael Driggs, "CCCT Working Group on Space Commercialization," November 29, 1984, and letter from James Beggs to Robert McFarlane, December 7, 1984, Box 12, Papers of Craig Fuller, RRL.
18. National Security Study Directive 3-85, "Shuttle Pricing," January 24, 1985, Folder 12772, NHRC.
19. "Shuttle Pricing Issue Paper," attached to Memorandum from John Poindexter to Members, Senior Interagency Group for Space, "Third SIG (Space) Meeting on the Shuttle Pricing Policy," July 16, 1985, CIA-RDP87M00539R000400520007-1, CREST.
20. Memorandum from Gil Rye to Robert McFarlane, "NASA Shuttle Pricing Proposal," May 1, 1985 with attached letter from James Beggs to the President, April 24, 1985, https://fas.org/irp/offdocs/nsdd/nsdd-144.htm.
21. "Industry Observer," *AWST*, April 1, 1985, 13. "Shuttle Pricing Issue Paper" prepared for Meeting of SIG (Space), May 28, 1985, CIA-RDP87M00220R000500530003-6, CREST, 6.
22. Ibid., 1–3.

23. Memorandum from Hayden Boyd to Beryl Sprinkel, "Sig (Space) Meeting on Shuttle Pricing," June 17, 1985, Box 5, Papers of Beryl Sprinkel, RRL.
24. Memorandum from Joe Wright to Beryl Sprinkel, Jim Burnley, Bud Brown, and Jim Beggs, untitled, July 8, 1985, and memorandum from Joe Wright to John Poindexter, July 9, 1985, Box 12, Outer Space Files, RRL.
25. "Shuttle Pricing Issue Paper," attached to Memorandum from John Poindexter to Members, Senior Interagency Group for Space, "Third SIG (Space) Meeting on the Shuttle Pricing Policy," July 16, 1985, CIA-RDP87M00539R000400520007-1, CREST.
26. Memorandum from Robert McFarlane for the President, "Shuttle Pricing for Foreign and Commercial Users," July 27, 1985, https://fas.org/irp/offdocs/nsdd/nsdd-144.htm.
27. A copy of NSDD 181 can be found at https://www.reaganlibrary.gov/sites/default/files/archives/reference/scanned-nsdds/nsdd181.pdf.
28. For an authoritative account of the nine shuttle missions during 1985, and indeed information on every aspect of the space shuttle program, see the impressive three-volume opus by Dennis Jenkins, *Space Shuttle: Developing an Icon, 1972–2013* (Specialty Press: Forest Lake, MN, 2016). Volume III details every shuttle flight.
29. For a selection of Reagan's letters, see Kiron K. Skinner, Annelise Anderson, and Martin Anderson, eds., *Ronald Reagan: A Life in Letters* (New York: Free Press, 2004).
30. Letters from Douglas Morrow to the President, February 11, 1985, and March 11, 1985; memorandum from Alfred Kingon to the President, "February 11, 1985 letter from Doug Morrow regarding his travel on the space shuttle," February 21, 1985; letter from Ronald Reagan to James Beggs, February 28, 1985; letter from Ronald Reagan to Doug Morrow, March 4, 1985, Box 11, Outer Space Files, RRL. Letters from Doug Morrow to the President, May 31, 1985, and June 17, 1985; letters from Ronald Reagan to Douglas Morrow, June 10, 1985, and June 26, 1985; note initialed RR to FFF (White House counsel Fred Fielding), December 16, 1985, Box 12, Outer Space Files, RRL.
31. Ronald Reagan: "Remarks to the Finalists in the Teacher in Space Project," June 26, 1985. Online by Gerhard Peters and John T. Woolley, APP, http://www.presidency.ucsb.edu/ws/?pid=38825.
32. Information on the teacher-in-space selection process comes from Dennis Jenkins, *Space Shuttle*, Volume III, III-115, III-116.
33. Memorandum from Alfred Kingon to Fred Ryan, "Ceremony to announce the teacher selected for a Shuttle flight," July 3, 1985, Box 12, Outer Space Files, RRL. Interview with Alan Ladwig, May 6, 2016.
34. Memorandum from Alfred Kingon to Donald Regan, "Space Shuttle," October 25, 1985, Box 15, Outer Space Files, RRL.

CHAPTER 18

Challenger

The loss of space shuttle orbiter *Challenger* and her seven-person crew on January 28, 1986, was a traumatic shock both to the American public and to the U.S. space program. Reacting to that shock occupied much of the Reagan administration's space policy attention during the rest of 1986. The space shuttle returned to flight only on September 29, 1988, just a few months before the end of the president's second term. Between the *Challenger* accident and the shuttle's return to flight, there were significant changes in U.S. space policy, in particular to the role of the shuttle as a central element in carrying out that policy. From the time it was approved in 1972 to the time of the accident, the space shuttle had been developed and operated on the basis of unachievable premises of frequent low-cost access to orbit and of its being able to be the sole U.S. means of access to space; after *Challenger*, those premises were discarded in favor of a more realistic role for the shuttle in future U.S. space efforts.

The *Challenger* accident also gave Ronald Reagan an unfortunate opportunity to display his best qualities. During the day of the accident and its immediate aftermath, Ronald Reagan was indeed the "great communicator," voicing the nation's anguish and providing whatever consolation was possible. His obvious grief and compassion helped the country accept the reality of what had happened without calling for scapegoats or demanding that the human spaceflight program be brought to an end.

The Shuttle Under Schedule Pressure

A tenth space shuttle mission during 1985 was scheduled for launch on December 18; 15 launches were scheduled for 1986. The National Aeronautics and Space Administration (NASA) shuttle team was operating under pressure to meet that schedule; the investigation into the *Challenger* accident was to

J. M. Logsdon, *Ronald Reagan and the Space Frontier*,
Palgrave Studies in the History of Science and Technology,
https://doi.org/10.1007/978-3-319-98962-4_18

reveal that persistent warning signs that all was not well with the shuttle program had been ignored in the push to meet an unrealistic target for frequently launching the shuttle.

One of the seven-man crew on the scheduled December mission was Representative Bill Nelson (D-FL). Nelson was chair of the Subcommittee on Space Science and Applications of the House of Representatives Committee on Science and Technology. He had been interested in making a shuttle flight since NASA had announced in 1983 the possibility of "ordinary citizens" flying on the shuttle, and soon after had written to NASA Administrator Beggs saying that he would like a flight opportunity. But as a junior congressman from the minority party, that was an unlikely prospect. Then, the 1984 election had resulted in a Democrat majority in the House of Representatives. Nelson competed with a more senior member for the space subcommittee chairmanship. By winning that competition, he put himself in a position for a chance to fly on a shuttle mission.

On September 6, 1985, Nelson got a letter from Beggs, saying "given your NASA oversight responsibilities, we think it appropriate that you consider making an inspection tour and flight aboard the shuttle." After Nelson accepted this invitation, Beggs told him that he would likely be part of the crew on the mission that also included "teacher-in-space" Christa McAuliffe; that mission was scheduled to launch aboard the orbiter *Challenger* in January 1986. Nelson indicated to Beggs that he could also be ready to fly aboard *Columbia* on its December mission. Hughes Aircraft employee Greg Jarvis, who had been scheduled to fly in December, was reassigned to the January *Challenger* mission after a Hughes satellite was removed from the manifest of the December mission. This opened up a seat on that flight, and on October 4, NASA informed Nelson that he "was now assigned" to the December mission.[1]

Even as NASA invited a second politician to fly aboard the shuttle, the space agency was also accepting applications for the second "citizen-in-space" mission, which would send a working journalist into orbit. That mission would reinforce the image that the shuttle was safe enough for nonastronauts. NASA mailed application forms to potential candidates in December 1985. The response was enthusiastic; 1703 applications were submitted by the January 15, 1986, deadline. Applicants included well-known journalists such as Walter Cronkite from CBS, Tom Brokaw from NBC, and Sam Donaldson, ABC's White House correspondent. Donaldson even asked Ronald Reagan for a letter of recommendation. His request was denied; Press Secretary Larry Speakes suggested that "since there are many other journalists making application to the same program, it would be unfair to single out Sam for such special treatment."[2]

As it turned out, there were seven separate launch delays associated with the December mission, and it was not launched on its six-day flight from pad 39A at Kennedy Space Center until January 12, 1986, increasing the already-high schedule pressure on the space shuttle team. Nearby, on pad 39B, *Challenger* was being prepared for its launch, at that point scheduled for January 23. The original schedule had called for launch on January 22, but a one-day postponement had been announced by NASA in December because of the initial delays in the preceding mission.

The uncertainty regarding the launch date complicated White House planning. There had been discussions at the White House since October about President Reagan and his wife Nancy attending the *Challenger* launch. A detailed timeline for a presidential trip to Kennedy Space Center on January 22 had been prepared. In early November, before the delays in the December launch, NASA had called the White House to say that it was thinking about moving the launch one day forward, to January 21. Even at this point, it was "unclear as to whether [the] President would attend on either day." It is not clear from the available evidence when a final decision was made that the Reagans would not attend the launch.[3]

As NASA pressed to launch the missions carrying Representative Nelson and then the teacher in space, William Graham was in his early weeks as NASA acting administrator. Veteran NASA employees resented his presence and were effectively shutting him out of day-to-day discussions. Graham was having problems getting his confidants appointed to NASA positions. Former administrator James Beggs, although on leave because of his indictment, was still occupying an office at NASA. He and Graham were not on speaking terms. NASA was clearly not ready to cope with what was about to happen.

A "Major Malfunction"

The *Challenger* launch had been postponed two more times in January, then on January 27 scrubbed with just over 14 minutes left in the countdown. It was rescheduled for the next morning. Overnight, temperatures near the launch pad dipped to unprecedented lows, and ice formed on the pad structures and on the shuttle itself. Even so, after two inspections of the icing, a "go for launch" clearance was given at 11:15 a.m. When *Challenger* lifted off at 11:38 a.m. EDT on January 28, with its seven-person crew of Dick Scobee, Mike Smith, Ron McNair, Judy Resnik, Ellison Onizuka, Greg Jarvis, and Christa McAuliffe aboard, the temperature near the launch pad was 36 degrees Fahrenheit, 15 degrees colder than at the time of any previous launch.

Videos of the launch later revealed small puffs of black smoke coming from the lowest joint in the right side solid rocket booster attached to the shuttle's external fuel tank; the first puff appeared less than one second after liftoff. The puffs indicated that the seal between two segments of the booster was being eroded by the burning of the solid fuel. The joint failure got much worse as the shuttle passed through several high wind shears during its ascent. By the 59-second mark, a plume of flame from the burning solid rocket fuel was escaping from the joint. The plume breached the skin of the large external tank, and 72 seconds after launch it caused a strut securing the solid rocket booster to fail; the booster pivoted into the shuttle's wing and then into the part of the external tank holding liquid oxygen. At the same time, the hydrogen tank ruptured, releasing a massive amount of highly flammable liquid hydrogen. Within milliseconds, the combination of hydrogen and oxygen spontaneously ignited, producing a fireball that enveloped the shuttle and placed it under severe aerodynamic loads, causing the orbiter to break apart. At this point, the shuttle was at an altitude of 48,000 feet. The sturdy crew compartment remained intact as the rest of the orbiter broke up, and, after climbing to 65,000 feet because of its upward momentum, it began a 2-minute, 45-second descent until striking the ocean surface traveling 207 miles per hour. While at least some of the crew were conscious during the descent, none survived the impact. The commission investigating the accident observed "during the period of the flight when the Solid Rocket Boosters are thrusting, there are no survivable abort options. There was nothing that either the crew or the ground controllers could have done to avert the catastrophe."

As was usual during a shuttle launch, as soon as the shuttle cleared the launch pad tower the public commentary describing the event had switched from Kennedy Space Center to NASA's Johnson Space Center in Houston, Texas. There it took public affairs officer Steve Nesbitt some forty seconds after the breakup to turn away from his console, which displayed only data, to look at a video screen which showed what had happened. His spontaneous comment was "obviously, a major malfunction."[4]

Initial Reactions

Ronald Reagan was scheduled to deliver his annual State of the Union Address on the evening of January 28, just hours after the liftoff. As the shuttle launch took place, he was meeting with several of his staff in preparation for discussing his speech with network news anchors. The meeting was interrupted at 11:43 a.m. as Vice President George Bush and National Security Adviser John Poindexter burst into the Oval Office to tell the president that the shuttle had exploded. Reagan, accompanied by Bush and Poindexter plus others of his staff, immediately went to the small study next to the Oval Office to watch replays of the accident. Reagan looked on in horror as he viewed the catastrophe (Fig. 18.1). As he watched the replays, he called his wife to alert her to what had just happened.

Just before 1:00 Reagan met for a few minutes with the news anchors. A young woman on the national security staff, Karna Small, took notes of what the president said in response to their questions. Reagan's remarks were later transcribed:

> **Q**. Mr. President, can you give us your comments on the tragedy so that we can tell the American people your words, your thoughts?
>
> **The President**. Well, what can you say? It's a horrible thing that all of us have witnessed it and actually seen it take place. And I just can't rid myself of the thought of the sacrifice and the families that have been watching this also, the families of those people on board and what they must be going through at this

Fig. 18.1 President Reagan looks in horror at a replay of the *Challenger* accident. With the president are (l to r) National Security Adviser Admiral John Poindexter, Director of Communications Pat Buchanan, Cabinet Secretary Alfred Kingon, Chief of Staff Donald Regan, and Deputy Press Secretary Edward Djerejian. (Photograph courtesy of the Reagan Presidential Library)

point. I'm sure all of America is more than saddened, feels the great weight of this, and wishes, as I do, that there was something we could do to make it easier for those who've suffered such a loss.

* * *

Q. Mr. President, do you want to see all systems halted until we find out explicitly what happened in this tragedy?

The President. Well, I'm not a scientist. I do have confidence in the people that have been running this program. And this is the first in, what is it, 56-some flights that something of this kind has happened. I certainly want everything done that can be done to find out how this could have happened and to ensure against its happening again. But there again, I have to say I'm sure that the people that have to do with this program are determined to do that right now. And I'm quite sure, also, when you look at the safety measures that sometimes those of us looking on have gotten a little impatient with when flights have been aborted, and it hasn't seemed as if the situation—well, it seems as if they were taking things too seriously. Now we know they weren't. And so, I'm confident that there will be no flight until they are absolutely as certain as a human being can be that it is safe.

Q. Mr. President, do you think it raises questions about having citizens aboard the space shuttle?

The President. Well, they're all citizens, and I don't think anyone's ever been on there that isn't a volunteer. I know I've heard many times from other people that have tried to give me reasons why they, or someone like them, should be included in flights of this kind. So, no, that is the last frontier and the most important frontier. We have to say that the space program has been most successful, most effective. And I guess we've been so confident of it that it comes as such a tremendous shock when something of this kind happens.

Q. Will you go ahead with your message tonight?

The President. Yes, I feel that things like that have to go on.

Q. Mr. President, are you afraid there'll be any public backlash against the space program because of this tragedy?

The President. I shouldn't think so, and I would certainly do everything I could to express an opinion the other way. You know, we have accidents in every line of transportation, and we don't do away with those things. They've probably got a better safety record than we have out on the highways.

* * *

Q. Mr. President, will you tell us exactly who brought you the news and exactly what you thought and said at that point?

The President. We were all sitting in there, and I was preparing myself for your questions on the State of the Union Address when the Vice President and John Poindexter came into the room. And all they could say at the time was that they had received a flash that the space shuttle had exploded. And we immediately went into the adjoining room where I have a TV set to get on this, because there was no direct word except that word that had been made public also. And there we saw the replaying and saw the thing actually happen. And it just was, as I say, a very traumatic experience.

* * *

Q.—so many children have, you know, been a part of this particular space shuttle because of the teacher, and they're doing classrooms. Can you say something that would help them to understand how this happened?

The President. I think people closer to them have got to be doing that. But as I say, the world is a hazardous place, always has been. In pioneering we've always known that there are pioneers that give their lives out there on the frontier. And now this has happened. It probably is more of a shock to all of us because of the fact that we see it happen now and—thanks to the media—not just hearing about it as if something that happened miles away. But I think those that have to do with them must, at the same time, make it plain to them that life does go on and you don't back up and quit some worthwhile endeavor because of tragedy.

* * *

Q. Sir, do you have any special thoughts about Christa McAuliffe, who, I think it was in this room, was named as the first teacher? What are your thoughts about her today?

The President. I can't get out of my mind her husband and her children. But then that's true of the families of the others. Theirs probably more so because the families of the others had been a part of this whole program and knew that they were in a hazardous occupation. But knowing that they were there and watching, this just is-well, your heart goes out to them.[5]

"Touch the Face of God"

Reagan speechwriter Peggy Noonan was in her office when she was told of the accident. She realized that President Reagan would likely want to address the nation, and Noonan wanted to be the person drafting his speech. Dick Darman, now at the Treasury Department, called Noonan, asking, "Is he going to speak? Are you writing it?" Noonan replied: "Don't know but assume so, and yes." Darman added: "The president has to speak to the children and reassure them that the world isn't ending and that there is both inherent purpose and danger in scientific exploration … It's very important."

Then Noonan got a call from a presidential assistant, saying that the president would not address the country until the search for survivors was officially suspended, but that the speech was needed as soon as possible. Karna Small brought Noonan her notes of Reagan's reactions, and they became "the spine of the speech." As she drafted the president's remarks, Noonan watched, "over and over," the videotape of the astronauts that morning as they prepared to travel to the launch pad: "Awkwardly, with heavy-gloved hands, they had waved goodbye." This reminded her of a poem she had learned in the seventh grade—*High Flight* by John Gillespie Magee, Jr., an American who had volunteered for the Canadian Air Force in 1940, before the United States joined what was to become World War II, and in 1941 had died in a midair collision. Noonan remembered that the poem "spoke of the joy of flying, of the sensation of breaking free from gravity, breaking 'the surly bonds of earth.'" With a "hunch" that Reagan would know the poem, she made an excerpt from it the last line of Reagan's remarks.

After Noonan completed her draft of Reagan's speech, it was reviewed, first by communications director Pat Buchanan and Donald Regan's assistants, and then in a larger meeting in Regan's office. Noonan later suggested that "no one is pleased but there is no time to rewrite" and that she was "depressed. I failed when the whole country needs something."[6]

While the speech was being prepared, a first decision, counter to Reagan's initial reaction, was to postpone the State of the Union Address. The White House staff quickly gathered information on how President Lyndon B. Johnson had reacted to the Apollo launch pad fire that had killed Gus Grissom, Ed White, and Roger Chafee on January 27, 1967, 19 years plus 1 day earlier. Acting press secretary Larry Speakes later in the day provided a timeline of postaccident events, noting that "the President [Johnson] issued a written three-sentence statement at the White House shortly after the launch pad fire." Johnson did not make a television or radio address. The Johnson statement had said: "Three valiant young men have given their lives in the Nation's service. We mourn this great loss. Our hearts go out to their families." She also noted that Johnson "did not order the flag flown at half-mast." After the *Challenger* disaster, flags around the country, including at the White House, were quickly lowered.

Acting NASA administrator Graham had been at the Kennedy Space Center the previous day, but when the launch was postponed he had returned to Washington. At the time of the accident, he was meeting with a congressman. Beggs was at NASA and went to look for Graham. Discovering he was on Capitol Hill, Beggs said to Graham's secretary: "As soon as he gets back, let me know because we need to do some things and do them quickly." When Graham returned to the NASA building, Beggs told him: "You should get on an airplane and get your ass down to Kennedy." Beggs also suggested that Graham immediately form an internal investigation board, and gave Graham a list of possible members of such a committee. Graham later remembered that "those were literally the first words" that Beggs had spoken to him since Beggs had introduced him at NASA in December. Graham went to the White House, where he briefly met with President Reagan before joining Vice President Bush and senators Jake Garn and John Glenn to fly to Florida for meetings with the family members of the *Challenger* crew.[7]

Ronald Reagan had only a few minutes to glance at Noonan's speech before, at 5:00 p.m. EDT, he went on nationwide television and radio. He said:

> Ladies and gentlemen, I'd planned to speak to you tonight to report on the State of the Union, but the events of earlier today have led me to change those plans. Today is a day for mourning and remembering. Nancy and I are pained to the core by the tragedy of the shuttle Challenger. We know we share this pain with all of the people of our country. This is truly a national loss.
>
> Nineteen years ago, almost to the day, we lost three astronauts in a terrible accident on the ground. But we've never lost an astronaut in flight; we've never had a tragedy like this. And perhaps we've forgotten the courage it took for the crew of the shuttle. But they, the Challenger Seven, were aware of the dangers, but overcame them and did their jobs brilliantly. We mourn seven heroes: Michael Smith, Dick Scobee, Judith Resnik, Ronald McNair, Ellison Onizuka, Gregory Jarvis, and Christa McAuliffe. We mourn their loss as a nation together.
>
> For the families of the seven, we cannot bear, as you do, the full impact of this tragedy. But we feel the loss, and we're thinking about you so very much. Your loved ones were daring and brave, and they had that special grace, that special spirit that says, "Give me a challenge, and I'll meet it with joy." They had a hunger to explore the universe and discover its truths. They wished to serve, and they did. They served all of us. We've grown used to wonders in this century. It's hard to dazzle us. But for 25 years the United States space program has been doing just that. We've grown used to the idea of space, and perhaps we forget that we've only just begun. We're still pioneers. They, the members of the Challenger crew, were pioneers.

And I want to say something to the schoolchildren of America who were watching the live coverage of the shuttle's takeoff. I know it is hard to understand, but sometimes painful things like this happen. It's all part of the process of exploration and discovery. It's all part of taking a chance and expanding man's horizons. The future doesn't belong to the fainthearted; it belongs to the brave. The Challenger crew was pulling us into the future, and we'll continue to follow them.

I've always had great faith in and respect for our space program, and what happened today does nothing to diminish it. We don't hide our space program. We don't keep secrets and cover things up. We do it all up front and in public. That's the way freedom is, and we wouldn't change it for a minute. We'll continue our quest in space. There will be more shuttle flights and more shuttle crews and, yes, more volunteers, more civilians, more teachers in space. Nothing ends here; our hopes and our journeys continue. I want to add that I wish I could talk to every man and woman who works for NASA or who worked on this mission and tell them: "Your dedication and professionalism have moved and impressed us for decades. And we know of your anguish. We share it."

There's a coincidence today. On this day 390 years ago, the great explorer Sir Francis Drake died aboard ship off the coast of Panama. In his lifetime the great frontiers were the oceans, and an historian later said, "He lived by the sea, died on it, and was buried in it." Well, today we can say of the Challenger crew: Their dedication was, like Drake's, complete.

The crew of the space shuttle Challenger honored us by the manner in which they lived their lives. We will never forget them, nor the last time we saw them, this morning, as they prepared for their journey and waved goodbye and "slipped the surly bonds of earth" to "touch the face of God."

Writing in his diary on the evening of January 28, Ronald Reagan called the date "a day we'll remember for the rest of our lives." He added "there is no way to describe our shock & horror."[8]

As he had finished his remarks, Reagan to Noonan "looked lost" and "stricken." That perception was confirmed as the president called Noonan the next morning to thank her for her words. He told Noonan: "I did the remarks. I read them and then at the end I just had the feeling that I'd failed. I thought I'd done badly and I hadn't done justice. And of course I was so sad about what had happened. And I got off the air and I thought 'Well, not so good.'" Noonan's "hunch" about Reagan knowing the Magee poem turned out to be correct. The president told her: "I hadn't heard that in years, but of course I knew it from years back, the war. And I think it was written on a sort of tablet or plaque outside [Reagan's daughter] Patti's school that I took her to when she was a young girl."

Reagan's unhappiness at his speech delivery did not last long. Overnight a "deluge" of telephone calls and telegrams reached the White House, thanking Reagan for his consoling remarks. Reagan told Noonan that even Frank Sinatra had called him, and "Frank didn't call after every speech." Reagan's *Challenger* speech would go down as one of the most memorable addresses in Ronald Reagan's eight years as president.

How to Respond?

As President Reagan's speech was being prepared, the White House staff was also discussing the other immediate actions needed, in particular the location of a memorial service for the *Challenger* astronauts and the organization of the investigation into the causes of the accident. With respect to the location of the service, the question was whether to hold it in Washington at the National Cathedral, the traditional venue for occasions of national mourning, or in Houston, the base for NASA's human spaceflight program and thus the home of most of the astronauts' families. With respect to the investigation, the issue was whether it should be carried out under NASA supervision, as had been the case after the 1967 Apollo 1 fire and the 1970 Apollo 13 near-tragedy, or by a group independent of NASA.

The White House consulted NASA's Graham about the location of the memorial service. His first recommendation was to hold the event in Houston, but later changed his advice to suggest holding it in Washington, both to reflect the worldwide character of the shocked reaction to the accident and to better accommodate President Reagan's schedule. Someone from the White House, most likely Reagan aide Dennis Thomas, contacted senior astronaut John Young, who was serving as NASA liaison to the *Challenger* families. Young's reaction was that a Washington service would be "nice," but that it would be "mighty hard" for all of the families to be taken away from their support system in Houston to make the trip to the capital so soon after the accident. Young suggested that "it would be a great honor" if Reagan would come to Houston. First lady Nancy Reagan was also consulted. She responded: "We have to go to the families." Ronald Reagan agreed, and the memorial service was set for the Johnson Space Center.[9]

Consoling the Families and NASA Employees

Even before heading to Houston, Reagan had what he described as a "tougher job" that his speech on the evening of the accident. On January 29, Reagan called each the families of the seven *Challenger* crew members to express his sympathy. "All of them," he wrote in his diary that evening, "asked that we continue the space program—that's what their loved ones would have wanted."[10]

On Friday, January 31, the president and his wife flew to Houston to attend the memorial service for the *Challenger* crew. They first met privately with the families, and then were among the 14,000 attending the service. He told the assemblage:

> We come together today to mourn the loss of seven brave Americans, to share the grief that we all feel, and, perhaps in that sharing, to find the strength to bear our sorrow and the courage to look for the seeds of hope. Our nation's loss is first a profound personal loss to the family and the friends and the loved ones of our

shuttle astronauts. To those they left behind—the mothers, the fathers, the husbands and wives, brothers and sisters, yes, and especially the children—all of America stands beside you in your time of sorrow.

What we say today is only an inadequate expression of what we carry in our hearts. Words pale in the shadow of grief; they seem insufficient even to measure the brave sacrifice of those you loved and we so admired. Their truest testimony will not be in the words we speak, but in the way they led their lives and in the way they lost their lives—with dedication, honor, and an unquenchable desire to explore this mysterious and beautiful universe.

* * *

We will always remember them, these skilled professionals, scientists, and adventurers, these artists and teachers and family men and women; and we will cherish each of their stories, stories of triumph and bravery, stories of true American heroes. On the day of the disaster, our nation held a vigil by our television sets. In one cruel moment our exhilaration turned to horror; we waited and watched and tried to make sense of what we had seen. That night I listened to a call-in program on the radio; people of every age spoke of their sadness and the pride they felt in our astronauts.

* * *

We think back to the pioneers of an earlier century, the sturdy souls who took their families and their belongings and set out into the frontier of the American West. Often they met with terrible hardship. Along the Oregon Trail, you could still see the grave markers of those who fell on the way, but grief only steeled them to the journey ahead. Today the frontier is space and the boundaries of human knowledge. Sometimes when we reach for the stars, we fall short. But we must pick ourselves up again and press on despite the pain. Our nation is indeed fortunate that we can still draw on immense reservoirs of courage, character, and fortitude; that we're still blessed with heroes like those of the space shuttle Challenger.

Writing in his diary that evening, Reagan said: "It was a hard time for all the families & all we could do was hug them & try to hold back our tears."[11]

The Rogers Commission

Within a few hours after the *Challenger* accident, NASA took initial steps toward forming an internal mishap board to investigate what had happened and why it had happened. This was the approach that NASA had followed in the aftermath of the January 1967 Apollo 1 fire and the April 1970 explosion as the Apollo 13 spacecraft was on its way to the Moon.

However, this was not the path followed in the wake of the *Challenger* mishap. Reagan's chief of staff Donald Regan recounted that his initial reaction as he and the president watched replays of the accident was "that no ordinary investigation would be adequate" and that he advised Reagan that "he must appoint a blue-ribbon group to probe every aspect of the situation." Regan noted that "NASA and its supporters on Capitol Hill, backed by John Poindexter, wished to leave the investigation in the hands of the space agency." The issue was debated for the next several days. Those supporting a NASA-led investigation suggested that it was a way of demonstrating White House "faith in NASA's integrity." NASA indicated its intent to appoint a number of distinguished non-NASA individuals to the panel, with only a minority of its members coming from those parts of NASA not directly connected to the space shuttle program. Those supporting creation of an independent panel suggested that the context of the space program had "changed radically" since the time of Apollo, and that a NASA-led panel would not be as credible as an independent investigation. According to Regan, the decision was not made final until January 31, when, "on the way back to Washington from Houston after the memorial service the two points of view were put to the President and he opted for an independent commission."

An executive order creating the commission was drawn up on February 1 and signed by the president two days later. It declared that the commission "shall investigate the accident to the Space Shuttle Challenger, which occurred on January 28, 1986." In carrying out that investigation, the commission would "(1) review the circumstances surrounding the accident to establish the probable cause or causes of the accident; and (2) develop recommendations for corrective or other action based upon the Commission's findings and determinations." The commission was to report to the president and the administrator of NASA within 120 days.[12]

President Reagan appointed William Rogers, Nixon-administration secretary of state, as chair of the investigative group, which soon became known as the Rogers Commission. Members of the commission were rapidly recruited and announced on February 3. They included Apollo 11's Neil Armstrong as vice-chairman, astronaut Sally Ride, Nobel Prize-winning physicist Richard Feynman from the California Institute of Technology, and eight other men from diverse backgrounds. The executive order establishing the commission anticipated that it could have had as many as 20 members, but only one other person, "father of the Boeing 747," Joseph Sutter, was added beyond the 12 individuals initially identified. On the afternoon of February 3, President Reagan announced their appointment to the White House press corps, saying, "As we move away from that terrible day we must devote our energies to finding out how it happened and how it can be prevented from happening again. It's time now to assemble a group of distinguished Americans to take a hard look at the accident, to make a calm and deliberate assessment of the facts and the ways to avoid repetition."[13]

Was There White House Pressure to Launch?

In the days after the accident, rumors surfaced that the White House had put pressure on NASA to launch *Challenger* no later than January 28 because President Reagan intended to mention the launch in his State of the Union Address that evening. One of those determined to discover whether there was any substance to these rumors was Senator Ernest Hollings (D-SC), the ranking minority member of the Senate Committee on Commerce, Science, and Transportation. Hollings wrote to the top White House lawyer, Counsel to the President Fred Fielding, on March 20, saying that in the two months since the accident he had been arguing that "it must be determined, without a doubt, whether or not there was any pressure on NASA to launch on January 28." Hollings, "to assist in the process," requested the telephone records of nine White House staff likely to have had conversations with NASA's Headquarters, Kennedy Space Center, and Johnson Space Center between January 20 and January 28. Hollings noted that "to date, I have not been particularly pleased with White House cooperation concerning the shuttle tragedy" and that "rumors of White House involvement have persisted too long—and the time has come to be more open and to put this issue to rest once and for all." Hollings had been requesting hearings about the accident, but his committee's chairman, Senator John Danforth (R-MO) had refused his requests.

Fielding responded to Hollings on March 27. He advised Hollings that the White House telephone system was not capable of keeping a record of every call, either outgoing or incoming, but that his office had conducted an internal review and that this review "disclosed no record, recollection or indicia" of communications with respect to the date of the shuttle launch. Fielding added: "You make note of rumors of 'White House involvement' in the Shuttle tragedy. Such rumors appear to be totally based in irresponsible statements and charges. Our review reveals no evidence to substantiate these charges; I am advised that the Presidential Commission [investigating the accident] has reached the same conclusion." Senator Hollings's office released Fielding's letter. On April 3, White House spokesman Larry Speakes reacted to that release, adding the information that several White House staff members had indeed telephoned NASA in the eight days before the *Challenger* launch, but that none of these calls dealt with the timing of the launch.[14]

The White House also made available the final version of the State of the Union Address as it existed on January 28. No part of the address mentioned NASA, but one of those "American heroes" Reagan intended to mention was 21-year-old Richard Cavoli. Reagan was to say that "today, the science experiment he began in high school was launched on the space shuttle *Challenger*." When Reagan actually delivered the State of the Union Address on February 4, that line was changed to: "Richard, we know that the experiment you began in high school was launched and lost last week, yet your dream lives."[15]

Rogers Commission Gets to Work

The initial operating concept for the Rogers Commission was spelled out in the memorandum to Ronald Reagan transmitting the executive order establishing the group: "The Commission would work closely with NASA on the investigation ... The Commission would also review NASA's report and findings on the Space Shuttle accident and submit its findings on the validity of NASA's report." Within a few days after the accident, the NASA investigative team had identified the failure of the joint in the right side solid rocket booster as the technical cause of the accident.

The Rogers Commission members then learned at a February 14 hearing at the Kennedy Space Center that NASA managers had overruled the recommendation of the manufacturer of the solid rocket booster, Morton Thiokol, that *Challenger* not be launched because of the anticipated unprecedented cold temperatures. They quickly decided to change their relationship with the NASA investigative team, which included some of the managers involved in the launch decision. "Visibly disturbed" at what they had heard, the commission broadened the scope of its inquiry to include NASA's organization and management practices, shifting the focus from a technical failure to NASA itself. By the end of an executive session later on February 14, the commission issued a statement saying, "NASA's decision making process may have been flawed."

The commission decided to exclude from the investigation anyone from NASA who had been involved in the *Challenger* launch decision and soon added its own investigators, many coming from the Federal Bureau of Investigation, to its staff. That way the commission could conduct its own investigation, not just oversee what the NASA investigative team was doing. By the time it finished its work and submitted its report to President Reagan, the commission's full-time staff had grown to 43, plus some 140 part-time support staff. The commission had held 35 investigative sessions and over 160 individuals were interviewed during its investigation.[16]

The Rogers Commission work was concluded by early June. The commission's report, made public on June 9, concluded that "*the cause of the Challenger accident was the failure of the pressure seal in the aft field joint of the right Solid Rocket Motor*. The failure was due to a faulty design unacceptably sensitive to a number of factors." In particular, said the report, the low temperature at the time of launch may have caused the rubber-like material from which the seal's O-rings were made, designed to seal any unanticipated gaps in the primary seal, to lose the flexibility needed to perform its function. The commission also concluded that "the decision to launch the Challenger was flawed. Those who made that decision were unaware of the recent history of problems concerning the O-rings and the joint and were unaware of the initial written recommendation of the contractor advising against the launch at temperatures below 53 degrees Fahrenheit and the continuing opposition of the engineers at Thiokol after the management reversed its position." Had those in charge of the decision to launch been made aware of the debate between NASA and Morton

Thiokol about the possible impact of low temperature on the pressure seal, they almost certainly would not have allowed the launch to proceed.

The Rogers Commission made nine recommendations. Those recommendations dealt only with the space shuttle and its management by NASA; they did not address broader questions with respect to the policy context within which the shuttle should operate in the wake of the *Challenger* accident. On June 13, new NASA administrator James Fletcher, who had replaced William Graham in May, met with the president to discuss the commission's findings. After the meeting Reagan wrote to Fletcher, saying, "I have completed my review of the report from the Commission on the Space Shuttle CHALLENGER Accident. I believe that a program must be undertaken to implement the recommendations as soon as possible." Reagan asked NASA to report within 30 days on its implementation plan.[17]

It would take 27 more months, until the shuttle's return to flight on September 29, 1988, to fully comply with that presidential directive. In the interim, the space shuttle's central role in the U.S. space program was significantly modified.[18]

Notes

1. Bill Nelson's account of his shuttle flight is Bill Nelson with Jamie Buckingham, *Mission: An American Congressman's Voyage to Space* (New York: Harcourt Brace Jovanovich, Publishers, 1988). The quotes are from 31–36, 52. "Washington Roundup," *AWST*, 17; "Washington Roundup," *AWST*, February 18, 1985, 15, and "Shuttle Marketing Approach Spurs Hughes Launch Buy," *AWST*, November 4, 1985, 18.
2. Dennis Jenkins, *Space Shuttle: Developing an Icon, 1972–2013* (Specialty Press: Forest Lake, MN, 2016), Volume III, III-17. Memorandum from Larry Speakes to Donald Regan, January 14, 1986, Box 12, Outer Space Files, RRL.
3. This discussion of planning for Reagan's attendance at the *Challenger* launch is rather speculative. It is drawn from two unsigned documents: "Proposed Itinerary for the Visit of President & Nancy Reagan to the Kennedy Space Center, FL," undated, and a March 5, 1986, discussion of "Calls from Fanseen (NASA's White House liaison James Fanseen) re Shuttle Launch." That document says, "AHK did not return call." This likely refers to Donald Regan's assistant Alfred Kingon, who had initially proposed that Reagan attend the launch. The documents are in Box 15, Outer Space Files, RRL.
4. Most of this account of the *Challenger* launch comes from the "*Report to the President on the Space Shuttle Challenger Accident*," June 6, 1986 (the Rogers Commission). The quoted passage is on p. 18. The report is available at https://spaceflight.nasa.gov/outreach/SignificantIncidents/assets/rogers_commission_report.pdf.
5. Ronald Reagan: "Exchange With Reporters on the Explosion of the Space Shuttle Challenger," January 28, 1986. Online by Gerhard Peters and John T. Woolley, APP, http://www.presidency.ucsb.edu/ws/?pid=37635.
6. This account of Noonan's speech preparation comes from Peggy Noonan, *What I Saw at the Revolution: A Political Life in the Reagan Era* (New York: Random

House, 1990), 253–255 and Peggy Noonan, "In the White House," *AARP: The Magazine*, December 2015/January 2016, 28–29.

7. Memorandum from Larry Speakes to Donald Reagan, "Events following launch-pad fire January 27, 1967," January 28, 1986; memorandum from Katherine Ladd to David Chew, "PRESIDENTIAL ACTIONS FOLLOWING APOLLO ACCIDENT IN 1967," January 28, 1986, both in Box 14, Outer Space Files, RRL. Joseph Trento, *Prescription for Disaster: From the Glory Days of Apollo to the Betrayal of the Shuttle* (New York: Crown Publishers, 1987), 292.
8. Ronald Reagan: "Address to the Nation on the Explosion of the Space Shuttle Challenger," January 28, 1986. Online by Gerhard Peters and John T. Woolley, APP, http://www.presidency.ucsb.edu/ws/?pid=37646. Ronald Reagan, *The Reagan Diaries*, Douglas Brinkley, ed. (New York: Harper Collins, 2007), 386–387.
9. This account is based primarily on handwritten unsigned and undated notes, clearly composed in the immediate aftermath of the accident, in the files of Donald Regan's deputy, Dennis Thomas. Box 14, Papers of Dennis Thomas, RRL.
10. Noonan, *What I Saw at the Revolution*, 257–258 and Noonan, "In the White House," *AARP: The Magazine*, December 2015/January 2016, 29. Reagan, *The Reagan Diaries*, 387.
11. Ronald Reagan: "Remarks at the Memorial Service for the Crew of the Space Shuttle Challenger in Houston, Texas," January 31, 1986. Online by Peters and Woolley, APP, http://www.presidency.ucsb.edu/ws/?pid=36402. Reagan, *The Reagan Diaries*, 387.
12. Donald T. Regan, *For the Record: From Wall Street to the White House* (New York: Harcourt Brace Jovanovich, 1988), 333. John M. Logsdon, "Return to Flight: Richard H. Truly and the Recovery from the *Challenger* Accident," in Pamela Mack, ed., *From Engineering Science to Big Science: The NACA and NASA Collier Trophy Research Project Winners*, NASA SP-4219 (Washington, DC: Government Printing Office, 1998), 347–348. Undated handwritten notes, Box 14, Papers of Dennis Thomas, RRL. Executive Order 12546—Presidential Commission on the Space Shuttle Challenger Accident, February 3, 1986, https://www.reaganlibrary.gov/research/speeches/20386a.
13. Ronald Reagan: "Remarks Announcing the Establishment of the Presidential Commission on the Space Shuttle Challenger Accident," February 3, 1986. Online by Peters and Woolley, APP, http://www.presidency.ucsb.edu/ws/?pid=36502.
14. Letter from Senator Ernest Hollings to Fred Fielding, March 20, 1986, and letter from Fred Fielding to Senator Ernest Hollings, March 27, 1986, Box 14, Outer Space Files, RRL. Gerald Boyd, "White House Says It Found No Pressure to Launch," *NYT*, April 4, 1986, D18. No evidence that there was White House pressure to launch was discovered in the course of research for this study.
15. Letter from Jay Stephens to Ray Molesworth, April 29, 1986, Box 14, Outer Space Files, RRL.
16. Memorandum from James Miller to the President, "Proposed Executive Order Entitled "Presidential Commission on Space Shuttle Challenger Accident," February 1, 1986, RRL. The two best accounts of the *Challenger* launch decision are Diane Vaughn's book with that title, published in 1996 by the University of Chicago Press, and Allan McDonald and James Hansen, *Truth, Lies, and*

O-Rings: Inside the Space Shuttle Challenger Disaster (Gainesville, FL: University Press of Florida, 2009). Information about the Rogers Commission comes from Logsdon, "Return to Flight," 348, 353.

17. *Report to the President on the Space Shuttle Challenger Accident*, 73, 83. Ronald Reagan: "Letter to the Administrator of the National Aeronautics and Space Administration on the Recommendations of the Presidential Commission on the Space Shuttle Challenger Accident," June 13, 1986. Online by Peters and Woolley, APP, http://www.presidency.ucsb.edu/ws/?pid=3744.
18. The lengthy process of returning the space shuttle to service is detailed in Logsdon, "Return to Flight."

CHAPTER 19

Recovering from the Accident

Prior to the *Challenger* accident, the space shuttle had been the unquestioned focal point of U.S. claims to space leadership. Between 1981 and 1985, it had demonstrated unique capabilities unmatched by any other country. The shuttle was a source of national pride in technological achievement and provided a sense of future achievement, a point repeatedly made by Ronald Reagan in his many space-related remarks. The National Aeronautics and Space Administration (NASA) was striving to make the shuttle, due to its purported reliability and competitive pricing, the launch vehicle of choice for all those wanting access to space and a key to developing space into an arena for new economic activity.

In the aftermath of the accident, and particularly after the reviews by the Rogers Commission and others assessing the reality of the shuttle's operations and NASA's less-than-stellar performance in managing them, it became clear that there was a high degree of self-delusion in the way the shuttle had been perceived. Rather than a vehicle delivering on the promises that NASA had made to Richard Nixon had made in 1972, claiming that the space shuttle would "revolutionize transportation into near space by routinizing it" and would "take the astronomical costs out of astronautics," the shuttle was revealed to be a vehicle that was expensive and difficult to operate safely.

Organizational sociologist Gary Brewer commented after the *Challenger* accident that NASA during Apollo "came close to being the best organization that humans could create to accomplish" the goal that had been set for it, but by the time of the accident NASA was "no longer a perfect place." NASA, struggling to meet the promises that had accompanied shuttle approval on a constrained post-Apollo budget, was not the excellent organization that had sent Americans to the Moon. As the response to the accident was being debated, *The New York Times* editorialized "the loss of NASA's Challenger shuttle was no accident, striking a well-run agency like a bolt from the blue. It sprang directly from 15 years of hidden mismanagement, waste and fraud that

J. M. Logsdon, *Ronald Reagan and the Space Frontier*,
Palgrave Studies in the History of Science and Technology,
https://doi.org/10.1007/978-3-319-98962-4_19

have become routine in NASA operations … NASA needs a new policy, a new direction, and renewed dedication and competence." The accident was the culmination of space policy developments that had been set in motion by Richard Nixon's post-Apollo decisions 15 years earlier and which had not yet been reversed by the Reagan administration.[1]

The stage was thus set for a debate over post-*Challenger* launch policy. It would take seven months, with "lots of blood on the floor" according to one participant in the process, for final decisions on that policy to be made. Those decisions were then embodied in a December 1986 directive on "United States Space Launch Strategy."[2] That directive set out a new pathway for assuring U.S. access to space, and changed the role of the space shuttle, from being the primary means of access to space for all U.S. government payloads and an active participant in the global competition for commercial launch contracts, to being a system devoted to launching almost exclusively NASA payloads, carrying out various operations in orbit, and eventually launching, assembling, and servicing elements of the space station. In this last role, the shuttle came full circle to its original *circa*-1969 rationale as an essential element of facilitating permanent human presence in Earth orbit.

Initial Response

To begin the process of reacting to the *Challenger* accident, the National Security Council (NSC) called a meeting of the Interagency Group for Space [IG (Space)] for February 5. The call to the meeting noted: "It is necessary to review the availability of capabilities and resources to ensure the continuity of the U.S. space launch program."[3] Participants at the February 5 meeting created a new IG (Space) working group, cochaired by NASA and the Department of Defense (DOD), to prepare a report that would "consolidate all pertinent data on impact of the loss of the orbiter has on the U.S. space launch program." In addition, the report would "review expendable launch vehicle programs and determine necessary programmatic changes which ensure an early and effective use of boosters to augment the Shuttle." Finally, it would identify "costs [and] timing milestones," design a "strategy ensuring that early expenditures and program changes efficiently coordinate with mid and long term goals," examine "the role of private sector involvement," and determine "legislative options and what role the aerospace plane could play."[4]

At its initial meetings, the working group reached general agreement on what if approved would constitute a major policy change; the recommended new policy would say that the "shuttle would not compete with private sector ELVs for commercial and foreign launch business." This statement represented a reversal of the central issue that had been so heatedly debated during 1985 in the context of setting the shuttle price beginning on October 1, 1988 (see Chap. 17). However, what this agreement would actually mean in practice was quite unsettled. NASA wanted, after the shuttle returned to flight, to honor all existing commercial and foreign launch commitments, only gradually phasing out of commercial launch activity. This was likely to mean that the shuttle would be flying reimbursable payloads well into the 1990s, casting in doubt the prospects for enough new business to convince the commercial sector to invest the large sums needed to keep the expendable launch vehicle (ELV) production lines in being. Advocates of commercial ELVs argued that the sooner the shuttle stopped flying such payloads, the better; ideally, all existing NASA commitments would be immediately abrogated, creating a pool of payloads needing a launch, mainly communication satellites, that could form the initial basis for a commercial ELV industry.[5]

Working Group Report

Over the next several months, the NASA and DOD co-chaired study would be the focal point of Reagan administration discussion on options for recovering from the *Challenger* accident. After working its way through several drafts, the working group's report was available for Senior Interagency Group for Space [SIG (Space)] review by mid-April. The report identified four possible options for proceeding:

- Option 1—"Return to pre-accident baseline (four orbiters—no additional ELVs)"
- Option 2—"Three-orbiters—expanded ELVs"
- Option 3A—"Four-orbiters by FY 1990-expanded ELVs"
- Option 3B—"Four-orbiters by FY1991-expanded ELVs."

All agencies participating in the working group except the Office of Management and Budget (OMB) preferred Option 3A; OMB wanted to relieve short-term budget pressures by delaying construction of a new orbiter, and thus preferred Option 3B. The 1983 decision to slowly develop "structural spares" for a fifth orbiter (see Chap. 6) would turn out to have been prudent. If a replacement for *Challenger* was to be approved, NASA would not be starting from scratch, since the orbiter component production lines were still open and some progress had been made in constructing parts of an orbiter. The report identified two "unresolved issues": "How can required funding be provided?" and "When and how should NASA discontinue flying commercial and foreign payloads that do not require a manned presence?" Those two issues would dominate discussions in the months to follow.[6]

The DOD, anticipating a lengthy period of shuttle grounding, did not wait for decisions on the future role of the shuttle to initiate remedial action. Pete Aldridge, by April, Acting Secretary of the Air Force while still also director of the National Reconnaissance Office, told a Senate committee that month that the service hoped to quickly get additional funds to procure a "substantial number" of medium lift boosters to launch military payloads such as Global Positioning Satellites. These boosters would be in addition to the already-authorized larger complementary expendable launch vehicles (CELVs). The

Air Force procurement would guarantee a market large enough to convince the supplier of whichever booster, most likely General Dynamics's *Atlas*, was selected to keep its production line open. These ELVs, said Aldridge, could not only serve military needs but also be able to compete with *Ariane*space and other foreign suppliers of launch services for commercial launch contracts. The United States by pursuing this path, he told the committee, had a "golden opportunity" to promote a U.S. commercial ELV launch industry. With this proposal, Aldridge allied himself and the Air Force with the advocates of commercial ELVs, even though he also suggested that the service also supported a replacement orbiter to provide insurance against a future shuttle accident.[7]

How to Pay the Costs?

A thorny issue that had been identified in the immediate aftermath of the accident was how to pay the multibillion dollar costs of whatever launch recovery program was adopted. On March 31, officials from the concerned agencies met with OMB to address that issue. One important complication in providing the required funding was the existence of the "Balanced Budget and Emergency Deficit Control Act of 1985," generally known as the Gramm-Rudman-Hollings Act after its Senate sponsors. President Reagan had signed the act into law on December 12, 1985. It set deficit reduction targets for federal spending and, most significantly, called for automatic across-the-board reductions in spending, called "sequestrations," if those targets were not met. Adding to the Federal budget the unanticipated billions of dollars required to recover from the shuttle accident could trigger the sequestration process, an outcome the White House very much wanted to avoid. Finding a way to finance the launch recovery plan in the face of this constraint would become a vexing policy and budgetary issue.

At the March 31 meeting, the OMB representative indicated that "full offsets [reductions in spending for other programs] are required if the President and the Congress are not to abrogate their policies and responsibilities under current law." If no offsets were identified, that would "signal to the Congress that fiscal responsibility no longer matters." But "if full offsets are required, space program needs must be traded against other priority Administration programs." NASA's budget was already under pressure. Reductions proposed in the Fiscal Year (FY) 1987 budget request that Ronald Reagan had sent to the Congress in February had already caused an 18-month slip in the space station schedule. In addition, NASA had been required to reduce its FY 1986 budget outlays by $223 million to meet Gramm-Rudman-Hollings requirements. Finding offsets within the reduced NASA budget in the current and future FYs to cover the estimated $2.8 billion cost of a replacement orbiter would be extremely challenging.[8]

SIG (Space) Meets

National Security Adviser John Poindexter called a meeting of SIG (Space) for April 18. Poindexter had replaced Robert McFarlane in the top NSC position in December 1985. He was familiar with space issues, having been involved with them since he joined the NSC staff in 1981 and chairing many of the meetings during the 1985 debate on shuttle pricing. Gil Rye had also left the NSC staff in September 1985. He had been replaced as the top space policy staff person by another Air Force colonel, Gerald "Jerry" May. May would turn out not to be the skilled bureaucrat that Rye had been. Between them, Poindexter and May would struggle to manage the launch recovery process.

In his call to the meeting, Poindexter said that it would "discuss three major issues related to the loss of Challenger." The first issue would be "identifying funding offsets ... We cannot submit an unfunded request to the Congress." The second issue would be "deciding policy changes to relieve NASA of the responsibility for launching all commercial and foreign payloads." Poindexter added: "While all SIG participants agree to privatization of a commercial ELV capability, how and when such systems will be made available is an issue." The third issue was the timing of the release of the IG (Space) working group report in terms of its relationship to the release of the Rogers Commission report.[9]

The first order of business as the SIG (Space) meeting was called to order was a report from Pete Aldridge on the explosion that morning, just five seconds after launch, of a Titan 34D ELV carrying a National Reconnaissance Office photo-intelligence satellite. This was the second failure in a row of the Titan; with the shuttle grounded and these two failures, the United States had lost its capability to launch critical reconnaissance satellites and other large payloads, emphasizing the urgency of agreeing on a recovery strategy.

The next 45 minutes of the meeting were spent discussing the ELV commercialization issue. Secretary of Transportation Dole told her colleagues that the heads of two ELV manufacturers, Martin Marietta and General Dynamics, were "very interested in entering the commercial ELV market, but they feel they need a clear signal that the Government will not compete with them for this business." The Department of Transportation felt that a "date certain" for ending the use of the shuttle was "imperative." Representing NASA at the meeting was Acting Administrator Bill Graham. Graham said that he agreed with the decision to eventually take the shuttle out of the business of commercial launches, but that "the commercial ELV capability does not exist now, and the transition is, therefore, a problem." Graham believed that NASA had "an obligation to support the US COMSAT industry. Setting a date certain [for withdrawing the shuttle from the commercial launch market without assurance that U.S. ELVs would be available] would not show U.S. support for this industry," since manufacturers of communication satellites need some guarantee with respect to when and how their satellites would be launched. Deputy Secretary of Commerce Bud Brown agreed that "the ELV industry needs assurance that the Government will not compete"; he noted with respect to crafting a recovery plan that "the problem is that nobody trusts anyone else."[10]

That the Reagan administration was having difficulty coming up with a viable approach to financing an accident recovery plan soon came under "blistering, bipartisan attack" from Congress. The chairman of the Senate Subcommittee on Science, Technology, and Space, Slade Gorton (R-WA), called the failure to request recovery funds "inexcusable." His Democrat counterpart, Don Riegle (D-MI), suggested that the nation's space policy had become enmeshed in a "classic bureaucratic tangle," with the result that "at the very highest level we're drifting."[11]

A New (Old) NASA Administrator

Correcting that drift would not be Acting Administrator Graham's responsibility. The White House on March 6 had announced that President Reagan would nominate James Fletcher as the new NASA administrator. Fletcher had already served in that position from April 1971 to May 1977, as the space shuttle was approved and developed.

There had apparently been some thought among those in the White House who had supported his nomination as deputy administrator that Graham should be promoted to administrator on a permanent basis. However, there was bad blood between Graham and James Beggs, who had resisted Graham's appointment, and this antagonism effectively blocked that possibility. Beggs told the White House that he would not leave his position as administrator until he was assured that Graham would not get the job. Beggs received that assurance, and on February 25 resigned as administrator, clearing the way for his replacement.[12]

As it waited for Beggs to resign, the White House had been evaluating candidates for the position. Former astronaut Neil Armstrong, at that point serving as vice-chair of the Rogers Commission, Apollo 10 commander General Thomas Stafford, Pete Aldridge from the Air Force, and General Lew Allen, director of NASA's Jet Propulsion Laboratory, were among those under consideration. By early March, Fletcher's name had risen to the top of the candidate list; he had strong support from the space industry and from members of Congress, in particular Senator Jake Garn. Garn was pressuring the White House to find a "white knight" to take over NASA, and Fletcher in his mind fit that description. Garn knew Fletcher well. Garn was from Utah, where Fletcher had been president of the University of Utah for eight years before coming to NASA in 1971, and Garn and Fletcher attended the same Mormon Church in the Washington area. As the White House checked potential nominees for the NASA job with key people in the Senate and House, they found that Garn had a strong preference for Fletcher and objections to several of the other candidates. Both because he chaired the Senate subcommittee handling NASA's appropriations and because he had a year earlier flown aboard the shuttle, Garn's congressional colleagues deferred to his views; the White House also accepted them.

Fletcher was thus asked to meet with Chief of Staff Don Regan to interview for the NASA position. He told Regan he did not want the job, but he also indicated that it would be difficult to turn down a direct request from President Reagan to take it. That request came in the form of a brief telephone call from the president on the afternoon of March 6.

Earlier on that day, Fletcher had told a reporter from *The New York Times* that he did not want the job and would take it only if no other suitable candidate could be found. Fletcher told the *Times* reporter that the NASA position was "not good for me" because he would have "to give up all the other activities I'm involved in." Fletcher had headed a study of the technical feasibility of Reagan's Strategic Defense Initiative proposal and was still involved in the undertaking, and he was chairman of a board overseeing the cleanup of a damaged nuclear reactor at Three Mile Island in Pennsylvania. He had previously said that he "would have to be dragged kicking and screaming" back to NASA. But a few hours after his telephone conversation with Reagan, the White House press office announced that Fletcher would be the president's nominee as NASA administrator.[13]

Fletcher was sworn in by President Reagan on May 12. Bill Graham returned to his position as deputy administrator. At the swearing-in ceremony, Reagan told those gathered in the Roosevelt Room of the White House: "Over 25 years ago the people of this great nation made a commitment to go where no man has gone before and to do what no man has ever done before ... Everyone knew this was a great nation, and great nations pursue great efforts.... In the 1600's another explorer with unlimited vision and an insatiable search for knowledge recognized the sacrifices of others. Sir Isaac Newton said, 'If I have seen further, it is by standing upon the shoulders of giants.' Well, the giants of the world's finest space program, we're proud to say, are Americans." Speaking directly to Fletcher, Reagan said: "Jim, you're coming on board at a time when NASA and the country need you. We need a steady hand on the tiller."[14]

Fletcher had already taken up residence at NASA in early April to prepare himself for taking over the space agency leadership, but he could not officially represent NASA until he was confirmed and sworn in. Until then, he worked with senior NASA staff, getting briefed on the various issues he would face once he took charge of the agency. One of those preparing Fletcher after his absence of nine years was Norm Terrell, NASA's associate administrator for policy. Terrell told Fletcher:

> We are faced with the political problem that DOT, supported by DOC, OSTP, OMB and possibly DOD, is trying to take advantage of the present situation to undo the 1985 pricing decision. In fact, they want to remove the Shuttle from the commercial market at any price.
>
> Bill Graham has already offered to quit marketing the Shuttle until existing customers have been taken care of. This would give the U.S. entrepreneurs a period of time to get themselves established ... However, this was not enough for DOT and their friends. Their proposal is that we announce that NASA will fly no more commercial communication satellites after October 1, 1988—period.
>
> I have been asking myself and others for two years what job DOT might be given in this business that would provide them useful work to do, give them a stake in the [space] enterprise and get them out of our hair. No one has come up with a good idea. In fact, DOT's approach—that Shuttle marketing is antithetical to their role—makes it awfully hard.[15]

As he prepared to enter the interagency fray as NASA's new top official, James Fletcher was quite aware of the conflict over the shuttle's future role that he would inherit. What he did not anticipate was the changed context for determining that role, with multiple new actors competing with NASA to influence space policy decisions.

Decisions Deferred

A meeting of the NSC with President Reagan in the chair to discuss "decisions to reconstitute the U.S. space launch capabilities" was called for the afternoon of May 15. Escalating the decision-making level to the NSC itself, rather than have a recommendation to the president originate at the SIG (Space) level, indicated the importance assigned to the question of how to proceed. Attached to the meeting call was a draft National Security Decision Directive (NSDD) on "U.S. Space Launch Strategy." The directive indicated that "NASA will procure a replacement orbiter," that "the STS will no longer provide launch services for commercial and foreign payloads exclusively for revenue-generating purposes," and that "the Commercial Space Working Group of the Economic Policy Council (EPC) will develop a transition plan that considers the method and timing to implement this policy." (As noted in Chap. 16, President Reagan in April 1985 had eliminated the Cabinet Council system, creating an Economic Policy Council chaired by Secretary of the Treasury James Baker that assumed the space policy role previously played by the Cabinet Council on Commerce and Trade.) The agenda for the NSC meeting called on John Poindexter to highlight the issues for decision, Jerry May to present the findings of the IG (Space) Working Group, and Randy Davis, Fred Khedouri's successor as OMB's associate director for natural resources, energy, and science, to discuss funding options (Fig. 19.1).[16]

The meeting was characterized by "heated debate." Apparently Poindexter and May had assumed on the basis of prior discussions that "everybody had already made up their minds to approve a replacement orbiter. This assumption was 'just not correct.'" In particular, White House Chief of Staff Don Regan, who had not previously been involved in the accident recovery discussions, "repeatedly asked pointed questions about the need for a new shuttle orbiter, suggesting that it might be outdated," since it would be based on 1970s technology but not be available until the early 1990s. There was also some discussion of the possibility of a privately funded replacement orbiter. There was at the meeting, "despite strong opposition from NASA," top-level agreement on taking commercial and foreign satellites off the shuttle. One press account of the meeting noted that "this would represent a major retreat" from "the administration's earlier goal to make the shuttle program finance itself." But there was no agreement on procuring a replacement orbiter, and the draft NSDD was not approved.

The meeting was also the first time that Ronald Reagan had been "directly presented with the results of a contentious internal administration review of the future of the space program after Challenger." As the meeting ended, mak-

Fig. 19.1 At a May 15, 1986, NSC meeting, NSC Director for Space Programs Col. Gerry May (standing) briefed the attendees on plans for recovering from the *Challenger* accident. Pictured in this photograph are (on near side of Cabinet Room table from l to r): Secretary of Defense Caspar Weinberger, President Reagan, Secretary of State George Shultz, NASA Administrator James Fletcher, and two unidentified individuals. On the other side of the table (l to r) are: National Security Adviser Admiral John Poindexter, Director of Central Intelligence William Casey, and Army Chief of Staff General John Wickham, Jr. (Photograph courtesy of Reagan Presidential Library)

ing no decisions, "the president asked for more information." Chief of Staff Regan directed the NSC to prepare a "new assessment" of the space launch issue. Writing in his diary that evening, Reagan noted that "getting NASA back in operation" was "a tough problem & a costly one. We arrived at no answer."

In the wake of the meeting, *The New York Times* reported "a feeling was said to be spreading throughout the Government, the science community and the aerospace industry that the future of the space program hung in the balance." The *Times* added: "However the issues are resolved … the shuttles will no longer play the dominant role they had enjoyed in the American space program. They may still be the flagships, but hardly the only ships, or necessarily the preferred ships, of the many and diverse users of space."[17]

As the results of the NSC meeting were publicly reported, Reagan's indefatigable correspondent Douglas Morrow wrote to the president, saying, "I've been hearing that the 'White House' is now leaning toward a delay regarding the construction of a new Shuttle." Morrow wondered whether "this is your position or someone else's," but at any rate would be "a big mistake." Reagan quickly responded, telling Morrow "my desire is to go forward if it is at all possible." He added: "We do have some problems, the financial one of course, but a number of others such as the backlog of various machines to be put into space. We've only been able to touch on that and whether a spurt in non-manned launchers should be used to reduce the backlog." The president told Morrow: "We haven't come together to make a final decision yet," but "I'm for going ahead with a shuttle unless the things I've mentioned here make it impractical and unwise." As had been the case in terms of the space station decision in 1983, Ronald Reagan knew the outcome he wanted, but would wait for the results of a full examination of the issue before he made his final choice.[18]

A Privately Funded Orbiter?

The concept of private sector funding to construct a space shuttle orbiter had first received attention within the Reagan administration in 1982, when SpaceTran, a firm founded by economist and entrepreneur Klaus Heiss, had proposed such an undertaking. With the White House decision that an additional, fifth, orbiter was not needed, the idea lost steam. In 1985, Astrotech International, a company founded by the retired head of Rockwell International, Willard "Al" Rockwell, had reversed the proposition, offering to lease or buy from the government one or more of the four existing orbiters and use it as the basis of a commercial space transportation business. Rockwell had created an Astrotech subsidiary called Space Shuttle of America for this purpose. That idea also went nowhere.[19]

Then, in the immediate aftermath of the *Challenger* accident, Rockwell through Astrotech offered to provide $1.5 billion of private funds to build a replacement shuttle. The *Wall Street Journal* reported on February 6 that "Astrotech would own the new shuttle, lease it to NASA and be responsible for getting business customers for the whole shuttle fleet. In return, Astrotech would receive 33% of all earnings from the shuttle fleet." Rockwell met with Don Regan on February 10 to call attention to his proposal, and on February 25 wrote Isaac Gillam, IV, NASA's assistant administrator for commercial programs, spelling out Astrotech's terms for the offer in some detail. Rockwell told Gillam: "We feel very strongly … that the use of willing investors' dollars, instead of taxpayers' dollars, is a preferable alternative." By this time, Astrotech had changed the name of Space Shuttle of America to General Space Corporation, and it was under that banner that Rockwell's proposition was put forward.[20]

The proposal had received little or no attention as IG (Space) and SIG (Space) debated how best to react to the shuttle accident, but apparently Regan raised the possibility of private financing at the May 15 NSC meeting, and the idea caught President Reagan's interest as a way of resolving the uncertainty regarding how to pay for a replacement orbiter. On May 28, Rockwell wrote to Regan, repeating "my offer to finance the replacement shuttle and lease it back to the government."[21]

NASA Deputy Administrator Graham wrote Poindexter on June 18, answering "some of the questions you asked of Dr. Fletcher concerning the opportunity for private funding of a Shuttle orbiter." Graham pointed out that "the viability of a privately financed orbiter is directly linked to the Shuttle fleet continuing to serve the needs of the commercial sector." With the decision then emerging from the interagency deliberations to prohibit the space shuttle from launching commercial payloads, noted Graham, a privately operated shuttle could not depend on a revenue stream from commercial customers, and thus "the private proposer would have to recover his capital investment, operating costs, and profit from revenues realized from serving the U.S. government exclusively."[22]

Even with this significant limitation on the viability of a privately financed orbiter, the idea apparently remained attractive to Regan. When he met with Fletcher on July 11 to discuss a planned briefing to President Reagan on NASA's response to the Rogers Commission, he asked again about the concept. Later that day Fletcher sent an impassioned letter to the chief of staff, saying that "I may not have made clear my strong position on this." He told Regan: "I have analyzed the question of private versus conventional financing of the replacement from all possible aspects. And I have concluded that private financing, at this time, would be a serious mistake." Fletcher suggested that "it is an unfortunate but inescapable fact of life that NASA, its program, and America's future in space is [*sic*] in a period of serious crisis ... We have already lost six precious months. To add to this the many months that the consummation of private financing would require would only seriously compound NASA's problems." He argued: "We are at war with the Soviets for preeminence in space ... It would be very short-sighted to allow a very dubious dollars-and-cents equation, necessarily accompanied by inordinate delays, to prejudice our ability to regain our forefront status in this space conflict." Fletcher suggested that "if we are asked to seek private financing ... I am afraid it will appear as if the President is giving NASA and its program a vote of dubious confidence." He told Regan: "I don't want, and I know you don't want, this President ... to be charged with a possible failure ... that might result from our now treating the financing of an essential fourth orbiter like a commercial business deal instead of as a national priority." This kind of complaining tone was to characterize many of the interactions between Fletcher and the White House over the next three years.[23]

Further Delaying the Decision

It took almost three months after the May 15 NSC meeting to finally decide that NASA would procure a replacement orbiter. In the interim, the various parties with interests in the launch recovery policy and plans made their views known, and NASA and OMB engaged in a continuing interchange with regard to the comparative benefits of an approach that placed primary future dependence on ELVs and one that was more balanced between the use of a four-orbiter shuttle fleet and ELVs. NASA argued strongly that depending on only three orbiters over their anticipated 15-year lifetime was too risky, given both the possibility of losing another orbiter and the likelihood that one of the orbiters could be out of service for extended periods for maintenance or upgrades. Procuring a fourth orbiter would provide insurance against these contingencies and thereby assure that the shuttle would be available to make its unique contributions to U.S. space capabilities and leadership.

At the May NSC meeting, NASA had suggested modifying the draft NSDD language on prohibiting the shuttle from launching "revenue-generating" payloads to say that the space shuttle would no longer "compete with a viable and competitive U.S. private sector launch services industry." This wording would have allowed NASA, until the commercial ELV industry was "viable and competitive," to continue to launch those commercial payloads already under contract and perhaps even to continue to compete with U.S. ELVs as well as *Ariane* for future launch contracts.

This suggestion was not acceptable to the advocates of a commercial ELV industry, who saw the NASA language as a ploy to discourage manufacturers from making the investments needed to get their production lines restarted and to market their launchers. Secretary of Commerce Malcolm Baldridge wrote to Regan on June 4, saying that if the NASA change was accepted, it "would doom the development of an ELV industry." He argued for a "clear and concise" statement that would provide "the concrete assurances that the ELV industry needs." Baldridge also told Regan that "I agree with your position in the NSC meeting that we are not getting much by purchasing a replacement shuttle," and thus "I oppose procuring a replacement orbiter at this time." Secretary of Transportation Elizabeth Dole also weighed in. Writing Regan, she made the point that the companies producing ELVs "have asked for a clear and unambiguous signal that they will no longer have to compete with the Federal government." She thus argued for retaining the language in the draft NSDD, with perhaps the phrase "phase out" substituted for "will no longer provide." Such a change would be acceptable, "if, and only if the Economic Policy Council is charged with expeditiously determining both the timing and method" for such a phase out.[24]

In the weeks after the May 15 NSC meeting, Poindexter and May attempted to develop a redrafted option paper and NSDD that responded to the questions raised at the NSC meeting and thus could form the basis for a presidential decision. *The Washington Post* on June 10 reported that a "National Security Council session is to be held soon" to deal with the replacement orbiter choice. But interagency disputes over the transition in the shuttle's future role and NASA-OMB disagreement over the financing of a replacement orbiter continued to block the level of agreement needed to once again bring the issue before the president for decision. The NSC staff was not able to forge enough of a consensus to move forward.[25]

Other parts of the government were taking more decisive action. The Congress was close to passing an emergency supplement of $526 million to the FY1986 NASA budget. These funds would pay for the steps needed to accommodate the recommendations of the Rogers Commission, including the redesign and testing of the solid rocket motor joint and modifications to the remaining three orbiters, but included no money for a replacement orbiter. The Congress also increased the DOD budget by $309 million so that the Air Force could order an additional 13 CELVs and 12 medium-size ELVs, as the national security community accelerated its move toward lessened dependence on the space shuttle. The Air Force also decided to place its California launch facility for the shuttle at Vandenberg Air Force Base in "mothball" status, since it would be a number of years, if ever, before there would be a shuttle launch from that facility. Before the Challenger accident, the first launch from the Vandenberg site had been planned for summer 1986. The Vandenberg facility had cost several billion dollars to develop; it was now uncertain whether there would ever be a payoff from that investment.[26]

NASA was also making changes that affected the replacement orbiter decision. On June 19, the space agency announced that it was canceling plans to use the powerful Centaur upper stage to launch heavy payloads from the shuttle to their final orbits. This step was taken primarily on safety grounds; the Centaur was fueled by highly flammable liquid hydrogen, and in the post-*Challenger* environment NASA decided that its use in the shuttle payload bay was excessively risky. This meant that other approaches to launching ambitious planetary spacecraft such as *Galileo* and *Ulysses* would have to be pursued. With the Rogers Commission report in hand, NASA, in order to have the time to respond to the commission's recommendations, changed its estimated downtime before the shuttle could return to flight from 12 months to at least 18 months, and more likely two years.

Another NSC meeting on the launch recovery policy was tentatively scheduled for June 24, but did not take place. *The Washington Post* reported that "although the White House had announced that Reagan might make a decision this week on the space program … it now appears that he will not act until next month at the earliest." White House spokesperson Larry Speakes was quoted as saying, "The president does favor a fourth orbiter … but he hasn't heard all the arguments. He may change his mind." Reagan himself had had the following interchange at a June 11 press conference:

> **Q.** Mr. President, NASA is awaiting your decision on how to replace the Challenger spacecraft. Could you tell us tonight how you would finance a fourth orbiter? And if you can't tell us that, could you explain what's holding up your decision?
>
> **The President.** Well, for one thing, we're studying the report that we've received, and there are many things that have to be decided. There is a backlog now of space cargo that is supposed to be up there. And we have the problem of determining whether we shouldn't increase the number of unmanned launchers, for many of those things, that could put them in space, and then see where we can come with the—believe me, I want to go forward, and I think we all do, with the shuttle program. But how soon we can get to that is a question, and in the meantime should we emphasize more of the unlaunched [unmanned] to move on that backlog that we have of cargo that needs to get into space. So, I don't have an answer for you on this except that, yes, I think we should go forward with another shuttle.

> **Q.** If I may follow up, sir: Would you insist that fourth orbiter incorporate all the recommendations of the Rogers commission when it is built?
>
> **The President.** Well, again, we're still in the midst of studying that now that we've just received it. So, I can't answer something as specific as that about that.[27]

Fletcher, Poindexter, and OMB Director Jim Miller met on June 23 to discuss the differences between NASA and OMB. Fletcher brought to the meeting a paper that described "our points of agreement and positions where we differ." Areas of agreement, the paper asserted, included that "the first priority task in reconstitution of national space launch capability is to restore the Space Transportation System, with its three remaining orbiters, to operational status at the earliest date consistent with safety" and that "the capability of the lost orbiter should be replaced." NASA and OMB still differed on how best to replace that launch capability. The NASA position was that "replacement of the fourth orbiter is the most economical way to provide space launch capability" and that procurement of that replacement orbiter should begin "in FY1986 for a FY1990 delivery." With respect to how to fund orbiter replacement, "NASA is unable to assume the total burden of this unanticipated funding impact without sacrificing the balance of the civil space program." The NASA paper characterized the OMB position as being "to defer replacement orbiter procurement until FY 1987 and to undertake this activity with additional funding from NASA offsets."

OMB Director Miller wrote Fletcher later the same day, saying that given the short time remaining in FY 1986, which would end on September 30, "the option of a commitment to a 1986 start" on a replacement orbiter "is no longer credible." Miller said that he was concerned about "the impact of any commitment now to orbiter replacement" and that "we do not yet have a sufficiently clear picture of the totality of these outyear needs" to determine whether "it would be wise to pursue the fourth shuttle option." Miller asked NASA to provide detailed "payload sequences under five scenarios." Providing this information was bound to take some days. Miller told Fletcher: "We expect Presidential decisions on the orbiter replacement issues will be considered in conjunction with your 30 day follow-up report to the President on the Rogers Commission recommendations."[28]

Another contributor to the delay, according to some involved in the decision process, was the way in which space policy decisions were being made. Earlier in June, *The Washington Post* reported that Don Regan was expressing "dissatisfaction ... with delays by a larger interagency group [SIG (Space)] studying the space program's future." Regan, with his "take charge" style, was making himself central to the decision process. The *Post* commented that "the U.S. space program is still a Humpty Dumpty that Washington policy-makers still don't know how to put back together. It's not just that they can't agree. Part of the problem is that there are few good choices." Another problem, according to another *Post* article, was the weak performance of National Security Adviser Poindexter in managing the space policy process. Poindexter was depicted as being "slow to resolve internal disputes, often overlooks important political considerations and has not mastered the art of explaining complex issues to a president accustomed to dealing in broad concepts rather than details." According to some of his colleagues, Poindexter had "allowed key decisions on the future of the U.S. space program to languish." The fact that Jerry May, Poindexter's staff person on space issues, was less effective than his predecessor, Gil Rye, was also likely a contributing factor to what was characterized as "gridlock" on how to move forward.[29]

The *Post* also noted that "in the view of many, the question of Challenger's replacement should be part of a much broader reassessment of the civilian space program." Instead, "the fourth orbiter has become an emotional symbol of support for the space program ... and all other actions wait on this decision." Also, "decision-makers have to come to grips with the notion that the country can't have the champagne space program that it has come to expect on the beer budgets it is providing."[30] The *Challenger* recovery process would drag on for several more months, with continuing conflict over both procuring a fourth orbiter and the character of the transition away from using the shuttle to launch commercial and foreign payloads.

Notes

1. Gary D. Brewer, "Perfect Places: NASA as an Idealized Institution" in Radford Byerly, ed., *Space Policy Reconsidered* (Boulder, CO: Westview Press, 1989), 169; "Escape the NASA Nightmare," *NYT*, April 25, 1986, https://www.nytimes.com/1986/04/25/opinion/escape-the-nasa-nightmare.html.
2. National Security Decision Directive 254, "United States Space Launch Strategy," December 27, 1986, https://www.reaganlibrary.gov/sites/default/files/archives/reference/scanned-nsdds/nsdd254.pdf.
3. Memorandum from William Martin, executive secretary, National Security Council, to members of the Interagency Group (Space), "IG (Space) Meeting – February 5, 1986," undated but early February 1986, CIA-RDP92B00181R0015014600-2, CREST.
4. Memorandum from Rodney McDaniel to IG (Space) Members, "IG (Space) Actions," February 6, 1986, CIA-RDP92B00181R001701590054-9, CREST.
5. Intelligence Community Working Paper, IG (Space) Meeting (3/6/86), undated but March 5, 1986, CIA-RDP92B00181R00150460002-2, CREST.
6. Interagency Group on Space, "Recommendations for the U.S. Space Launch Program," April 14, 1986, CIA-RDP88G01117R000300780001-2, CREST., ES-4-ES-6, ES-8, NSDD-1, NSDD-2.
7. "USAF Suggests Ending Use of Shuttle for Commercial Satellite Launches," *AWST*, April 21, 1986, 29.
8. Norman Terrell, "Post-Challenger Interagency Group Facts and Figures," undated but early April 1986, Box 10, Papers of James Fletcher, NARA. Memorandum from Randall Davis to (redacted), Deputy Director for Programs and Budget Staff, Intelligence Community Staff, "Funding Options for U.S. Launch Program Recovery," April 1, 1986, with attached draft memorandum on "Funding for National Space Launch Program Recovery," CIA-

RDP88G01117R000300780001-2, CREST. "Washington Roundup," *AWST*, January 20, 1986, 17.
9. Memorandum from John Poindexter to members, Senior Interagency Group for Space, "SIG (Space) Agenda," April 17, 1986, https://fas.org/irp/offdocs/nsdd/nsdd-144.htm.
10. Author's name redacted, deputy director for Policy, Policy and Planning Staff, Intelligence Community, Memorandum for the Record, "SIG (Space) Meeting, 18 April 1986," April 24, 1986, CIA-RDP88G01117R000300780001-2, CREST.
11. Laurie McGinley, "Senators Assail Reagan for Failure to Set Budget Plan to Replace Lost Space Gear," *The Wall Street Journal*, May 9, 1986.
12. Gaylord Shaw, "Feud Blocks New NASA Leadership," *Los Angeles Times*, February 22, 1986, 19.
13. This account of Fletcher's choice is taken from Joseph Trento, *Prescription for Disaster: From the Glory of Apollo to the Betrayal of the Shuttle* (New York: Crown Publishers, 1987); Phillip Boffey, "Fletcher, Reluctant to the End, Picked to Head NASA a Second Time," *The New York Times*, March 7, 1986, https://www.nytimes.com/1986/03/07/us/fletcher-reluctant-to-the-end-picked-to-head-nasa-a-2d-time.html; Storer Rowley, "Reagan Recalls Fletcher to Head NASA," *Chicago Tribune*, March 7, 1986, http://articles.chicagotribune.com/1986-03-07/news/8601170439_1_james-beggs-james-fletcher-head-nasa; and an interview with Jeff Bingham, a long-time member of Garn's staff, on December 7, 2017.
14. "Washington Roundup," *AWST*, April 28, 1986, 15; "Escape the NASA Nightmare," *NYT*, April 25, 1986; Stuart Diamond, "NASA Wastes Billions, Federal Audits Disclose," *NYT*, April 23, 1986, 1. Reagan's remarks can be found at https://reaganlibrary.gov/archives/speeches/51286a.
15. Memorandum from Norm Terrell to Dr. Fletcher, May 2, 1986, Box 10, Papers of James Fletcher, NARA.
16. Memorandum from Rodney McDaniel, "National Security Council Meeting on U.S. Space Launch Capabilities," May 12, 1986, with attached draft NSDD, CIA-RDP88GO1117R000501600003-6, CREST.
17. David Hoffman and Michael Isikoff, "Regan Questions Need for New Space Shuttle," *WP*, May 16, 1986, A1; John Noble Wilford, "President is Reported to Be Near Approval of a New Space Shuttle," *NYT*, May 25, 1986, A1. Ronald Reagan, *The Reagan Diaries*," Douglas Brinkley, ed. (New York: Harper Collins, 2007), 411.
18. Letter from Douglas Morrow to R.R. (Ronald Reagan), May 21, 1986, and letter from "Ron" (Ronald Reagan) to Douglas Morrow, May 29, 1986, both in Box 14, Outer Space Files, RRL.
19. Letter from James Bayless to Craig Fuller, January 17, 1985, with attached background memorandum, Box 12, Outer Space Files, RRL. "Astrotech's Plan to Lease Shuttle Looks Dead for Now," *Defense Daily*, February 1, 1985, 171.
20. Bryan Burrough, "Astrotech Offers Financing to Build New Space Shuttle," *Wall Street Journal*, February 6, 1986, 4; letter from James Cicconi to Donald Reagan, February 11, 1986, and hand-delivered letter from Willard Rockwell to Isaac Gillam, IV, February 25, 1986, Box 14, Outer Space Files, RRL.
21. Letter from Al Rockwell to Donald Regan, May 28, 1986, Files of Valerie Neal.

22. Letter from William Graham to John Poindexter, June 18, 1986, Box 4, Papers of James Fletcher, NARA.
23. Letter from James Fletcher to Donald Regan, July 11, 1986, Box 14, Outer Space Files, RRL. "Fletcher Cites Soviet Threat in Asking for New Orbiter," *AWST*, July 28, 1986, 23.
24. Letter from Malcom Baldridge to Donald Regan, June 4, 1986, and letter from Elizabeth Dole to Donald Regan, undated, Box 14, Outer Space Files, RRL.
25. Kathy Sawyer, "Decisions for Reagan," *WP*, June 10, 1986, A15.
26. Memorandum from Jim Miller for Don Regan, "Further Information on Shuttle/ELV Funding in Urgent Supplemental," June 17, 1986, Box 14, Outer Space Files, RRL.
27. David Hoffman, "Reagan Likely to Delay Space Program Decisions," *WP*, June 24, 1986, A7. Ronald Reagan: "The President's News Conference," June 11, 1986. Online by Gerhard Peters and John T. Woolley, APP, http://www.presidency.ucsb.edu/ws/?pid=37431.
28. Letter from James Fletcher to James Miller, June 23, 1986, Files of Valerie Neal, and letter from James Miller to James Fletcher, June 23, 1986, Files of Darrell Branscome. Dr. Branscome is a veteran leader in the space community, working for NASA, the Congress, and industry. In 1986, he was a senior adviser to NASA Administrator Fletcher dealing with the post-*Challenger* recovery efforts. I am grateful for his willingness to share material from his files.
29. Kathy Sawyer, "Decisions for Reagan," *WP*, June 10, 1986, A15; Kathy Sawyer, "U.S. Policy-Makers Struggle to Rechart NASA's Course," *WP*, June 22, 1986, A1; Lou Cannon and David Hoffman, "Reagan Adviser Poindexter Under Criticisms by Colleagues," *WP*, June 24, 1986, A1.
30. Kathy Sawyer, "U.S. Policy-Makers Struggle to Rechart NASA's Course," *WP*, June 22, 1986, A1.

CHAPTER 20

Correcting a Policy Mistake

The debate regarding producing another shuttle orbiter to replace *Challenger* continued through most of summer 1986, as did the conflict over the future role of the shuttle. If the shuttle were to be removed from its role in launching commercial and foreign payloads, after losing its planned monopoly in 1985 on national security launches, the basic premise upon which the shuttle program had been approved and carried out before the *Challenger* accident—that the vehicle could provide regular and affordable access to space for a wide variety of payloads, not only for the U.S. government but also for commercial and foreign users—would be abandoned. The decision to make the shuttle the sole U.S. means of access to space would have been recognized as a very significant policy mistake.[1]

Steps Toward a Decision on a New Orbiter

On July 14, 1986, Ronald Reagan met with Don Regan, John Poindexter, and James Fletcher to discuss the National Aeronautics and Space Administration's (NASA) plans for implementing the recommendations of the Rogers Commission report. By this time, NASA was forecasting that the shuttle would return to flight only in spring 1988, more than two years after the *Challenger* accident.

Reagan reported on this meeting to his correspondent Doug Morrow, telling him "money isn't the big delay on the shuttle, safety is." He added that "Jim Fletcher was in yesterday with a full report. He has declared that the next shuttle launch can't be scheduled until the 1st quarter of 1988. The study of and testing of the solid rocket boosters are part of the problem. Added to this are safety features on the shuttles themselves." Reagan said: "Yes, the money problem has to do with several hundred million to make a start, but it's my understanding from Jim that this does not change the 1988 date." He added:

J. M. Logsdon, *Ronald Reagan and the Space Frontier*,
Palgrave Studies in the History of Science and Technology,
https://doi.org/10.1007/978-3-319-98962-4_20

"It's true I wanted to look into this private funding because it was coming up from several directions. As it turns out, there would be a conflict with the private sector which is moving toward commercial launching of satellites."[2]

The National Security Council (NSC) scheduled a one-hour meeting on "U.S. Space Launch Capabilities" on July 29, 1986. Just before the meeting, the president met with a smaller group of his advisers to discuss the issues to be addressed (Fig. 20.1). The NSC meeting was attended by a very high-level group of officials, including the secretaries of state, defense, transportation, and treasury, the attorney general, the deputy secretary of commerce, the chairmen of the Joint Chiefs of Staff and the Council of Economic Advisers (CEA), and, from the White House, Don Regan and John Poindexter, among others. Ronald Reagan was in the chair.

NASA administrator Fletcher told the group that NASA's "first priority is to begin Shuttle flights as close to possible to the first quarter of CY [calendar year] 1988." The second priority was "to keep the space station on schedule," and the third priority was "to replace the *Challenger* orbiter," which should be "not just a NASA priority but a national priority." Fletcher said that NASA had looked at the idea of private funding for the orbiter, but "we support a Government-sponsored plan … We will soon need $500 million to keep it on track." Fletcher closed his remarks by lamenting to the president "NASA is in

Fig. 20.1 Just before a July 29, 1986, NSC meeting on whether to approve a new shuttle orbiter to replace *Challenger*, President Reagan discussed the issue with Senator Jake Garn (R-UT). Also present at the meeting were NASA Administrator James Fletcher, Secretary of Defense Caspar Weinberger, National Security Adviser John Poindexter, and Chief of Staff Donald Regan (not shown). Garn was instrumental in finding a way to fund the replacement orbiter. (Photograph courtesy of Reagan Presidential Library)

terrible shape in a number of areas … Our morale is bad … Without your support everything will unravel." Fletcher's comments continued his pattern of complaining about the lack of White House support for NASA during the post-*Challenger* recovery period.

James Miller, Office of Management and Budget (OMB) director, suggested that "we wait until 1988 to [decide to] replace the orbiter." Miller argued that "no additional launch capacity is needed by NASA … Since it is not needed, and no clear funding is identified, then we recommend you wait." In rebuttal, Secretary of Defense Weinberger noted: "In 1972 I was at OMB and the nation would not have a STS if we had listened to the OMB [staff] arguments … We need a fourth orbiter, SDI and the Space Station. We need them all." Weinberger noted that "European and Soviet capabilities are moving forward. They don't have a public opinion or an OMB." Secretary of State Shultz added: "I agree with Cap. And the sooner we step up to this decision the better."

Secretary of Transportation Dole told the group: "The private sector is ready to move out to provide launch capability using ELVs. They need the signal that the Government will off-load the commercial satellite business from the Shuttle. This is an important signal that will have a major impact … I say make no decision on a fourth orbiter until we see what the commercial industry can do." Deputy Secretary of Commerce Brown added: "We should repair the design of the STS … Redesigning an advanced STS and letting private industry ELVs pick up the slack is the way to go." Attorney General Meese remarked: "I don't think there is much difference between the positions of Transportation, Commerce or NASA"; but Shultz responded, "Oh no. I disagree." Clearly, there was no consensus among the president's most senior advisers on how best to move forward.

As was often the case, Ronald Reagan listened to the discussion without giving a sense of his preferences. Rather, he asked, "What is the status of the space plane? [the national aerospace plane that Reagan had announced in his 1986 State of the Union Address]" and was told that as a research program the plane would not be ready for regular flights until well after 1995. Then Reagan made some rather rambling remarks:

> There has been no mention made of a return from our investment or spin-off. We have fishing nets made from materials of spaceage technology. Sufferers from diabetes are using pharmaceuticals derived from the space programs. We have repaired or salvaged satellites on orbit. If we do not move forward with the procurement of a fourth orbiter, how much will we delay the Space Station? How much further will other nations move ahead of us? We wonder what the Soviets are cooking up in space and how far ahead of us are they.

Later in the session, Reagan told the group "I have one question about recent space activities that has nothing to do with dollars … Prior to the Shuttle *Challenger* launch, Soviet trawlers were seen speeding away at flank speed from the launch areas. Is there any possibility that sabotage could have played a role in the *Challenger* accident?" Reagan was told by Poindexter that there was "no evidence to support any assertion of sabotage."[3]

The meeting ended without a decision. Writing in his diary that evening, Reagan noted that "the problem is now mine & it's a tough one."[4]

Fletcher had been reluctant to return to NASA as administrator, and he in the subsequent months had become very frustrated by the interagency and OMB opposition to replacing *Challenger* with a new shuttle orbiter. The lack of a decision at the July 29 meeting added to his frustration. On August 6, he wrote a letter to Reagan, using unusually strong language for a communication from an agency head to the president. Fletcher suggested that at the meeting he might "not have sufficiently emphasized how strongly I feel regarding the urgent need for an immediate decision." He added: "It is not my style too come on too strong, even when I'm angry." Closing the letter, he repeated; "I am not a table-pounder or a shouter, but I hope I won't mislead you as to how strongly I feel."

In his letter, Fletcher told Reagan: "Mr. President, believe me, the space program is desperately in need of your help right now. Morale throughout the Agency is the lowest it has ever been." NASA's employees "have not been supported or motivated. Quite the opposite … a clear, basic situation is being constantly clouded and stalled by the intrusion of an unending series of irrelevant peripheral issues." He added: "To be honest, Mr. President, I sometimes wonder if I can really pull the program together without your strong support … I am certain that there is nothing … in deeper present crisis or more in need of immediate help than the space program." He asked Reagan to announce a start on the replacement orbiter in Fiscal Year (FY) 1987 while "not compromising NASA's other programs—especially the Space Station"—and proceeding with a commercial expendable launch vehicle (ELV) program "in an orderly way, not by demanding that NASA immediately cancel all foreign and commercial launch contracts." Fletcher suggested that "what seems to be happening is that the replacement orbiter … is being held hostage" by advocates of commercial ELVs "until NASA agrees to cancel all foreign and commercial contracts. I think this is wrong."[5]

Removing Commercial Payloads from the Shuttle

Fletcher was correct in his assessment that a decision on a replacement orbiter was closely linked to a parallel decision to take commercial payloads off of the shuttle. The May 15 NSC meeting on recovering from the *Challenger* accident had deferred a decision on a proposed National Security Decision Directive (NSDD) that had said that, among other provisions, "the STS will no longer provide launch services for commercial and foreign payloads exclusively for revenue-generating purposes," and "the Commercial Space Working Group of the Economic Policy Council (EPC) will develop a transition plan that considers the method and timing to implement this policy." On June 25, Cabinet Secretary Alfred Kingon wrote to Chief of Staff Regan, noting that the EPC working group "has held back from developing an implementation plan, anticipating that the NSDD directing its mandate to do so would be approved in short order." Kingon suggested that "there does not seem to be a reason to hold up further." He added: "The sooner we send a signal to the private sector that NASA will no longer compete for launches of commercial and foreign nondefense payloads, the sooner private sector work can begin in earnest." Regan agreed, and tasked the EPC "to immediately begin work on developing a space commercialization implementation policy."[6]

Developing a Transition Strategy

Developing that policy would be the EPC Working Group on Space Commercialization. This group had replaced the former Cabinet Council on Commerce and Trade working group of the same name after the cabinet councils were abolished in 1985. The group was still chaired by the Department of Commerce. The department's new deputy general counsel, Bob Brumley, had replaced Richard Shay as working group chair. Brumley, like his boss Malcom Baldridge, had come not to trust NASA and was determined to get NASA out of its monopoly position with respect to U.S. civilian space activity. Brumley was assisted in his work at the Department of Commerce by Courtney Stadd, who had joined the secretary's staff in 1984. Other active members of the working group included Ken Pedersen, NASA associate administrator for external relations, assisted by Peggy Finarelli, and Madeline Johnson, who in November 1985 had replaced Jenna Dorn as the head of the Office of Commercial Space Transportation at the Department of Transportation.[7]

The working group had a draft of its report ready by late July. The report identified three possible transition strategies. Option A was "termination of services"; in this option, "NASA shall no longer provide launch services for foreign and commercial payloads." NASA's existing launch contracts would be "terminated in the national interest immediately by Executive Order." This option would allow exceptions only "for special circumstances such as shuttle-dependent payloads." Option B was labeled "deferred services." In this option, NASA would "accept no new orders and enter into no new contracts." However, NASA could honor its existing commitments, but only beginning four years after the shuttle returned to flight, at that point planned for early 1988. These four years would provide a window of opportunity for U.S. ELVs to compete for those and other payloads. Option C was "phase out"; it reflected NASA's position that the shuttle, once it resumed operation, should continue to launch commercial payloads until the U.S. ELV industry had demonstrated its viability. In this option, "government launch systems will not compete with a viable and competitive U.S. private sector launch services industry." The government would enter "into no new contracts to provide commercial launch services, except for commercial payloads that can only be launched on the Shuttle or as directed by statute." However, this option would not "require the termination of existing U.S. Government commitments" and would permit "the government to fulfill its contractual obligations to those users who chose to wait for the extended period required for launch." Under this option, the shuttle could still be launching commercial payloads well into the 1990s.[8]

Advocates of the "termination of services" approach had to contend with the reality that it would be several years until commercially-provided U.S. ELVS able to launch communication satellites would be available. Because preaccident policy had anticipated that the shuttle would replace government-procured ELVs for providing launch services for commercial customers, the manufacturers of those vehicles, McDonnell Douglas, General Dynamics, and Martin Marietta, had been in the process of shutting down their production lines, and it would take both time and investment to return them to operation. The 1985 NASA-Air Force agreement to allow procurement of ten complementary expendable launch vehicles (CELVs) had allowed Martin Marietta to stop the shutdown process for the *Titan* booster, but the manufacturers of the *Delta* and *Atlas* launch vehicles were waiting either for a clear signal from the government with respect to banning shuttle launches of commercial payloads or for possible Air Force orders for one or both of the launchers before they would commit to the sizeable investments required to reactivate production capabilities. It would thus be several years before the first newly produced commercial ELVs could be available; in the interim, it would be non-U.S. ELVs, particularly *Ariane*, which might profit from a decision to immediately remove the shuttle from providing launch services to commercial and foreign customers.

As it became known that the White House was considering barring the shuttle from launching commercial payloads, one leader of the communications satellite manufacturing industry made his worried views known. Allen Puckett, chairman of the board and chief executive officer of Hughes Aircraft Company, wrote Poindexter at the NSC on July 23, saying that "we understand serious proposals are being considered to exclude commercial communication satellites from future use on the Space Shuttle." Such a step, he suggested, would result in "killing the American communications satellite industry." Hughes had made a major commitment to the use of the shuttle to launch its satellites, and Puckett argued that, in the absence of U.S. ELVs capable of launching those satellites, "*Ariane* is the only viable launch alternative." Because Arianespace could give attractive "package deals" to European satellite manufacturers to launch their payloads on *Ariane*, Puckett suggested that it would be likely "that Arianespace pricing policy will eventually force U.S. satellite makers out of business." He ended his letter by saying, "It would be tragic to preclude commercial users from the Shuttle and so condemn the only American commercial success in space that is likely to emerge in our time." Puckett's position was at one extreme among satellite manufacturers; the EPC working group report noted only that "communications satellite industry representatives present diverse views, ranging from preference for a mixed STS/ELV fleet to exclusive reliance on ELVs."[9]

Options Evaluated

The working group report was discussed at a July 30 meeting of the Economic Policy Council, the day after the indecisive NSC meeting on the replacement orbiter. The meeting was chaired by Secretary of Treasury James Baker. The main topic of discussion was which of the three options identified in the working group's report should be recommended for presidential approval. Secretary of Transportation Dole told those in attendance that U.S. industry would not invest in commercializing its ELVs without a "crisp, clean assurance that they will not have to compete with NASA." She thus supported Option A, as did representatives of the Departments of Energy, Commerce, and Defense and of OMB and the CEA. The Department of Treasury did not take a position on a preferred option; only NASA supported Option C, which was characterized by as having "lots of fuzzy edges."

The Washington Post the next day reported that "President Reagan's top advisers yesterday endorsed a far-reaching plan to stop NASA from launching commercial and foreign satellites on the shuttle," a step which "would effectively end the space agency's long-standing efforts to promote the shuttle as an economically self-sufficient space transportation system." This endorsement came "despite strong objections" from NASA. The *Post* report noted that "the debate over how to handle commercial and foreign satellites has received less attention than a new orbiter, but administration officials and some aerospace experts say it has equally significant consequences for the future of the space program."[10]

Finally, a Replacement Orbiter

A meeting of the White House Economic Policy Council with the president present was called for August 7, primarily to present to Reagan options for taking commercial satellites off of the shuttle. At the start of the meeting, Ronald Reagan announced his decision that "he wanted a fourth orbiter started in FY1987." The new orbiter would utilize the structural spares approved for development at the end of the 1982–1983 debate over whether to build a fifth orbiter; having those structural elements already in development proved very fortuitous.[11]

The White House Press Office on August 15 issued a presidential statement that "the United States will, in FY 1987, start building a fourth space shuttle to take the place of Challenger, which was destroyed on January 28th. This decision will bring our shuttle fleet up to strength and enable the United States to safely and energetically project a manned presence in space … A fourth orbiter will enable our shuttles to accomplish the mission for which they were originally intended and permit the United States to move forward with new, exciting endeavors like the building of a permanently manned space station." Reporting Reagan's decision, *Aviation Week* noted that while there was relief that the replacement orbiter decision had finally been made, "the six-month delay in the decision … has generated significant criticism of White House space leadership."[12]

The statement's reference to the shuttle's original intent was a recognition that the concept of a reusable shuttle had originated as a means of constructing and supplying a space station, and that this would be a major future mission for the vehicle. Other reasons given for Reagan's decision to go ahead with the replacement orbiter included U.S. commitments to its international partners to launch the Spacelab and eventually foreign contributions to the space station; the realities that there were still critical national security payloads that could only be launched on the shuttle; that some versions of the Strategic Defense Initiative plans depended on shuttle use; the hope that shuttle-based research could accelerate the development of commercial space activities in advance of the availability of the space station; and the recognition that the Soviet Union was developing its own shuttle as part of an overall buildup in its space capabilities.

During the lengthy debate on whether to approve a replacement orbiter, OMB insistence that its cost be offset by reductions elsewhere, either internal to the NASA budget or elsewhere, had been a major stumbling block. Reagan's decision to go ahead with the orbiter and to provide the modest down payment needed to get started did not resolve this question. *The New York Times* commented that where funds would come in future years was "less clear." This was indeed the case. Regan staff assistant Richard Davis had prepared a memorandum to prepare White House Press Secretary Larry Speakes to answer questions about the presidential statement. He told Speakes that while the president intended to request an additional $272 million in new FY 1987 budget authority for getting started on the new orbiter, "we will face much more difficult problems in finding the financial resources to continue our space recovery in the years subsequent to 1987."[13]

It turned out that the funds needed to build the replacement orbiter were provided from an unexpected source. In early September, President Reagan submitted to Congress his proposal to supplement the NASA budget request for FY 1987 to support an early start on the replacement orbiter. This year-by-year approach did not sit well with space-oriented members of Congress. In particular, Senator Jake Garn (R-UT), who had flown aboard the shuttle in April 1985, was one of the members of Congress calling the administration approach to financing the replacement orbiter "inadequate and indefensible." Garn was chairman of the Senate subcommittee which appropriated funds for NASA.

Over the summer, as the executive branch debate went forward, Garn had been in conversations with his fellow senator Ted Stevens (R-AK), who chaired the defense appropriations subcommittee, about financing the launch recovery program. They and their subcommittee staff members, Stephen Kohashi for Garn and Sean O'Keefe for Stevens (O'Keefe in 2001 became NASA administrator), had decided that there was enough unallocated money in the Department of Defense (DOD) budget to enable a transfer of several billion dollars to the NASA budget to cover the costs of a new orbiter. They justified the transfer on the basis that the shuttle was a national capability used both for NASA and for national security missions.

When the NASA appropriations bill came to the floor, Kohashi got the final go-ahead from Stevens and Garn (who was in the hospital, the previous day having donated a kidney to his daughter) to go ahead with the transfer; no other senator objected. Eventually $2.4 billion of DOD funds were transferred to the NASA FY 1987 budget. That action removed any financial barriers to constructing the new orbiter, later named *Endeavour*, on a rapid schedule. The transfer resulted in the NASA FY 1987 appropriation being $10.5 billion, a 40 percent increase over its FY 1986 budget of $7.8 billion. This proved to be a one-time jump reflecting the budget for the new orbiter, not a permanent increase.[14]

No More Commercial Launches

As noted earlier, the meeting of the Economic Policy Council to present to President Reagan its recommendation of a transition strategy for shuttle payloads took place on the afternoon of August 7. Leading off the meeting, NASA's Fletcher stressed the threat of litigation if NASA had to cancel existing contracts, and argued that NASA should be allowed to honor those agreements. Commerce Secretary Baldridge and Transportation Deputy Secretary James Burnley spoke in opposition to Fletcher, saying it was important to make clear to U.S. industry that NASA was out of the commercial satellite launching business. Secretary of Defense Weinberger, OMB Director Miller, and Chairman of the CEA Beryl Sprinkel agreed with Baldridge and Burnley.

The Departments of Commerce, Treasury, and Transportation, OMB, the U.S. Trade Representative (USTR), and the Office of Science and Technology Policy (OSTP) supported allowing the use of the shuttle to launch payloads uniquely designed for such launch. Commerce, Treasury, Transportation, CEA, and OSTP also supported allowing shuttle launch of payloads with foreign policy or national security implications; they were joined by State, Defense, and NSC. Treasury, USTR, and OSTP also supported allowing the launch of payloads that would be costly to retrofit. Only NASA supported allowing it to make a "best effort" to honor all of its existing 44 contracts.

After hearing the opposing points of view, President Reagan did not make a choice among the alternatives, hoping that a compromise could be reached. He directed the NSC and EPC to work together to prepare within 30 days a shuttle manifest—the schedule of future launches—that would emphasize NASA, STS-unique, and national security payloads but still include a "minimal commercial" aspect. He also asked the two groups to prepare a policy statement for his approval that would reflect "NASA's commercial launch role as one of exception based on foreign policy [and] STS uniqueness" and "provide strong support for an industrial commercial launch industry." Reporting on the meeting, *The Washington Post* noted that "while the President made it clear he wants the shuttle program to reduce its commercial role and encourage a private rocket industry to develop, the advisers still must decide how much to reduce it and over what period."[15]

Eugene McAllister, the White House executive secretary of the EPC, was not pleased with the outcome of the August 7 EPC meeting. A few days later, he made "another pitch for a Presidential decision on space commercialization" that went beyond announcing a revised manifest for future space shuttle launches. He hoped for a statement of policy, asking, "How do we resolve any conflicts? How does the President decide between Secretary Baldridge and Dr. Fletcher if they disagree whether one of the 44 payloads is shuttle unique or costly to retrofit?" He suggested: "If we get a top-down decision, specifying which exceptions are desired, the President is not forced into making micro-decisions. Rather we can make the decision … farther down in the bureaucracy." He added: "Looking at only the 44 contracts as a commercialization decision doesn't provide a signal for the future." Rather, McAllister proposed, "if we are going to make an announcement on privatization later this week when the decision on the fourth shuttle is announced, what could be better than: 'NASA shall no longer provide launch services for commercial or foreign payloads, except those that are shuttle unique or have national security or foreign policy considerations.'"[16]

McAllister's argument carried the day. A presidential statement announcing the replacement orbiter decision was issued on August 15; it also said: "It has been determined, however, that NASA will no longer be in the business of launching private satellites. The private sector, with its ingenuity and cost effectiveness, will be playing an increasingly important role in the American space effort. Free enterprise corporations will become a highly competitive method of launching commercial satellites and doing those things which do not require a manned presence in space." The statement added: "NASA will keep America on the leading edge of change; the private sector will take over from there." A memorandum preparing Press Secretary Larry Speakes for questions related to the statement gave two reasons "for getting NASA out of the commercial satellite launch business." One was that "the private sector is capable of providing commercial satellite launch services, and is eager to do so." However, "the private sector will not make the investments necessary to provide launch services unless they are assured they will not have to compete with NASA." The other reason was that "once out of the commercial launch business, NASA can concentrate on what it does best, and what no one else is doing: science and research."[17]

Final Decisions

At the August 7 EPC meeting, President Reagan had asked for a shuttle manifest that reflected the outcome of that day's discussions. As a step in preparing such a manifest, by September 11 the Economic Policy Council had prepared a memorandum for presidential decision that presented Ronald Reagan with two options:

> Option 1: NASA shall no longer provide launch services for commercial and foreign payloads subject to exceptions for payloads that: (1) are Shuttle-unique; or (2) have national security and foreign policy implications.

> Option 2: NASA shall no longer provide launch services for commercial and foreign payloads subject to exceptions for payloads that: (1) are Shuttle-unique; or (2) have national security and foreign policy implications; or (3) are essentially complete and costly to retro-fit.

The president was told that Option 1 was supported by the Departments of Defense, Commerce, Labor, and Transportation, plus OMB, CEA, and NSC. Option 2 was preferred by the Department of State, OSTP, and NASA. The two options "differ only in the treatment of up to eleven purely commercial payloads, which have already been constructed." Twenty of the existing 44 contracts would qualify for a shuttle launch under Option 1. Under Option 2, because there were likely to be only 25 flight opportunities for non-U.S. government payloads between the shuttle's return to flight in early 1988 and 1995, when existing contracts expired, NASA would make "best efforts" to launch five of the 11 "expensive to retro-fit" payloads on the shuttle. NASA preferred this option "out of concern that the Congress and domestic commercial satellite companies may severely criticize the Administration if NASA is not allowed to fly any U.S. communication satellites while flying ten satellites for foreign and international use." Reagan was also advised that "both options one and two carry the threat of suits by customers unable to fly on the shuttle." But "White House Counsel believes that option two ... poses a higher risk of litigation with greater liabilities at stake" because "it is not based on as clearly an articulated distinction as option one."[18]

Reagan made his final decision between the two options in the following days. Apparently at first he initialed his "RR" approval opposite Option 2. When that mistake was discovered, Chief of Staff Regan took the decision document back to the president, who crossed out his initial "RR" and wrote the initials opposite Option 1. On September 25, 1986, NASA Administrator Fletcher was informed that "the President has approved Option 1." NASA was directed to "revise its manifest to include only those payloads that are either Shuttle-unique or have national security and foreign policy implications. The manifest will then be made public."[19]

A Changed Role for the Space Shuttle

In order to develop a new schedule for the space shuttle once it returned to flight, NASA had to juggle the need to launch its own payloads, including major scientific missions like the Hubble Space Telescope, the *Galileo* mission to Jupiter, and the *Magellan* mission to Venus, the critical Tracking and Data Relay Satellites need to relay communications from the shuttle and various orbiting spacecraft, the remaining DOD payloads that had been shuttle-optimized, other shuttle-unique payloads, and spacecraft of other countries and the European Space Agency (ESA). Sponsors of all these payloads were eager to get them into space as soon as possible and so pressured NASA for an early slot on the shuttle schedule. NASA had to base its manifest on a realistic forecast of shuttle flight rate, once all the recommendations of the Rogers Commission had been incorporated and the shuttle gradually reintroduced to regular operations. Finally, NASA also had to give formal notification to those entities holding existing launch service contracts that the shuttle would not be available to launch their satellites.

Foreign Payloads

Particularly sensitive were the proposed launch dates for various non-U.S. payloads, since not only technical but also foreign policy considerations were involved. On September 23, Assistant Secretary of State for Oceans and International Environmental and Scientific Affairs John Negroponte wrote to Poindexter's acting deputy at the NSC, Alton Keel, saying that "we are very concerned that foreign payloads of significant foreign policy interest" be placed as high on the manifest as possible. In particular, he said, "representatives of Japan and the Federal Republic of Germany have intervened with us concerning the launch priority of their Spacelab missions. Both have linked that issue to the prospects for the continued support of the space station project." Negroponte asked that these two missions be launched "no later than 1990." He sent a similar letter to Fletcher at NASA.

Secretary of State Shultz was even stronger in his pressure. He wrote to Fletcher on September 13, saying that the Department of State "has informed the NSC that, from a foreign policy point of view, INSAT D-1 [an Indian multipurpose geostationary satellite] is our highest priority … An early launch is very important to our bilateral relationship with India … We will therefore be following with great interest the preparation of a new manifest for shuttle launches."[20]

NASA responded to Shultz's request by deciding to launch INSAT D-1 on a Delta ELV rather than on the shuttle. There were three unassembled Deltas available, and NASA thought that it would be more efficient to assemble all three, use two to launch the Indonesian communication satellite *Palapa* and INSAT D-1, and use the third to launch a Western Union satellite that had been in the shuttle launch queue. This move caught Chief of Staff Regan's attention and ire. On September 26, he sent a handwritten note to Cabinet Secretary Kingon, saying, "Spoke to Poindexter about this. NSC reminded NASA two wks [weeks] ago of priority for Indonesia & India, no mention of WU [Western Union]. This is an outrage—I told P [President Reagan]—OK on the two I's but no to WU—a back door attempt by NASA to get into commercial." In dealing with NASA in the post-*Challenger* period, Regan had also come to distrust the agency's behavior.[21]

New Manifest Announced

NASA's Fletcher sent a revised shuttle manifest to Don Regan on October 1. He told Regan that he planned to announce the new manifest at an October 3 press conference. NASA also sent to its "commercial & foreign customers" an announcement of the new shuttle schedule, saying, "We sincerely regret the agonies of the past several months caused by schedule and policy uncertainties and look forward to a time of renewed commitment and stability." *Aviation Week* commented that the new manifest was "forcing fundamental changes in shuttle flight operations." The magazine reported a belief among NASA managers that "the manifest would have been ready by last July but was held up by the White House Economic Policy Council," which included "members unfamiliar with significant space flight issues." There was a sense that the White House decision process "had introduced political concerns, delays and complexity into decisions that many NASA managers said should have been left for space program officials." Used to controlling its activities, NASA was having a hard time adjusting to shared responsibilities for space decisions.[22]

The new manifest anticipated that the shuttle would return to flight on February 18, 1988, almost 25 months after the *Challenger* accident. It projected five shuttle flights in 1988 and 10 launches in 1989, with the launch rate gradually increasing to 16 per year. (The actual return-to-flight date was nine months later, on September 29, 1988. The shuttle was launched twice in 1988, five times in 1989, and six times in 1990.) NASA also provided an explanation of its manifest decisions, noting:

- Because of the two-year downtime and a reduced flight rate, a substantial backlog of payloads will exist for some time.
- The priorities for payload assignments had been: (1) national security-related payloads; (2) shuttle operational capability and dedicated science payloads; and (3) other science and foreign and commercial payloads.
- Because of the "intense competition of many payloads for limited flight opportunities," most payloads would be delayed more than the two-year shuttle downtime, "with some payloads being delayed three or more years."[23]

A New U.S. Space Launch Strategy

Somewhat after the fact, on December 27, 1986, the White House announced that Ronald Reagan had signed a NSDD on a "United States Space Launch Strategy." The directive, NSDD-254, incorporated the policy decisions made over the past three years, and especially in the wake of the *Challenger* accident. The directive declared that "the U.S. national space launch capability will be based on a balanced mix of launchers," including both the shuttle and ELVs. The national security space sector would use both vehicles "as determined by specific mission requirements," with "selected critical payloads" designed so that they could be launched by either system. NASA would use the "unique" shuttle capabilities "to provide manned access to space" and to provide "other unique STS capabilities." Funding for "procurement of a replacement fourth orbiter" would begin in FY 1987, and NASA would establish "sustainable" flight rates.

The policy statement referred to the May 1983 NSDD-94 on a commercial ELV industry, saying that the "principles and policies" enunciated in that directive "remain valid." Reflecting the heated debates of the preceding months and to emphasize their outcome, the policy stated that the shuttle "will be phased out from providing launch services for commercial and foreign payloads that do not require a manned presence or the unique capabilities of the STS" and added that "NASA shall no longer provide launch services for commercial and foreign payloads unless those spacecraft have unique, specific reasons to be launched aboard the shuttle. Those reasons are: the spacecraft must be man-tended or the spacecraft is important for national security or foreign policy purposes." The directive indicated that it was essential that "the reconstituted U.S. space launch assets provide a balanced, robust, flexible space launch capability which can function independently of failures in any single launch vehicle system, allow a return to regularly scheduled launch operations, meet continuing requirements, help make up for lost launch opportunities and reassert global space leadership."

NSDD-254 was thus the death knell of what had been perhaps the most significant policy mistake in the history of the U.S. space program—planning to make the space shuttle the primary, and eventually the only, means of U.S. government access to space. From the time it returned to flight in September 1988 until it was retired from service in July 2011, the space shuttle would be launched 110 more times. One of those launches resulted in the 2003 *Columbia* accident, once again killing a seven-person crew. Eight of the missions were dedicated to national security payloads. The remaining 101 missions advanced NASA's objectives; in particular, 37 missions were dedicated to assembling and outfitting the space station. The shuttle was a remarkable technological achievement, but it was never the reliable and affordable launch vehicle that had been promised at its inception. As former NASA Administrator Mike Griffin wrote in 2010, "What the shuttle does is stunning, but it is stunningly less than what was predicted." It was while Ronald Reagan was president that this reality became evident.[24]

Conclusion

Aviation Week during the post-*Challenger* debate had observed that "the United States is at a historic juncture with its space program and is in critical need of positive action to end the indecision blocking definition of a coherent policy that will reassert U.S. leadership in space … Saying that the U.S. space program is in shambles overstates the case, but not by much." In fact, "a malaise pervades the U.S. space program … What one sees is continuing indecision." The publication placed a large share of the blame for this situation on the Reagan White House and the president himself, saying that while Reagan should be in charge of setting the nation's space goals, "top White House advisors who have direct access to the President either are too technically unsophisticated to grasp the importance of rekindling an aggressive, well-balanced space program, or are preoccupied with other issues they deem to have higher political quotients." The fact that it took so long to make a decision on a replacement orbiter was a "national disgrace."[25]

After several months of delay, Ronald Reagan on July 22, 1986, had been formally presented with the report of the National Commission on Space that had been established in October 1984. That report set forth a bold vision for the next half-century of U.S. space activities. The presentation was made by commission chair Thomas Paine and vice-chair Laurel Wilkening; it consisted only of a five-minute photo opportunity that was intended to "shut down the Commission's activity on a high note." After the photo opportunity, Paine held a press conference and the White House issued a statement, saying, "The President informed the Commissioners that he continues to support a strong space program that will maintain U.S. leadership in space."[26]

These were basically empty words. The ambitious recommendations of the National Commission fell on deaf ears in the Reagan White House. Rather than use the occasion of the *Challenger* accident to set forth new space goals, the White House during 1986 had focused on the narrower issues of restoring U.S. launch capabilities and fixing the problems in the space shuttle program. An opportunity to provide an exciting future for the U.S. space program had been squandered.

Whether it could have been seized, and whether Ronald Reagan would have been willing to approve in the wake of the *Challenger* tragedy a major redefinition of space goals requiring significant increases in the NASA budget, is questionable. Reagan was a pro-space president, but he gave higher priority to fiscal responsibility, even as the Federal deficit burgeoned. There was continuing conflict between NASA seeking to control its own destiny and those who wanted to end its monopoly on U.S. civilian space activities by getting American business more involved in space. Still, there would be a continuing attempt in the final two years of the Reagan presidency to set a new space leadership goal for the United States, even as the space station partnership with U.S. allies was finalized and cooperation with the Soviet Union was renewed.

Notes

1. For an immediate post-*Challenger* discussion of this issue, see John M. Logsdon, "The Space Shuttle: A Policy Mistake?" *Science*, May 30, 1986. I amended the conclusions of this article, but only slightly, in John M. Logsdon, *After Apollo? Richard Nixon and the American Space Program* (New York: Palgrave Macmillan, 2015), Epilogue.
2. Letter from "Ron," Ronald Reagan to Douglas Morrow, July 16, 1986, Box 14, Outer Space Files, RRL.
3. Minutes of National Security Council Meeting, July 29, 1986, Jason Saltoun-Ebin, *The Reagan Files*, http://www.thereaganfiles.com/19860729-nsc-134-us-space.pdf.
4. Ronald Reagan, *The Reagan Diaries*, Douglas Brinkley, ed. (New York: Harper Collins, 2007), 428.

5. Letter from James Fletcher to the President, August 6, 1986, Box 4, Papers of Richard Davis, RRL.
6. Memorandum from Alfred Kingon to Donald Regan, "Space Commercialization Policy," June 25, 1986, Box 14, Outer Space Files, RRL.
7. Interview with Bob Brumley and Courtney Stadd, March 9, 2016.
8. Memorandum from Eugene McAllister for the Economic Policy Council, "Agenda and Paper for July 30 Meeting," July 29, 1986, with attached Memorandum from Commercial Space Working Group to the Economic Policy Council, "Transition Plan to Encourage Commercial ELVs," CIA-RDP88G01117R000702250002-3, CREST.
9. Letter from Allen Puckett to John Poindexter, July 23, 1986, Box 14, Outer Space Files, RRL.
10. Note from James Hirsch to deputy director of Central Intelligence Robert Gates (DDCI), "Economic Policy Council Meeting, 30 July 1986," July 30, 1986, CIA-RDP88G01117R000702250002-3, CREST; Michael Isikoff, "Plan Cuts NASA Use of Shuttle," *WP*, July 31, 1986, A1.
11. Note from James Hirsch to the director of Central Intelligence, "Space Commercialization," CIA-RDP88G0111R000702260002-2, CREST.
12. Ronald Reagan: "Statement on the Building of a Fourth Shuttle Orbiter and the Future of the Space Program," August 15, 1986. Online by Gerhard Peters and John T. Woolley, APP, http://www.presidency.ucsb.edu/ws/?pid=37769. Craig Covault, "Reagan Authorizes Orbiter to Replace Challenger," *AWST*, August 18, 1986, 18.
13. Phillip Boffey, "Commercial Launching by NASA Ordered Shifted to Private Sector," *NYT*, August 16, 1986, 1; memorandum from Richard Davis, through Alfred Kingon, for Larry Speakes, "Friday Space Briefing," August 14, 1986, Box 4, Papers of Richard Davis, RRL.
14. Phillip Boffey, "Commercial Launching by NASA Ordered Shifted to Private Sector," *NYT*, August 16, 1986, 1. NASA Oral History Interview with Jeff Bingham, who in the 1980s was on Senator Garn's staff, by Rebecca Wright, November 9, 2006, https://www.jsc.nasa.gov/history/oral_histories/NASA_HQ/Administrators/BinghamJM/BinghamJM_11-9-06.pdf; interview with Jeff Bingham, December 7, 2017.
15. Memorandum from Eugene McAllister for the Economic Policy Council, "Agenda and Paper for the August 7 Meeting," August 7, 1986, with attached memorandum from the Economic Policy Council for the President, "Commercializing Satellite Launch Services," August 5, 1986; note from James Hirsch to the Director of Central Intelligence, "Space Commercialization," CIA-RDP88G0111R000702260002-2, CREST; Kathy Sawyer, "Reagan Pushes Advisers to Set Shuttle Plan," *WP*, August 8, 1986, A4.
16. Memorandum from Eugene McAllister to Alfred Kingon, "Space Commercialization," August 12, 1986, Box 4, Papers of Richard Davis, RRL.
17. Ronald Reagan: "Statement on the Building of a Fourth Shuttle Orbiter and the Future of the Space Program," August 15, 1986. Online by Peters and Woolley, APP, http://www.presidency.ucsb.edu/ws/?pid=37769; memorandum from Richard Davis through Alfred Kingon for Larry Speakes, "Friday Space Briefing," August 14, 1986, Box 4, Papers of Richard Davis, RRL.

18. Memorandum for the President signed by James Baker for the Economic Policy Council, "Space Commercialization," September 11, 1986, Box 14, Outer Space Files, RRL.
19. The anecdote with respect to Reagan initialing the wrong option is based on an interview with Courtney Stadd, May 16, 2016; memorandum from Alfred Kingon to James Fletcher, "Space Commercialization," September 25, 1986, Box 14, Outer Space Files, RRL.
20. Letter from John Negroponte to Alton Keel, September 23, 1986; letter from George Shultz to James Fletcher, September 13, 1986, both in Box 6, Papers of James Fletcher, NARA.
21. Note from Don Regan to Al [Kingon], September 26, 1986, Box 14, Outer Space Files, RRL; memorandum from Director of International Affairs to Administrator, September 30, 1986, "Consultations with International Partners on the Shuttle Manifest," September 30, 1986, Box 7, Papers of James Fletcher, NARA.
22. Craig Covault, "New Manifest for Space Shuttle Generates Payload Sponsor Debate," *AWST*, October 13, 1986, 22–23.
23. Letter from James Fletcher to Don Regan, October 1, 1986, with attached "National Space Transportation System Manifest," Box 14, Outer Space Files, RRL. Telegram from Richard Truly to Commercial and Foreign Customers, "NASA Space Shuttle Manifest," October 3, 1986, Box 7, Papers of James Fletcher, NARA.
24. Ronald Reagan, National Security Decision Directive 254, "United States Space Launch Strategy," December 27, 1986, https://www.reaganlibrary.gov/sites/default/files/archives/reference/scanned-nsdds/nsdd254.pdf. Griffin's quote is in his essay "The Legacy of the Space Shuttle" in Wayne Hale, General Editor, *Wings in Orbit: Scientific and Engineering Legacies of the Space Shuttle*, NASA SP-2010-3409, available at https://ntrs.nasa.gov/search.jsp?R=20110011792.
25. Craig Couvalt, "Fletcher Cites 'Turf Battles' in Space Program Decision Delays," *AWST*, September 15, 1986, 77–78; Donald Fink, "Who's in Charge?" *AWST*, July 28, 1986, 11.
26. Memorandum from Alfred Kingon and John Poindexter to Donald Regan, "National Commission on Space," July 17, 1986, and Office of the Press Secretary, The White House, July 22, 1986, Box 4, Papers of Richard Davis, RRL.

CHAPTER 21

The Home Stretch

As 1987 began, with the Reagan administration starting its final two years in the White House, much of the optimism about the ability to put the United States on a new course in space had dissipated. The *Challenger* accident and the long and conflictual process of developing a launch recovery strategy had diverted administration attention from the possibility, as recommended by the National Commission on Space among others, of setting long-range space objectives. The National Aeronautics and Space Administration (NASA) Administrator James Fletcher, uncomfortable in his job, was complaining that "he has no patron or even a willing ear in the White House."[1] Reagan's signature space initiative, the space station, was in budget trouble, and, with the grounding of the shuttles, there was little progress in space commercialization. During 1987–1988, the interagency rivalries that had emerged over the roles of the shuttle and expendable launch vehicles shifted their focus to the NASA space station and a potential commercial alternative. As one White House staffer commented in September 1987, there was a "long standing issue: Should NASA have a monopoly on space? The President has said no. NASA has said yes. The issue continues."[2]

There were several reasons why White House attention to space issues had diminished. By this time almost all of those Reagan intimates who had come to Washington with the president-elect to promote the "Reagan Revolution" either had returned to private life or had moved out of the White House to other jobs in the administration. Those who had departed included several individuals who had been particularly interested in a revitalized U.S. space program. None of those in the top levels of 1987 White House staff had the level of interest in space affairs as Ed Meese, Craig Fuller, Richard Darman, or Bud McFarlane had had during Reagan's first term. Fletcher was correct in perceiving the space

J. M. Logsdon, *Ronald Reagan and the Space Frontier*,
Palgrave Studies in the History of Science and Technology,
https://doi.org/10.1007/978-3-319-98962-4_21

program had no advocate in Reagan's inner circle; indeed, his constant complaining about the need for the president to give more support to NASA was becoming an irritant.

Ronald Reagan himself had become mired down in what had come to be called "Iran-Contra," an affair in which arms were secretly sold to Iran in return for efforts to release U.S. hostages held by an Islamist militant group, with some of the proceeds from the sales used to support the Contra rebels trying to overthrow the left-leaning government of Nicaragua. After denying in November 1986 that such a transfer had taken place, Reagan in March 1987 went on national television to admit that transfer had indeed happened. This reversal and the subsequent investigations significantly diminished the president's political clout during much of his remaining time in office, especially as the Senate came under Democrat control after the 1986 midterm election, resulting in Democratic control of both houses of Congress.[3]

As he recovered from the impact of the Iran-Contra scandal, Reagan focused much of his attention in 1987–1988 on foreign policy and national security issues and, in particular, on his interactions with new Soviet leader Mikhail Gorbachev. Space issues were a secondary focus of the Reagan-Gorbachev agenda, which concentrated on arms control; there was no agreement on a major new joint U.S.-Soviet space initiative. With the shuttle grounded and no new space initiatives to announce, Reagan was far understandably less visible on space issues than had been the case in the earlier years of his administration.

Ronald Reagan asked former Senate majority leader Howard Baker to become his chief of staff after February 1987 firing of Donald Regan, who in addition to being a poor fit to the requirements of the position had earned the enmity of Nancy Reagan. Before he would accept the position, Baker wanted to make sure that the rumors were unfounded that Reagan had lost his mental acuity and might no longer be fit to carry out his presidential duties, necessitating the invoking of the 25th amendment to the Constitution to remove him from office. Baker had a long meeting with Reagan, discussing a range of issues "in depth," and "came away … totally reassured." Baker thus agreed to become White House chief of staff.

Another newcomer to the inner circle was Frank Carlucci, who became Reagan's national security adviser in December 1986. Carlucci replaced John Poindexter, who was one of those most involved in the Iran-Contra affair and who had been forced to resign a few weeks earlier. Even Baker and Carlucci did not stay in their positions until the end of Reagan's second term. Baker resigned in mid-1988 to care for his ailing wife and was replaced by his deputy Ken Duberstein. Carlucci was named secretary of defense in November 1987 and was replaced by his deputy, General Colin Powell. None of these individuals showed particular interest in NASA or civilian space matters.[4]

Space Station Under Review

While officials in Washington during 1986 were focusing space policy discussions on issues related to recovering from the shuttle accident, NASA engineers at the Johnson and Marshall Space Flight Centers had been working on a detailed design for the space station. They were paying limited attention to program costs, focusing instead on creating what they judged to be the needed capabilities for a productive facility. At the start of 1987, top officials at NASA headquarters were just learning that the estimated cost of the space station design emerging from this process had increased dramatically, from the $8 billion in Fiscal Year (FY) 1984 dollars that had been the basis for the president's approval of the program, to over $14 billion. This discovery of an over 80 percent budget increase precipitated an urgent review of how to deal with the unanticipated cost growth, with President Reagan, while sustaining his basic commitment to the program, approving a modified, less fast-paced approach to developing the space station.

Office of Management and Budget (OMB) director James Miller on February 10, 1987, told President Reagan: "NASA has informed us that the cost to complete the Space Station program will be sharply higher than the figures presented to you by NASA when you approved the program." He alerted the president that "a lengthy debate on your civil space program and priorities is likely to ensue once the revised cost estimates are released to the Congress." On February 19, White House Chief of Staff Don Regan, who would be fired eight days later, indicated to his White House associates that the increased cost estimates had "disturbed" and "startled" Ronald Reagan. He added that "while the President fully supports the idea of a Space Station … we probably should have some serious discussions of where we are going in this project."[5]

A major reason for the cost increase was that those engineering the station had changed the design presented to the president in late 1983 to double the on-board electrical power; increase the crew from six to eight astronauts, including up to three international crew members; and provide a facility for satellite servicing. The station design was modified into a substantially larger "dual keel" configuration with a number of trusses supporting station elements. NASA had also added the ground-based support for the station to the project's costs; this was a departure from prior practice. Apparently, no one in a senior level at NASA Headquarters, preoccupied with the post-*Challenger debates*, had been paying close attention to the station design activity. NASA's Fletcher commented: "We screwed up … I should have gotten into the whole space station cost analysis much earlier."[6]

The White House staff soon came up with a budget-focused approach to dealing with the cost increases. Because the emphasis was on controlling program costs rather than broader policy issues, neither the Senior Interagency Group for Space [SIG (Space)] nor the Economic Policy Council (EPC) were involved. William Graham had left his position as NASA deputy administrator on October 1, 1986, to move to the White House as presidential science adviser and director of the Office of Science and Technology Policy (OSTP). Graham had departed with a substantial degree of bitterness regarding his treatment by NASA's career senior staff, and was reported to be "increasingly skeptical of the station program" as OSTP worked with OMB to craft a response to the budget problem.

The elements of the new approach were outlined in a March 19, 1987, memorandum to the president signed by Miller, Graham, and Carlucci from the White House and NASA's Fletcher. They told Reagan "we believe that the [space station] program should not be terminated" but that "the Administration should examine more intensively lower cost alternatives." To do this, they recommended that the program move forward in phases, starting with a descoped "revised baseline" which would lead to an "initial manned capability in 1995, leading to permanently manned operations by early 1996." This represented a two-year delay in station availability. The descoped configuration would use a single truss rather than the more elaborate "dual keel" configuration. It would consist of two U.S. modules for research and habitation, four equipment-docking modules, and a separate instrument-carrying platform in polar orbit. This design anticipated that Europe and Japan would each provide a research module and Canada would provide a robotic arm. There would be 50 kilowatts of power, rather than the 87.5 kilowatts in the original design.

Another element of the White House plan would be "an independent technical and cost review of the Space Station program," to be completed by September 1987. At that point, the program would be reviewed with the president, and if the results were acceptable, NASA would be authorized to issue contracts for station construction and OMB would propose a "rolling three-year Congressional commitment and a total cost ceiling" for the station program.

President Reagan on April 3 approved this approach. *The Washington Post* reported that Reagan had "approved a $12.2 billion plan to save the proposed space station by splitting its construction into two phases in order to defer some of its skyrocketing costs." The administration plan did not glean total approval. For example, one congressional staffer was quoted as suggesting "that the station is following the pattern of underfunding and overpromising established by the space shuttle program."[7]

National Academy Review

On April 10, OMB director Miller wrote to Frank Press, president of the National Academy of Sciences, asking that the National Research Council (NRC), the operating arm of the National Academies of Sciences and Engineering, undertake a technical and cost review of the space station program. Press accepted that assignment. Robert Seamans, former top NASA official during Apollo and Richard Nixon's secretary of the Air Force, now an MIT professor, was named to head the review committee.

The NRC committee completed its work by early September and briefed its findings to White House and NASA officials on September 10. The study concluded that the phased approach to station development was "reasonable" and "a good compromise among the needs of early users." There was low confidence in NASA's cost estimates, alerting the White House to the potential for future cost increases. The committee also "raised concerns about the management structure for the program." The committee recommended that the United States should "clarify its long-term goals in space" before deciding on a particular design for the second phase of space station development.[8] Even with these reservations, the White House saw the NRC report as enough of an endorsement of the Phase 1 station design for the project to proceed, and NASA was authorized to select the industrial contractors to begin development of that design.

Problems in Congress

Even as the administration developed a new approach to the station program, the program was running into trouble in the Congress. With a Democrat majority in the Senate, Senator William Proxmire (D-WI), rather than Jake Garn, became the chair of the appropriations subcommittee with jurisdiction over NASA. Proxmire was a long-time skeptic of the value of human spaceflight, and in 1987 led an effort to cancel the space station, moving to delete all funding for the program. He was overruled by the full Appropriations Committee, even though it approved station funding at a level $200 million less than the president's request. NASA's Fletcher had indicated that if the station budget were reduced by that amount, the program could continue with a several-month delay, but "any significant reduction below that level would lead me to … a reevaluation of the future of the program."

Internal to NASA, Deputy Administrator Dale Myers, another veteran of the beginning of the space shuttle program in early 1970s, when he had served as head of human spaceflight, was also expressing concern. Myers had replaced Graham in October 1986. Myers asked Fletcher "whether we are entering the same situation that we entered with the shuttle in 1972?" Then, reflected Myers, "we got starved for budget from the first year on [after shuttle approval], contributing to design decisions and operational stringencies that might have contributed to the *Challenger* accident." Myers, like Fletcher, thought that without adequate budget NASA might have to reevaluate its approach to the space station program, and perhaps even cancel the effort.[9]

Congress in a voluminous continuing resolution ultimately appropriated $425 million for the station during FY 1988, more than $300 million less than the president had requested. Even so, NASA and the White House decided to continue the project. Despite short-term budget issues and the anticipated increased costs in future years of the space station, the United States at the end of 1987 thus remained committed to the undertaking, as 15 years earlier it had remained committed to space shuttle development in the face of budget constraints. The pattern of NASA "straining to do too much with too little" that had first appeared during the space shuttle program was being perpetuated.[10]

Trying to Jump Start Commercialization: The Industrial Space Facility

Budding momentum behind space commercialization had been a victim of the *Challenger* accident and the subsequent lack of access to space to try out new ideas. In a 1987 editorial titled "The Broken Promise of Space Commercialization," *Aviation Week & Space Technology* suggested that "senior NASA managers have chosen to dismantle the commercial space program rather than expend the effort to constructively solve commercialization problems." The editorial noted "short-sighted policy decisions by senior NASA managers are destroying the foundation laid for commercial space ventures over the last five years."[11]

This neglect of commercialization did not sit well with those who believed that NASA was trying to preserve its monopoly on civilian space activity. Trying to recapture a focus on facilitating private sector space efforts became during 1987 the focus of efforts by the Working Group on Space Commercialization of the EPC, the same group that had led the 1986 struggle to get commercial payloads taken off of the space shuttle. The working group was supposed to have been dissolved after that struggle, but instead had been continued, with a charter that gave it, according to one NASA official, "authorization to 'invade' every aspect of NASA activity in the near-term and beyond."[12]

As the focus of its pro-commercialization efforts, the majority of the working group decided to advocate the Industrial Space Facility (ISF) that had initially been proposed in 1982 as the first project of a new company, Space Industries, Incorporated (SII). The Houston-based firm was started by retiring NASA "chief designer" Max Faget and several of his associates; initial investors were Houston businessmen, some with close ties to Vice President Bush and then White House chief of staff James Baker. The ISF was described as a "mini space station"; its 10-meter-long core module was to provide a very low-gravity environment for research and eventual in-orbit manufacturing. The ISF would most of the time operate without astronauts present; it would be only crew-tended during space shuttle visits. The original business plan developed by SII had anticipated that there would be a number of commercial firms eager to use the facility.

The post-*Challenger* collapse of the enthusiasm for early payoffs from microgravity activity in orbit changed SII's plans. A high level of government-funded business became essential to the viability of the ISF. The company now argued that the ISF would be a valuable precursor test bed for hardware and operations associated with NASA's space station, which would not be available until the mid-1990s. In October 1986, SII had entered into a partnership with Westinghouse Electric to complete ISF design and development, suggesting that the ISF could be ready for launch as early as 1991. The partnership hoped to raise $706 million to finance the facility; of that amount $475 million would be debt and $231 million in equity participation. To raise that amount of capital, Space Industries saw as essential a government guarantee that NASA would act as the "anchor tenant" for the ISF.[13]

By the end of 1987, SII lobbying efforts had been able to garner "strong support for this proposition from the White House, OMB, Congress, and the Departments of Commerce and Transportation." This support was reflected in the December 1987 report of the Working Group on Space Commercialization to the EPC, which proposed that "the Administration will announce a Federal commitment to the Industrial Space Facility (ISF) developed by the commercial sector." Elements of that commitment would include "a minimum $140 million lease agreement per year for five years. The lease would be awarded on a sole-source basis to Space Industries, Incorporated."[14]

Fletcher Objects

Although some in NASA saw the ISF as potentially valuable, they notably did not include James Fletcher. As a January 7, 1988, EPC meeting approached at which the nature of the government commitment to the ISF would be discussed, Fletcher made his strong reservations about the commitment to the ISF known. He wrote a December 31 letter to EPC chair James Baker, with copies to the other EPC members, saying that "our principal concern with the current proposal rests with its timing and not with its spirit or intent." He told Baker that "NASA cannot support a major Administration commitment to an early (1991–1993) lease by the Government of large portions of a space facility for which, in our view, commensurate firm requirements have not been identified either within or without the Federal government." Fletcher was also "deeply troubled" by the notion of "proceeding on a noncompetitive, sole-source basis."

Also on December 31, Fletcher wrote a more politically oriented letter to a White House colleague, saying that he was "writing to alert you that in its enthusiasm for possible space privatization initiatives, the Economic Policy Council may recommend an action which could be seriously embarrassing to the Administration." Fletcher said that his concern "involves the sole source aspects of the proposed action, the dollar amount involved and the directed schedule." He said that "we have identified some potential uses for the ISF in the early 1990's but no requirements that would warrant commitment of anything like $700 million." Fletcher was "seriously concerned that a directed sole source award of a large contract to a company with recent connections within the Administration … could result in an appearance of influence which could prove embarrassing to the President." Representing SII was James Baker's former law firm in Houston. SII's financial consultant was Shearson, Lehman, Hutton; Richard Darman had joined the Wall Street firm after leaving his Treasury position earlier in 1987, and was one of those advising SII on how to finance the ISF.

Fletcher had a long-time habit of writing notes on small pink pieces of paper in his difficult-to-decipher handwriting to his colleagues and even to himself. In one of those notes, Fletcher observed: "The more I dig into it [the ISF lease], the more farcical it becomes … It appears that the Commerce Dept.

(with OMB support) wants NASA to spend money which they do not have to buy services which they don't need for some unknown reason, which could be to bail out a company composed of recently departed NASA (and Administration officials!)." The parenthetical phrase had then been crossed out.[15]

ISF Commitment Debated, Approved

Fletcher's concerns did not convince supporters of the ISF to modify their basic course, although they did force a modification in the procurement approach. In preparing Baker for the January 7 meeting, EPC Executive Secretary Gene McAllister laid out the arguments for making a commitment to the ISF, which he described as "essentially a scaled down version of the Space Station." Favoring the commitment was the fact that the ISF was "a cheap way to get the private sector started in space: $700 million over five years compared to the $30 billion Space Station." The ISF "may be ready five years before the Station" and "is something tangible—a flagship for the private sector expansion into space." He suggested that "the crux of the matter is how the ISF and the Space Station fit together … NASA is concerned that a commitment to the ISF signals diminished interest in the Space Station … This is a real concern."

The EPC meeting deferred approval of the ISF initiative. The working group report that formed the basis for the meeting was strongly biased in favor of the ISF, suggesting that the "ISF lessens the chance of national reliance on a single facility [the space station. This is] perhaps analogous to the U.S. decision 15 years ago to rely on the Shuttle as the sole national space transportation system." The report noted that "the Treasury Department estimates that the annual cost to the Federal Government of committing to ISF … would amount to less than one-twentieth the annual Federal interest cost of a deployed Space Station," and that "while it is difficult to state a firm demand estimate for the ISF … existing demand estimates are no less firm than what existed in 1983 when the Administration gave a go-ahead to the Space Station."

The sole-source character of the proposed commitment, however, was bothersome to some other EPC members in addition to Fletcher. The council asked the working group "to examine and report back to the Council on procurement and funding issues related to the proposed ISF." A revised report was prepared for a January 14 EPC meeting. Among the changes intended to make the initiative more palatable to NASA were the stipulations that "the funding for the commitment to the ISF would be added to NASA's budget," rather than having to be offset by NASA reductions elsewhere, and that "the proposed commitment is not intended to diminish the Administration's commitment to the Space Station." The report added an alternative to the sole-source award to Space Industries; another possibility was "beginning the process for a competitive bid."[16]

It was this modification in the procurement approach that broke the log jam. Cabinet Affairs Secretary Nancy Risque, who had replaced Alfred Kingon in that role after Don Regan was fired in February 1987, in a memorandum to chief of staff Howard Baker noted that "although most agencies have a strong policy preference for a noncompetitive bid, we have not found absolute and unimpeachable justification for sole sourcing. In an attempt to break the deadlock on policy, EPC and NSC [National Security Council] staff have come up with a compromise competitive bid formulation that they believe they can sell to the majority of Council members who supported the noncompetitive bid." Baker approved the new formulation, it was quickly ratified by the EPC members, and it was included in a February 9 decision memo for President Reagan. The memo said: "The Federal Government would pursue a competitive bid process for microgravity services in a private space facility on an accelerated basis, with a mid-summer target for awarding a contract. This facility should be available by the end of FY 1993." Reagan was told that this option "has the unanimous endorsement of the Council members"; he initialed his approval. The decision memo noted that James Baker, chair of the EPC, "has recused himself on this issue."[17]

Commercial Initiatives Announced

On February 11, 1988, the White House released a "fact sheet" on the "President's Space Policy and Commercial Space Initiative to Begin the Next Century." Ronald Reagan on January 5 had approved a new National Space Policy that would replace the July 4, 1982, policy statement; the development of that new policy will be discussed in Chap. 23. Announcing the new National Space Policy had been delayed until the EPC could reach agreement on the newly designated Commercially Developed Space Facility (CDSF) and 14 other incentives to spur space commercialization. At a press conference following the release of the fact sheet, NASA's Fletcher was placed in the uncomfortable position of defending an action he had strongly opposed. He was asked: "You've said that NASA has absolutely no use identified for the industrial space facility and now you're suggesting that we're going to spend $700 million?" Fletcher responded: "We had to reorient our thinking … and in that mode we think that we can make very good use of that industrial space facility."[18]

Space Station and CDSF as Competitors?

Congressional skepticism regarding the wisdom of a government commitment to the new facility delayed issuing a Request for Proposals (RFP). Because the name ISF was specific to the SII proposal, the program was newly designated the CDSF. NASA's Fletcher by June 1988 was expressing "grave concerns on the prospects for securing early congressional approval of the CDSF." Fletcher was worried that some in Congress were "perceiving the actions of the Administration … in support of CDSF as indicating a higher Administration priority for CDSF than for the Space Station … We need all the help we can get to ensure that the Congress understands that the Space Station must take priority over CDSF, if a choice between the two should have to be made."

The White House legislative affairs office was aware of the situation, noting that "the two major Administration initiatives, the Space Station and the Commercially Developed Space Facility (CDSF) are in trouble on the Hill." They were "caught in the squeeze for funds and played off against each other." While the House Appropriations Committee had funded the space station at close to the president's request of $967 million for FY 1989, "in the Senate Chairman Proxmire is expected to propose only enough money ($250 million) to sustain the program until January 1989 when the next Administration can resurrect the program with a supplemental should they choose to do so … There is no enthusiasm or support for the CDSF from Space Station advocates in either House … They perceive quite correctly that the CDSF draws support away from the Space Station." Given this situation, NASA, the White House, and the Department of Commerce agreed that "it might make sense to 'lay low' on CDSF since at that time funding for the Space Station was at a critical juncture on the Hill."[19]

President Reagan Supports Space Station

Cabinet Secretary Risque reported to Chief of Staff Howard Baker on June 8 that "a great deal of frustration has developed both at NASA … and at Commerce and Treasury … over the fact that the President has not been more visible on the space issue … These agencies are seeking your support for having the President become much more active." At about the same time, OMB's Miller told Baker: "I wanted to let you know of the very serious situation facing the President's budget proposal" for NASA. Miller noted that "NASA has made every effort to convince Congress of the need to support the President's budget for space … Despite NASA's best efforts, the Congress has not been willing to appropriate the budget necessary for a strong and vital space program." He added: "I am very concerned that the actions of Congress will seriously imperil all of the president's space initiatives, but particularly the Space Station. This would not only be a major domestic embarrassment for the President, but an international one for the U.S. … The 7-year record of leadership that the President has shown in the space program could be lost." Miller suggested that "the President, either in a letter to the Congressional leadership or in a public statement, express his strong support for the space program … and indicate that these programs are vital to America's world leadership in the 21st century."

These suggestions produced quick results. On June 16, President Reagan sent a letter supporting the full range of NASA's programs to key members of Congress. To NASA's Fletcher, however, this letter was too general; he wrote to Chief of Staff Baker, saying that "because of the perceptions of many in Congress of the lack of Presidential interest," more specific presidential attention to the space station was needed. Fletcher attached a letter to the president for Baker to pass on to Reagan; in it, he told Reagan that the Senate action on the space station, if not reversed, could "force us to terminate the space station program. Without more help from you, we will not be able to turn this action around." He reminded Reagan that "it is 'your' space station." He asked the president to "intervene personally with the Senate to save the space station."

The White House reaction to this plea was to draft a second letter from Reagan, this one focused only on the space station. There was a debate within the White House regarding the wisdom of such a letter, since the Senate was considering shifting funds from the budget allocation for the Department of Defense to NASA to enable adequate space station funding, and the letter would seem to endorse such a switch. Alan Kranowitz, the top White House congressional lobbyist, suggested that such a shift "would concern the Secretary of Defense greatly. In deciding whether to ask the President to sign the proposed Space Station letter, the President's senior advisers should focus on where the money comes from as much as on where it goes."

The White House decided to go ahead with the letter, but have it also object to the potential transfer of funds. Reagan was told on the morning of June 22 that "affected agencies—NASA, Commerce, Treasury, OMB, and the NSC—believe that your involvement is crucial to save the Space Station program." He quickly signed letters to senators John Stennis (D-MS) and Mark Hatfield (R-OR), the chair and ranking minority of the Senate Appropriations Committee; the letters were delivered before the appropriations committee budget markup that afternoon.

The president wrote: "The appropriation bill as marked by the HUD-Independent Agencies subcommittee would ... effectively force termination of the Space Station program. This is unacceptable." Reagan argued that the country needed the space station as a means of assuring "our Nation's leadership role in the peaceful exploration and use of space."[20] Whether or not Reagan's letter had an impact, the conference committee reconciling the House and Senate versions of the NASA appropriations agreed to the House funding level of $900 million for the space station. A threat to the project's survival during the Reagan administration had been averted.

The CDSF Fades Away

By early August, with the NASA appropriation agreed upon, OMB suggested that "it is now time to consider what the next steps should be in implementing the CDSF." NASA was proposing to go ahead with initiating congressionally mandated independent studies of the CDSF, but to delay the RFPs until those studies were completed, thus deferring a CDSF contract until the next administration. OMB's concern was that "this delay could, when combined with NASA's penchant to absorb everything in its path, turn CDSF from a privately-designed, developed and financed project into a NASA-designed, developed and financed project." Still, there were "some significant drawbacks to issuance of the RFP now ... The private sector is already nervous, given the government's vacillation, and, except for one company [presumably SII], may not even respond to a RFP at this time."

By September, the White House decided to accept NASA's plan, delaying the release of the RFPs until spring 1989, after the independent studies were completed. Kranowitz, White House legislative director, had suggested that releasing the RFP "would convert existing Senate doubts about the CDSF into full-fledged opposition for institutional reasons. To release the RFP now would be to take an illusory step forward toward realization of President Reagan's CDSF goal."

NASA on September 19 contracted for the two external studies of the CDSF, to be delivered in April 1989, but did not release the RFPs for the facility. With this delay, the CDSF had been transformed from the highest profile commercial space initiative of the final year of the Reagan administration to an issue for decision by the next occupant of the White House. Then, almost as an anticlimax, a National Academy of Sciences study in April 1989 concluded that "the committee does not foresee a need for a U.S. human-tended free-flyer in the period prior to the Space Station to meet microgravity research or manufacturing requirements." This conclusion effectively ended the CDSF initiative.[21]

The ability of NASA and its congressional allies to undercut the CSDF left a bitter taste in the mouths of commercial space advocates within the government. Secretary of Commerce William Verity, who had taken that position after Malcolm Baldridge on July 25, 1987, had died after a rodeo accident, voiced his unhappiness in a speech a few days before the end of the Reagan administration, telling a space business audience that the incoming president, George H.W. Bush, needed to "take a good look at the future of NASA," suggesting that it had become a "government bureaucracy that runs a glorified trucking operation," rather than an agency focused on science and exploration. He added: "As we all know, private dollars tend to be more wisely managed … than those picked from the taxpayers' pockets … Government is just not stepping aside fast enough."[22] The rivalry between NASA and commercial space advocates that from 1983 on had become a central theme in Reagan administration space policy debates persisted even as Ronald Reagan left office.

Freedom: The Space Station Gets a Name

In early June 1988, NASA's Fletcher suggested to White House Chief of Staff Howard Baker that "we are at the point in the Space Station program where it is appropriate for us to give it a name which will help heighten public awareness and public appreciation for the program." Fletcher proposed that "the President himself select the name … It is fitting for him to select a name for this centerpiece of his civil space policy, a name that will project the image he desires." He told Baker that "we have conducted a fairly exhaustive study of possibilities," and that over 700 suggestions had been received. Those suggestions had been reviewed by both NASA and its potential space station partners, and narrowed down to a list of 90, and finally to 3 candidates. NASA's selection criterion was that "the name must be simple and easily pronounced and that it must be translatable without ambiguous or offensive meanings in the languages of our partners."

This process had resulted in the NASA recommendation that the station be named *Freedom*, with *Orion* and *Aurora* as alternatives. Fletcher urged that the president make his choice expeditiously, since "the Space Station is encountering extremely serious difficulties on the Hill" and that a presidential announcement of the station's name would demonstrate "his unflagging support for the program." He suggested that the president announce the name at the Economic Summit to be held during June 19–21 in Toronto, Canada. The next day, Fletcher amended this suggestion, proposing instead that the "preferred scenario" would be to announce the name at a White House Rose Garden ceremony with "those members of Congress critical to space station support" in attendance.[23]

Managing the naming process was assigned to the NSC, and on July 5, NSC deputy for defense policy William Cockell wrote to National Security Adviser Colin Powell, recommending that the name *Freedom* be approved and that President Reagan "devote a portion of an upcoming Saturday radio address" to announcing its selection for the space station. The following day, Powell sent a decision memorandum to the president, saying that "NASA has recommended, and we agree, that your selection of a name for the Station will heighten public awareness and appreciation for the program ... It also provides you with another opportunity to reaffirm your continued support for the program and explain its importance for America's future." Powell noted that "*Freedom* was the top choice of a selection team ... It conveys the appropriate image for the West's space station, and complements nicely the Soviet name for their space station, *Mir*, which translates as 'Peace.'"

On July 13, Ronald Reagan initialed his approval of Powell's recommendation that "you approve the name *Freedom* for the Space Station." The only dissent from that name had come from science adviser Graham, who suggested that "the President get other options for names that emphasize the forward-looking objectives of the American space program."

Despite Fletcher's suggestion, there was no high-profile announcement of Reagan's decision. On July 18, the White House press office simply issued a statement by Marlin Fitzwater, Reagan's assistant for press relations, saying, "The President today announced that the permanently-manned Space Station being developed by the United States, Canada, Europe and Japan will carry the name *Freedom*." He added: "The yearning for Freedom is a basic human motivation, and Freedom of the individual is a value shared by all the nations that will work together to build and use the Space Station. In a literal sense, the Space Station will provide Freedom from the confines of Earth's gravity."[24]

Notes

1. Memorandum from Robert Gates, deputy director of Central Intelligence, to director of Central Intelligence, "Meeting with NASA Administrator Fletcher, November 28, 1986," CIA-RDP88G01116R000600610001-8, CREST.
2. Memorandum from Michael Driggs to Gary Bauer, "U.S. Role in the Commercial Use of Space," September 24, 1987, OA 19224, Papers of Gary Bauer, RRL.

3. There are many books about the Iran-Contra affair. One of the best is Theodore Draper, *A Very Thin Line: the Iran-Contra Affairs* (New York: Hill and Wang, 1991).
4. On Howard Baker's hesitation in accepting the chief of staff position, see Chris Whipple, *The Gatekeepers: How the White House Chiefs of Staff Define Every Presidency* (New York: Crown, 2017), 148–149. On Poindexter's firing and his replacement by Carlucci, see John P. Burke, *Honest Broker? The National Security Adviser and Presidential Decision Making* (College Station, TX: Texas A&M Press, 2009), 218–225.
5. Memorandum from James Miller for the President, "Revised Cost Estimates for the Space Station," February 10, 1987, Box 14, Outer Space Files, RRL. Memorandum from Donald Regan to William Graham, "Revised Cost Estimates for the Space Station," February 19, 1987, and memorandum from Charles Cockell to Frank Carlucci, "Space Station Funding," February 17, 1987, Box 14, Outer Space Files, RRL.
6. Article by Charles Petit, *San Francisco Chronicle*, March 7, 1987, Box 17, Papers of Nancy Risque, RRL.
7. Theresa Foley, "White House Delays Actions on New Station Cost Estimates," *AWST*, March 2, 1987, 26; memorandum from Frank Carlucci, James Fletcher, William Graham, and James Miller for the President, "Space Station New Cost Estimates," March 19, 1987, Box 15, Outer Space Files, RRL; Kathy Sawyer, "Reagan Approves Two-Part Plan in Bid to Rescue Space Station," *WP*, April 4, 1987, A3.
8. Letter from James Miller to Frank Press, April 10, 1987, Folder 12463, NHRC; National Research Council, News Release, "Panel Approves Basic Space Station Design; Urges More Dependable Space Transportation," September 14, 1987, Box 15, Papers of Nancy Risque, RRL.
9. Letter from James Fletcher to James Baker, November 16, 1987, Box 5, Papers of James Fletcher; memorandum from Dale Myers to Jim Fletcher, November 24, 1987, Papers of Dale Myers, NARA.
10. The quoted words are from the *Report of the Columbia Accident Investigation Board*, August 2003, 209.
11. Theresa Foley, "The Broken Promise of Commercial Space," *AWST*, September 14, 1987, 15.
12. Memorandum from Isaac Gillam IV to Administrator, December 17, 1986, Box 5, Papers of James Fletcher, NARA.
13. Information on the Industrial Space Facility is drawn from Marcus Lindroos, "Industrial Space Facility," www.astronautix.com/industrialspacefacility.html; James Rose, "Industrial Space Facility (ISF) White Paper," December 1987, Box 5, Papers of James Fletcher, NARA, and unsigned, "Potential Demand for the Industrial Space Facility (ISF)," December 28, 1987, OA19828, Papers of Michael Driggs, RRL.
14. Eugene McAllister, Memorandum for the Economic Policy Council, "Agenda and Paper for the December 17 Meeting," December 16, 1987, with attached "Report from the Working Group on Space Commercialization," CIA-RDP92B00181R001000010017-1, CREST.
15. Jack Anderson and Dale van Atta, "A Space Contract That Almost Was," *WP*, April 27, 1988, D15. Letter from James Fletcher to James Baker, December 31, 1987, and letter from James Fletcher to Tom [no last name provided; probably

Thomas Griscom], December 31, 1987, Box 16, Outer Space Files, RRL. James Fletcher, undated handwritten note, Box 5, Papers of James Fletcher, NARA.

16. Memorandum from Eugene McAllister to James Baker, "One Hundred and Thirty-Fifth Meeting of the Economic Policy Council—January 7, 1988," January 6, 1987; memorandum from the Working Group on Space Commercialization, "Commercial Space Initiatives," January 7, 1988, Box 17, Papers of Nancy Risque, RRL; memorandum from Eugene McAllister for the Economic Policy Council, "Agenda and Papers for the January 14 Meeting," Box 5, Papers of James Fletcher, NARA.
17. Memorandum from Nancy Risque to Senator Baker, "Space Commercialization," February 5, 1988; memorandum from Eugene McAllister on Behalf of the Economic Policy Council to the President, "Space Commercialization," February 9, 1988, Box 17, Papers of Nancy Risque, RRL.
18. Office of the Press Secretary, The White House, Fact Sheet, "The President's Space Policy and Commercial Space Initiative to Begin the Next Century" and "Press Briefing by Administrator of NASA Dr. James Fletcher, Commerce Secretary William Verity, and Transportation Secretary James Burnley," February 11, 1988.
19. Letter from James Fletcher to Nancy Risque, June 7, 1988, Box 4, Papers of James Fletcher, NARA; memorandum from Gerald McKiernan and Larry Harlow to Alan Kranowitz, "Hill Outlook on Space Funding," June 10, 1988, Box 7, Papers of David Addington, RRL; memorandum from Robert Dawson, OMB, to Director and Deputy Director, "Current Status of the Commercially Developed Space Facility (CDSF)," August 8, 1988, Box 17, Papers of Nancy Risque, RRL.
20. Memorandum from Nancy Risque to Senator Baker, "Space Policy Initiatives," June 8, 1988; memorandum from James Miller to the Chief of Staff, "Funding Situation for the National Aeronautics and Space Administration," undated but early June 1988, Box 7, Papers of David Addington, RRL. Letter from Ronald Reagan to Jake Garn, June 16, 1988; letter from James Fletcher to Howard Baker, June 16, 1988; letter from James Fletcher to the President, June 16, 1988; memorandum from Alan Kranowitz to Rhett Dawson, "Propose Presidential Letters on Space Station," June 21, 1988; memorandum from Nancy Risque for the President, "Letter to Congress in Support of Space Station," June 22, 1988; "letter from Ronald Reagan to Mr. Chairman (John Stennis)," June 22, 1988; memorandum from Nancy Risque to Ken Duberstein and B. Ogelsby, "Space Policy Initiatives," July 8, 1988, Box 15, Papers of Nancy Risque, RRL.
21. Memorandum from Nancy Risque to Ken Duberstein and B. Ogelsby, "Space Policy Initiatives," July 8, 1988, Box 15, Papers of Nancy Risque; memorandum from Robert Dawson, OMB, to the Director and the Deputy Director, "Current Status of the Commercially-Developed Space Facility (CDSF)," August 8, 1988, Box 17, Papers of Nancy Risque; note from Alan Kranowitz to Nancy Risque, "Commercially Developed Space Facility," undated, Box 7, Papers of David Addington, RRL. National Research Council, *Report of the Committee on a Commercially Developed Space Facility*, April 1989, 53.
22. Vincent del Guidice, "Verity Urges Bush to Examine NASA's Role," January 10, 1989, https://www.upi.com/Archives/1989/01/10/Verity-urges-Bush-to-examine-NASAs-role/6418600411600/ph; "Washington Roundup," *AWST*, January 16, 1989, 15.

23. Letter from James Fletcher to Howard Baker, June 9, 1988; letter from James Fletcher to Thomas Griscom, June 10, 1988, Box 16, Outer Space Files, RRL.
24. Memorandum from William Cockell to Colin Powell, "Space Station Name," July 5, 1988; memorandum from Paul Schott Stevens to Rhett Dawson, "Press Release on Space Station Name," July 14, 1988, Box 16, Outer Space Files; memorandum from Colin Powell for the president, "Naming the Space Station," Box 15, Papers of Nancy Risque, RRL. Office of the press secretary, The White House, Statement by Marlin Fitzwater, July 18, 1988.

CHAPTER 22

Together in Orbit: Round Two

In the post-*Challenger* period, there was unfinished business with respect to international space partnerships. Increasing U.S. cooperation with other space-faring countries had been one of the four principal objectives of the national space strategy approved by Ronald Reagan in August 1984. Between 1986 and 1988, there was significant progress in achieving that objective.

One element of the strategy was to convince U.S. "friends and allies" to accept Ronald Reagan's invitation to participate in the space station program. The effort to turn that 1984 invitation into a multilateral space station partnership initially had moved forward quickly. By mid-1985, the National Aeronautics and Space Administration (NASA) had signed memorandums of understanding (MOUs) with its partner space agencies in Canada and Japan and with the European Space Agency (ESA) with respect to cooperation during the Phase B-detailed design period of the space station program. During that phase, the basic design of the core U.S. space station was being defined, as were the potential contributions of Europe, Canada, and Japan to the overall station complex. Europe and Japan each proposed to contribute a laboratory module attached to the U.S. core station; Canada would provide a mobile servicing system based on the technology it had contributed to the space shuttle. Europe would also contribute two free-flying research platforms and potentially a small space plane to the undertaking.

The next step after signing the Phase B MOUs would be crafting the agreements needed to underpin the space station partnership during the facility's development and multidecade operation. Taking that step turned out to be a contentious process that extended for more than two years and several times teetered on the verge of failure.

Ronald Reagan on October 30, 1984, had signed a congressional joint resolution calling for increased U.S.-Soviet space cooperation. One barrier to increasing that cooperation was the 1982 decision to allow the overall

J. M. Logsdon, *Ronald Reagan and the Space Frontier*,
Palgrave Studies in the History of Science and Technology,
https://doi.org/10.1007/978-3-319-98962-4_22

U.S.-Soviet space cooperation agreement, first signed in 1972 and renewed in 1977, to lapse. The 1984 congressional resolution called for renewing the agreement, but the administration's earlier hope had been first to get Soviets to participate in a spaceflight rescue project as a follow-on to the 1975 Apollo-Soyuz Test Project. The State Department had informally asked the Soviet Union in January 1984 whether it was open to that possibility; the response was negative, linking any space cooperation to arms control talks. Once again in January 1985 the inquiry was repeated, with the same response. With new leadership in the Soviet Union in the form of Mikhail Gorbachev, who took office as general secretary of the Communist Party in March 1985, the Soviet attitude would change, and eventually steps toward enhancing U.S.-Soviet space cooperation would follow.

Negotiating Space Station Cooperation

Between November 1985 and February 1988, difficult discussions regarding the terms and conditions of cooperation in the development and operation of the space station took place. Because the station program would involve a multidecade commitment by the participants, Europe took the lead in insisting that, in addition to separate MOUs between NASA and its space agency partners along the lines of those that had been in effect during Phase B, the basis for the partnership had to be a formal and binding agreement among participating governments, not just their space agencies. An immediate question was what form that government-to-government agreement should take. The partners indicated their preference for a formal treaty, which in the U.S. system of government would require ratification by the Senate after executive branch agreement. The Department of State, which had the lead in the government-to-government talks, argued that obtaining the two-third majority of the Senate required for treaty ratification might be difficult and could involve significant delays. The U.S. position was that an executive branch intergovernmental agreement (IGA) signed by the president or someone authorized to act on his behalf would provide an adequate guarantee of U.S. government commitment to the station program. The partners reluctantly accepted the U.S. position.[1]

The formal process of negotiating a single IGA among the United States, the nine governments of the member states of the ESA that decided to participate in the space station program, Japan, and Canada, as well as crafting three MOUs between NASA and each of its space agency partners, began in 1986. Leading the IGA negotiations was Robert Morris, a deputy assistant secretary in the State Department's Bureau of Oceans and International Environmental and Scientific Affairs (OES). Peggy Finarelli, by now the director of policy for the space station, was the NASA lead in the MOU negotiations, which took place separately from the IGA discussions but in close coordination with them.

Europe Demands a "Genuine Partnership"

On November 14, 1986, NASA Administrator James Fletcher reported to National Security Adviser John Poindexter on the status on initial negotiations on the space station agreements. Fletcher told Poindexter: "Our negotiations with Japan and Canada are going very well. We have not yet reached agreement, but the will of all parties to do so is clear. On the other hand, our negotiations with the Europeans … are not going well at all. I cannot rule out the possibility that the Space Station program may proceed … without European participation." Fletcher added: "Four months into the negotiations, the European Space Agency (ESA) has tabled a totally unacceptable counter proposal to our draft Memorandum of Understanding."

The May 1985 declaration at the end of the Economic Summit among the leaders of the major democratic countries had said that space station cooperation would take place "on the basis of a genuine partnership." Underpinning the unacceptable European position, first expressed at an October 30, 1986, NASA-ESA meeting, was a belief that if the space station cooperation was indeed to be a "genuine partnership," the United States, and especially NASA, had to treat ESA as a mature agency, fully capable of managing complex space efforts such as human spaceflight. ESA claimed that it was qualified to be an almost equal partner in the managing the station, even though its contribution to the undertaking was valued at some 25 percent of NASA's contribution. Fletcher observed that the "European approach is inequitable. In measurable ways, the Europeans want to take far more out of the Station than they are bringing to it." Most unacceptable to Fletcher was the approach to managing the station suggested by ESA; that approach would "not provide for strong centralized management [by NASA] … This objection is fundamental. NASA must retain the responsibility for overall program coordination and direction."

Fletcher noted that the ESA Director General Reimar Luest was flying to Washington on November 21 for an "urgent meeting." He believed that "the Europeans must be dissuaded from their proposed approach" and said that he would make "my thinking clear" to the ESA head, adding that "NASA is working hard to achieve the US goal of cooperation, but not at any price." The Fletcher-Luest meeting resulted in agreement to try to resolve the differences between the U.S. and European positions, and a November 24–25 negotiating session on the IGA had "a more cooperative tone." Even so, the issue of management control was to persist through the subsequent negotiations.[2]

The Department of Defense Intervenes

In addition to Fletcher's problems with the ESA negotiating position, there was another, unanticipated, and very disruptive intervention into the process from within the U.S. government. In November 1986, Frank Gaffney, the deputy assistant secretary of defense who was overseeing Department of Defense (DOD) involvement in the space station negotiations, called an "urgent meeting at the Pentagon of representatives of all U.S. agencies taking part in the space station talks." Secretary of Defense Caspar "Cap" Weinberger in 1983 had been a vocal opponent of the space station, and had made it clear that the DOD had no plans to use the station. But by 1986 the DOD position had changed, to one that insisted on its future right to make use of the station, even if there were no current plans to do so. Gaffney made it clear at the November 1986 meeting that "from that point forward, Defense would be determined to ensure that its potential interests be fully taken into account." In fact, according to one participant in the meeting, Gaffney gave the impression that "Defense would prefer that any space station be a solely U.S. national asset" and that he "would not be disappointed if an attempt at international cooperation in building and operating a space station failed." Gaffney and his associate Philip Kunsberg, assistant deputy secretary of defense for policy, who would become the DOD member of the negotiating team, were nationalistic in outlook and extremely skeptical of the benefits of international partnerships.[3]

Likely at the instigation of Gaffney and Kunsberg, Weinberger on December 1 sent Alton Keel, who had been National Security Adviser Poindexter's deputy and who had temporarily replaced Poindexter as he resigned on November 25 in connection with the Iran-Contra scandal, a provocative memorandum. Weinberger suggested that the positions being taken by potential U.S. partners in the space station could be "prejudicial to our future ability to use this asset in support of vital U.S. defense and intelligence functions." He warned that "a real danger exists" that the United States "may see an effective foreign veto over the U.S. use of the station for such activities as SDI experimentation." Weinberger urged Keel to "promptly convene" the Senior Interagency Group for Space [SIG (Space)] "to consider these policy issues and to make recommendations about any adjustments in our negotiating posture and positions, which may be deemed necessary." He added: "Obviously, until such a review is completed, we would be well-advised to suspend further negotiations."

This recommendation was not immediately accepted. At a previously scheduled December 2 meeting, the Interagency Group for Space [IG (Space)] decided "to quickly review a new draft agreement in hopes of permitting NASA to table it" in upcoming talks with ESA. This decision was not acceptable to DOD. On December 4, Undersecretary for Policy Fred Ikle wrote Poindexter, saying that "upon preliminary review of the new draft we continue to have important reservations regarding U.S. control and management of this major national resource." Ikle once again called for convening SIG (Space), adding that "in the meantime, negotiations should be deferred."[4]

The DOD objections resulted in the space station negotiations being suspended for almost three months. In the interim, an interagency working group convened by IG (Space) developed a new set of guidelines for the negotiators. That working group was chaired by Richard Smith, the OES principal deputy assistant secretary, who in the months to follow would be a crucial contributor to bringing the negotiations to a successful conclusion.

New Negotiating Guidelines

On February 3, 1987, President Reagan approved new principles and guidelines for space station negotiations that had been developed by Smith's working group. The guidance was issued as National Security Decision Directive (NSDD)-257. The NSDD included 10 statements of "general policies" and 17 specific negotiating guidelines. Among the general policies was a directive to "build a permanently manned space station consisting of a core U.S. Space Station, which, with the Canadian-provided Mobile Servicing Center, is capable of reliable autonomous operation by the U.S.," while also attempting "to secure the participation of U.S. friends and allies in the Space Station program." Negotiators were told to "promote world recognition of the Space Station as a national achievement of the United States," one which would "promote U.S. national interests."

Among the guidelines for the negotiating team was to "maintain the initiative" and "focus discussions on U.S. draft texts." Also, "ensure that the U.S. can at all times select the Commander and can control and exercise authority over all Space Station activities ... necessary to ensure safety and to enforce physical and information security procedures" and "establish management arrangements that ensure necessary U.S. control" and "explicitly provide for U.S. ability to make unilateral decisions where necessary." The most controversial provision was to "ensure that any foreign participants recognize and agree that the United States may use the U.S. elements of the Space Station and the Canadian-provided Mobile Servicing Center for national security purposes, consistent with U.S Law and U.S. international obligations, without their consent or necessarily their review."[5]

Space Station Negotiations Resume, Succeed

The provisions of NSDD-257 were quickly incorporated into a new draft of the space station IGA. That draft laid out the preferred U.S. positions "in stark, hard-line fashion." On February 11, 1987, potential partners in the space station met in Washington for an U.S. explanation of the new draft. They were "dismayed at its uncompromising nature, particularly the direct and specific reference it contained to national security uses of the space station. They went home angry."[6]

Recognizing that future negotiations were likely to be very argumentative, the State Department decided to replace Robert Morris as head of the U.S. government negotiating team with one of the department's most experienced negotiators, Richard Smith. Smith had chaired the working group that developed the principles and guideline that had been the basis for NSDD 257; that working group would continue in being to act as a U.S. interagency "backstop" for the negotiating team. Smith stayed on as chair of the backstopping group as well as becoming the lead U.S. negotiator.

Smith initially adopted a strategy of holding separate negotiating sessions with Japan, Canada, and Europe. The first of these sessions, with Europe, took place in Paris on February 25–27. Heading the European negotiating team was Reinhard Loosch, whom Smith described as a "seasoned and loquacious diplomat." Loosch was an official of the German foreign ministry and had led the European team since the start of discussions in 1985; according to Smith, "his positive attitude and his ability to keep the conversation going ... were helpful throughout the negotiating process."[7]

The Paris negotiations signaled that the path to agreement would be difficult. The State Department cable reporting on the meetings said that

> three days of intensive talks averted a threatened collapse of negotiations with European Space Agency (ESA) member states on space station cooperation. Perceived hardening and lack of flexibility on key issues in U.S. draft had led to uniform view in European countries that in absence of significant progress at this session there would be reconsideration of ESA States' acceptance of U.S. President's offer to cooperate on this project.[8]

Subsequent negotiating sessions took place with Japan on March 12–13 and with Canada on March 17–18; they also evidenced partner resistance to some of the provisions in the U.S. draft. Reflecting on these sessions, one State Department official observed that "many of the changes which the United States has introduced into the draft agreement [the IGA] over the last few months, taken cumulatively, appear to our partners to have changed the character of the program the President invited them to join." He added: "Our partners have expressed concern that the Space Station is becoming less a civil program and appears to be taking on the quality of a U.S. military facility. Many of the problems which the United States faces in the negotiations reflect this profound concern among our partners." While all partners "agree that the United States must retain the right to use its elements for national security purposes," they have "stated unequivocally that the explicit reference to 'national security purposes' [in the draft IGA] is unacceptable."[9]

Based on the results of this initial round of negotiations, Smith and the U.S. backstopping team developed a revised version of the IGA which removed the reference to national security uses in the agreement itself. Instead, there would be a "separate 'agreed minute' ... that would make clear that the negotiators envisioned national security uses." An "agreed minute" was a record of the negotiating history of an agreement that was not a formal part of the agreement itself; using such an approach to avoid introducing controversial language into an agreement text was a well-established tactic.

Weinberger Intervenes Again

Removing the national security use requirement from the text of the IGA by using the "agreed minute" approach was "extremely controversial ... Defense vigorously resisted any show of flexibility." Smith was in a difficult position, finding himself between the demands of his negotiating partner governments, on one hand, and his need to satisfy the positions of key U.S. agencies on the other. Then, the situation quickly got worse. On April 7, in what Smith described as a "bombshell" and *Aviation Week* characterized as a "haymaker swing at NASA" and "giving the international partners ... a solid punch in the nose," Weinberger sent a letter to Secretary of State George Shultz. In it, he suggested that "we may be in danger of paying too high a price for international cooperation" in the space station program. He added: "We will have paid too high a price if we:

- Fail explicitly to reserve the right to conduct national security activities on the U.S. elements of the Space Station, without the approval or review of other nations;
- Accede to multilateral decision-making on matters of Space Station management, utilization, or operation;
- Permit a one-way flow of U.S. space technology to participating nations, which are also our competitors in space;
- Allow the concept of "equal partnership" to displace either the reality or the symbol of U.S. leadership in the Space Station program.

Weinberger ended his letter by telling Shultz, "We must be prepared to go forward alone if the price of cooperation is too high."

Weinberger's letter was quickly leaked to various media outlets, including *The New York Times* and *The Washington Post*; the leaks presumably came from supporters of DOD's hard-line position. Deputy National Security Adviser Colin Powell sent a note to his boss Frank Carlucci on April 10, the day that the two papers published stories about the letter, saying, "This letter ... has spooked the allies even more. DOD & State are about to gridlock." It would be up to the National Security Council (NSC) staff to try to resolve the dispute; if that attempt was not successful, the dispute would go to the NSC and then the president for a decision. *Aviation Week* reported that "President Reagan is facing a choice between two space station policies that either would continue current international participation in a largely civilian project or transform the station into a purely NASA/Defense Dept. enclave." The report suggested that the timing of Weinberger's letter was "suspicious," and that his intent "was not only to alienate the international partners, but also to kill the station program outright." One European diplomat was quoted as saying, "To us, it was an obvious attempt to sidetrack the negotiations." Given the public knowledge of the conflict inside the Reagan administration, the State Department had no choice but to once again defer schedule negotiating sessions until the dispute was resolved.[10]

James Fletcher offered NASA's views on the situation in an April 13 letter to Carlucci. He told the national security adviser that making "the right decision" on the approach to negotiations "will ensure that the President continues to reap the financial, technical and political benefits of his far-sighted invitation to our friends and allies. The wrong decision will jeopardize the Space Station itself." Fletcher noted that "DOD use is *not* at issue here ... All of our partners have agreed that DOD will be able to use the Station. The issue here is *how* to reflect this in the agreements."[11]

As the dispute was being addressed inside the U.S. government, the space station negotiating team, "using our contacts at every level … undertook a damage control effort aimed at reassuring our partners that our offer of cooperation on a civilian space station for peaceful purposes was still in play." A meeting among the Departments of Defense and State, NASA, and the NSC staff was set for April 16 to try to resolve the disagreement. The NSC staff supported the State Department and NASA position. According to Smith, the parties "came within hours of taking the issue to the president before Defense agreed to go along with the NSC decision." The DOD representatives at the meeting "were not confident that Reagan would side with them" if the issue went to him for resolution. The participants at the meeting ratified the text of a one paragraph "agreed minute" to be attached to a revised IGA that said: "The partners recognize and agree that the U.S., Europe and Japan each will be the judge of what activities on their elements of the international space station complex fall within the requirement that all utilization will be for peaceful purposes in accordance with international law. Such utilization may include national security use." A revised draft of the IGA with this agreed minute was quickly prepared and sent to potential partners.[12]

Negotiating the Station Agreements

Once the DOD had withdrawn its insistence that the text of the station agreements explicitly allow national security uses of the facility, Smith and his negotiating team could resume their interactions with the potential space station partners. A round of talks in late spring and early summer 1987 were "difficult," with Japan being concerned about how station operations would be managed, and Europe continuing to complain that the United States was not offering "genuine partnership." Based on this round of negotiations, the U.S. team proposed to draft "a new negotiating text that would seek to respond to some of the concerns" that had been raised. Smith was convinced that "without a new text, the negotiations were headed toward certain failure." The DOD objected, and according to Smith it took a presidential decision to allow a new draft to be prepared. For the first time, the draft took a multilateral form, suitable for negotiations in which with all the partners would participate.

That new text was discussed with the partners separately in September, and a multilateral meeting was then scheduled in Washington on October 13–16, 1987. That meeting was an "intense negotiating session" that result in yet another new draft of the IGA. The U.S. delegation ended the meeting "with the understanding that Canada, Europe, and Japan would now reflect on whether we had achieved an acceptable basis for continued cooperation." The outcome of that reflection was uncertain. While the U.S. negotiating team described the discussions as "constructive and productive," some members of other negotiating teams disagreed, with one European describing the sessions as "the dialogue of the deaf."

Informal discussions among the partners continued throughout the fall. The issue of how best to acknowledge agreement that the United States (and the other partners) could use the station for national security purposes remained controversial; both Europe and Japan suggested that an agreed minute was "too formal a way" to express that agreement. Both were trying to avoid a public acknowledgment in their countries of acceding to the U.S. position. It was finally decided that the best way to express the agreement was through an exchange of letters between U.S. chief negotiator Richard Smith and the chief negotiators from Europe and Japan that recognized each participant's right to decide what activities could take place in the station elements they provided.

Another area of continuing controversy was how to express the reality that NASA would have the last word on the management of station operations, should the partners fail to reach consensus in a particular situation. European negotiators objected to language in the IGA stating that reality, since it would call attention to the fact that Europe would not have total control over its elements. Ultimately there was agreement that such authority would be vested in the Space Station Control Board, on which NASA had the right to make decisions in the absence of consensus; this avoided an explicit acknowledgment of the U.S. management role.[13]

Discussion among the negotiators on the IGA and the three MOUs continued into early 1988, with adjustments being made in the language and content of the agreements. The goal was to have the IGA contain only the top-level provisions required, with more detailed provisions on all aspects of the station program contained in the MOUs. The negotiating teams met once more in Washington on February 4–6, 1988, to review the documents and make sure they were ready to send to their governments for domestic review and approval. That process would take several months.

Agreements Signed

Representatives of the countries that had decided to participate in the space station program gathered at the Department of State on the afternoon of September 29, 1988, to sign the "Agreement among the Government of the United States of America, Governments of the Member States European Space Agency, the Government of Japan, and the Government of Canada on Cooperation in the Detailed Design, Development, Operation, and Utilization of the Permanently Manned Civil Space Station." Finarelli, who had led the NASA team negotiating the MOUs, told the press that "all sides getting a good deal … I think all four partners feel that they've won in this negotiation." By a fortuitous coincidence of timing, that morning the space shuttle *Discovery* had lifted off its launch pad at Kennedy Space Center; this was the first shuttle launch since the *Challenger* accident.[14]

With the signing of the space station agreements, the largest ever international technology project was formally initiated. The station partnership would stand as a major part of Ronald Reagan's space legacy.

Resuming Space Cooperation with the Soviet Union

In signing, on October 30, 1984, Senate Joint Resolution 236, "Cooperative East-West Ventures in Space," Ronald Reagan had said: "We are prepared to work with the Soviets on cooperation in space in programs which are mutually beneficial and productive."[15] This statement stimulated at least one person in the State Department's Soviet Bureau to begin "to look more seriously at what might be done to breathe some life into bilateral space cooperation" between the United States and the Soviet Union. On January 16, 1985, Soviet desk officer John Zimmerman examined the pros and cons from both the U.S. and the Soviet perspectives of increased cooperation in addition to the simulated space rescue mission that had already been proposed. This memorandum was followed on February 5 by a second suggestion from one of Zimmerman's associates that "a proposal to the Soviets to expand space cooperation presents us with a foil for anti-SDI/'militarization of outer space' propaganda by publicly demonstrating the Administration's peaceful intentions in space … A new program would not only offer symbolic value but have definite scientific merit." The author of this memorandum, Thomas Simons, Jr. of the Soviet Bureau, commented that "considering the well-known interest of Bud McFarlane, Gil Rye (NSC staff), and Jim Beggs, we expect that a Department proposal … would be favorably received in the White House."

Rye and Jack Matlock, the NSC Soviet expert, were aware of this thinking at the State Department. On February 28, Rye told Matlock that he was in favor of the State Department's initiative, which "could break some of the ice at Geneva," where the United States and Soviet Union were engaged in arms control talks that seemed to be going nowhere. Rye was drafting a speech for the president to give at a March 29 award ceremony, and told Matlock: "I'd like to insert some words on a U.S./Soviet space agreement." Matlock agreed that this was a good idea, and the State Department was asked to prepare such an insert.

This suggestion led to a March 22 State Department memorandum to National Security Adviser McFarlane that included draft language for the president's speech and a "non-paper," that is, not a formal proposal, "for immediate forwarding to the Soviet Government through Ambassador Dobrynin," alerting the Soviet leadership to what President Reagan would be proposing. The draft speech language would have had the president saying, "I have proposed to the Government of the Soviet Union that they meet with us in the coming months to begin discussions to establish a firm basis for future space cooperation." A quick interagency review revealed that the DOD and the Central Intelligence Agency "would prefer additional time to consider the initiative." On that basis, the proposal did not move forward; Reagan did not mention U.S.-Soviet space cooperation in his March 29 remarks. But the seeds of enhancing cooperation had been planted.[16]

Then, during a July 31, 1985, meeting between Secretary of State George Shultz and Soviet foreign minister Eduard Shevradnadze to prepare for a first summit meeting between President Reagan and new Soviet leader Mikhail

Gorbachev, Shultz gave the Soviet diplomat another "non-paper." The document said that "the United States would like to suggest that the U.S. and the U.S.S.R. enter into discussions aimed at improving cooperation on the peaceful uses of space ... The U.S. would also be interested in exploring the potential for renegotiation of the broader intergovernmental space agreement which expired in 1982." For the next year, there was no Soviet response to this suggestion.[17]

Ronald Reagan and Mikhail Gorbachev met for the first time at a November 19–20 summit gathering in Geneva. Civilian space cooperation was not an item on the summit agenda, although the need to avoid the weaponization of space through deploying anti-satellite weapons and aspects of the U.S. Strategic Defense Initiative (SDI) were discussed. In a toast at a dinner hosted by the Soviet Union at its Geneva mission, Reagan startled Gorbachev, among others, by his after-dinner remarks, saying that he had been telling Soviet Foreign Minister Shevradnadze that "if the people of the world were to find out that there was some alien life form that was going to attack the Earth approaching on Halley's Comet [which was to pass through the inner solar system in early 1986], then that knowledge would unite all the peoples of the world." Reagan called this concept of humanity uniting against an alien invasion his "fantasy"; he would occasionally repeat some version of his remarks during his final years in office.[18]

Gorbachev, Sagdeev, and Space Cooperation

Beginning in mid-1986, rapid progress on reestablishing a U.S.-Soviet cooperative relationship began. Returning from an international space meeting, leading U.S. space scientist Tom Donahue reported to NASA on a conversation with his Soviet counterpart Roald Sagdeev. Sagdeev was acting as a science and arms control adviser to Soviet leader Gorbachev, and told Donahue that Gorbachev agreed with him that renewal of the U.S.-Soviet space cooperation agreement "should go forward soon and not be tied to either (1) SDI or (2) a Reagan/Gorbachev summit meeting." Sagdeev also said that he understood this message had been communicated by the Soviet Embassy in Washington to the State Department. This communication may have been the much-awaited Soviet response to the July 1985 "non-paper." If what Sagdeev told Donahue was correct, the path to negotiations regarding space cooperation had been cleared.[19]

Sagdeev was a physicist who since 1973 had been director of the Space Research Institute of the Soviet Academy of Sciences, known in the West by the acronym IKI after its Russian name. As a young scientist, his work in physics won international recognition; he became one of the few Soviet scientists trusted to travel abroad. A cosmopolitan personality, after taking over IKI he had developed contacts among the global space and security communities. (In 1988, he even married Susan Eisenhower, the granddaughter of U.S. president Dwight D. Eisenhower.) He became active in discussions about controlling the spread of armaments, participating as a Soviet representative in arms control discussions in the United States and elsewhere.

Sagdeev's institute had developed the *Vega* mission to observe Halley's Comet, and Sagdeev had been active in the collaboration among Europe, Japan, the Soviet Union, and the United States to coordinate activities associated with the comet's once-every-76-years visit to the inner solar system. The United States had decided not to send a mission to Halley, but NASA's Jet Propulsion Laboratory (JPL) had initiated an International Halley Watch concept and offered the services of the U.S. Deep Space Network to support the European, Japanese, and Soviet missions to Halley. These activities were coordinated through the Interagency Comet Group with its members being NASA and the three space agencies with Halley missions. Through this and other interactions with JPL, Sagdeev became friendly with well-known U.S. scientist Carl Sagan and Sagan's organization The Planetary Society. That society had become a committed advocate for resuming U.S.-Soviet cooperation in space.

Sagdeev had known Mikhail Gorbachev casually since they both were students, but their careers had gone in very different directions and his professional association with Gorbachev began in connection with the November 1985 Reagan/Gorbachev summit. Sagdeev was invited to be one of a group of Soviet arms control experts advising the Soviet leader during the summit. The two developed a relationship of mutual trust, and so when Sagdeev in July 1986 reported that the Soviet Union was ready to renew the space cooperation agreement, it seemed to his U.S. counterparts that he could speak with authority. Sagdeev was to be a major actor in discussions of space cooperation during the remaining years of the Reagan presidency and afterwards.[20]

Renewing the Space Cooperation Agreement

The State Department sometime in mid-1986 had indeed received notice from the Soviet Union that it was open to talks about renewing the general space cooperation agreement, and anticipated that its renewal would become an agenda item for the next Reagan/Gorbachev summit. The department had developed a plan that called for a first step being a meeting between Soviet and U.S. space experts prior to government-to-government negotiations. By early August, NASA had selected Lew Allen, a retired general who was director of JPL, to lead the U.S. delegation. The purpose of the initial expert discussion was to gather information on Soviet programs and plans as a basis for the government-level negotiations.

A ten-person U.S. delegation traveled to Moscow for September 12–13, 1986, meetings, which were held at Sagdeev's institute. As he returned to the United States, Allen reported that "the discussions went well in a cordial and cooperative atmosphere." The two sides "identified areas of shared interest in five major space science disciplines: planetary exploration, life sciences, astronomy and astrophysics, earth sciences and solar-terrestrial sciences." Allen and Sagdeev had agreed "to keep the discussion of initiatives low-key," seeking "non-controversial initiatives" that would not "stress the system" and would be "most likely to lead to a 'general' agreement to cooperate in the five science disciplines." Potential high-profile initiatives such as the space rescue demonstration or other joint human missions were not discussed, but Allen reported to NASA administrator Fletcher that "if you feel that the President should be given a more dramatic option," he and Sagdeev "discussed the possibility of a cooperative Mars sample return mission for launch in 1996."

Reporting on the Moscow talks, *Aviation Week* suggested that "support is building in the White House and Kremlin to revive space program ties ... but the move is generating serious debate within the Reagan Administration ... The space initiative faces opposition from the Defense Dept. and some State Dept. officials concerned with technology transfer. The Central Intelligence Agency and other intelligence organizations have been supportive." The report noted that "the Administration has drafted a proposal for a new U.S./Soviet space cooperation pact to be discussed with the Soviets at the Oct. 11–12 summit planned between President Reagan and General Secretary Mikhail S. Gorbachev."[21]

That October summit was held in Reykjavik, Iceland. Going into the meeting, Reagan and Gorbachev had hoped to find ways of advancing the deadlocked arms control talks that had been taking place in Geneva since their first meeting in November 1985. At the Reykjavik summit, they came close to a historic agreement to eliminate all of their nuclear weapons, but instead the talks broke down in bitterness as Reagan refused to give up SDI atmospheric and space testing as a price for such an agreement. Before this breakdown, the two leaders did apparently endorse plans to negotiate a new space cooperation agreement, although no discussions of this issue appear in the official U.S. transcript of the meeting.[22]

Government-to-government negotiations on the new agreement took place in Washington during the week of October 26, 1986. Heading the U.S. negotiating team was John Negroponte, assistant secretary of state for OES; his team included representatives from NASA and the DOD. Negroponte decided that the best way to defuse DOD concerns about technology transfer was to include one of those most concerned, DOD's Stephen Bryen, as a member of the negotiating team. Bryen was surprised at Soviet willingness to accept stringent technology controls in the agreement. To Negroponte, this acceptance showed how much the Soviet Union wanted the agreement. Agreement on the text of a new agreement was reached on October 30. *Aviation Week* reported that "U.S. space officials expressed surprise at reaching such a significant new agreement on civil space when there is such sharp disagreement about the military aspects of the Strategic Defense Initiative." There were 16 specific areas of U.S.-Soviet space science cooperation called out as part of the agreement; however, cooperation in human spaceflight or an ambitious robotic Mars sample return mission were not included. Even Secretary of Defense Weinberger applauded the outcome, telling Secretary of State Shultz: "I am quite pleased that the recent US-USSR negotiations on space cooperation came out so well."[23]

The plan had been to have Reagan and Gorbachev sign the space cooperation at their next summit meeting, but after the collapse of discussions in Reykjavik, it was not clear when that meeting might take place. So Secretary of State Schultz and Soviet Foreign Minister Shevradnadze signed the pact as they met in Moscow on April 15, 1987. The agreement established five joint working groups. Those groups would begin their work over the next year. Resumed cooperation was off to a modest, non-urgent start.

Space Cooperation and the Reagan/Gorbachev Summits

Even though the October 1986 Reagan/Gorbachev summit in Iceland had ended in disagreement and disappointment, there had been enough progress on eliminating short- and medium-range ballistic missiles to allow the negation during 1987 of an intermediate nuclear forces treaty. The United States and the Soviet Union agreed to a December 1987 summit meeting in Washington to allow Reagan and Gorbachev to sign the treaty with appropriate ceremony. As preparations for the meeting progressed, the State Department sought other possible areas in which the two leaders could agree. NASA proposed as "a significant civil space initiative … a joint program using space-based resources to understand Global Change." NASA suggested that such a program "could be characterized both as a major bilateral initiative and a major U.S. initiative for the International Space Year (1992), one that could grow into a broad multilateral effort."

This suggestion made it onto the December summit agenda. The joint statement at the conclusion of the meeting noted that "the two leaders approved a bilateral initiative to pursue joint studies in global climate and environmental change through cooperation in areas of mutual concern, such as protection and conservation of stratospheric ozone, and through increased data exchanges … In this context, there will be a detailed study on the climate of the future. The two sides will continue to promote broad international and bilateral cooperation in the increasingly important area of global climate and environmental change." There was no mention of cooperation in the other space science areas that had been part of the renewed space cooperation agreement.[24]

There was some disappointment in the modest character of the proposed cooperation. In the United States, The Planetary Society, led by Carl Sagan, Bruce Murray, and Lou Friedman, had developed a "Mars Declaration" that advocated "a systematic process of exploration and discovery on the planet Mars—beginning with robotic roving vehicles and sample return missions and culminating in the first footfall of human beings on another planet." Signatories to the declaration would "endorse the goal of human exploration of Mars and urge that initial steps toward its implementation be taken throughout the world." The Society noted that "after hearing The Planetary Society's arguments, the Soviet Union has repeatedly asked the United States and other nations to join in making this an international endeavor." By November 1987, signatories of the declaration included "liberals and conservatives; Democrats and Republicans; high-ranking Army, Navy, Air Force and Marine officers, and leaders of peace groups; astronauts and religious leaders; politicians and poets; Nobel Laureates and football coaches; ambassadors, university presidents and former presidential science advisors; former Cabinet and sub-cabinet officers;

leaders of industry and labor; and every Administrator of the National Aeronautics and Space Administration since its founding, except for the present incumbent."[25]

These arguments had found a responsive audience in the editorial board of *The New York Times*. The newspaper in December 1987 published an editorial calling for U.S.-Soviet cooperation in going to Mars, saying that Ronald Reagan had "lost interest [in space] since the *Challenger* exploded," but "it's not too late for him to leave a larger legacy than the militarization of space. He could put a joint mission to Mars on the agenda of his next meeting with Mr. Gorbachev in Moscow."[26]

This suggestion reflected the fact that Reagan and Gorbachev at their Washington summit also agreed to meet again, this time in Moscow in the first half of 1988. But it would be Mikhail Gorbachev, not Ronald Reagan, who took the lead in suggesting a joint U.S.-Soviet mission to Mars. In advance of the May 29–31 meeting, Gorbachev in a wide-ranging interview with *The Washington Post* said:

> We are going to invite the president ... to cooperate on a flight to Mars.
> Why shouldn't we try to work together? We have great experience, you have great experience—let us cooperate to master the cosmos, to fulfill big programs ... This is a field for cooperation that would be worthy of the Soviet and American peoples. And I will make that proposal to President Reagan.[27]

The Gorbachev statement sent summit planners "into a panic," and "sparked a series of emergency meetings ... on how Reagan should respond to the overture." Officials were reported as wishing that Gorbachev had saved his proposal for private meetings or diplomatic channels, since Reagan's "standing could suffer if he is forced to turn down a specific offer." In the buildup to the summit, next steps in space cooperation were already on the agenda, but they were relatively modest in comparison to a dramatic announcement of a joint mission to Mars.[28]

According to Roald Sagdeev, Gorbachev's interest in a human mission to Mars was a result of the advocacy by the leaders of the Soviet space industry "as an immediate goal to serve as a form of conversion for the military-industrial complex." The new Soviet leader's economic reform proposals threatened the privileged position of the aerospace industry, and it was seeking a way to maintain that position in the face of a defense build down. In addition, Gorbachev "was still emotionally involved in the anti-SDI rhetoric and thought that Mars could divert the American military-industrial complex from SDI." Also, "space and rocketry still seemed to remain the stronghold of the country. So why not make it the role model for the rest of the national industry?"[29]

Reacting to the news that Gorbachev would call for a joint mission at the summit, on May 26, as Reagan arrived in Finland on his way to the Moscow summit, The Planetary Society bought a full page in *The Washington Post* to publish a copy of the Mars Declaration. Advocates of greatly increased cooperation were hopeful that a breakthrough in U.S.-Soviet space relations was in the cards.[30]

Gorbachev followed through on his intention to ask Reagan to approve a joint Mars mission. At his first one-on-one meeting with Reagan on the afternoon of May 29, Gorbachev suggested Reagan "give thought to opening up even greater cooperation in space between the two countries. If that came out of this meeting as a common desire, that would be a good result … As he had already said to the *Washington Post*, now the Soviets would like the U.S. to begin cooperation on a joint mission to Mars." Reagan, in a noncommittal response, said that "the U.S. program had been set back by the *Challenger* tragedy. But he had asked his people to look into the General Secretary's suggestion." Gorbachev gave Reagan his proposed text for a paragraph on Mars as part of the joint statement issued at the close of the summit. It read:

> The two sides noted that preparation and implementation of a manned mission to Mars would be a major and promising bilateral Soviet-American program, which at subsequent stages could become international. It was agreed that experts from both countries would begin joint consideration of various aspects of such a program.[31]

Sagdeev recounts that as Gorbachev and Reagan walked through the Kremlin's "courtyard of the czars" before a state dinner on May 29, "while passing by the memorable relics of ancient Russian 'super' projects that were never used, Gorbachev passionately spoke about another grandiose project capable of matching the achievements of our great Russian ancestors." This project was the joint mission to Mars. At the dinner, Gorbachev introduced Sagdeev to Reagan, saying, "This is the man who is promoting the flight to Mars." Sagdeev suggests: "I had the funny feeling that Gorbachev's words stuck some chord of curiosity in Reagan." Then, "as if to underscore his apparently successful start on his Mars public relations campaign … Gorbachev added: 'Academician Sagdeev has friends and colleagues in America who share the vision of a joint flight,'" such as Carl Sagan. Sagdeev recounts: "In a fraction of a second I could tell that something had clicked the wrong way. The guest of honor appeared to lose interest immediately. Gorbachev apparently did not understand that there was not a great deal of political compatibility between Ronald Reagan and Carl Sagan"[32] (Fig. 22.1).

The mention of Sagan was not a deciding factor. Reagan had been advised before the summit meeting not to agree to any discussions of a joint human mission to Mars, on the grounds that the United States had not itself decided whether it wanted to undertake such a mission. Reagan's encounter with Sagdeev may have reinforced this advice. At any rate, there was no such agreement at the summit. The space-related paragraph in the joint postsummit statement said only that:

> Recognizing the long-standing commitment of both countries to space science and exploration, and noting the progress made under the 1987 U.S.-USSR Cooperative Agreement in the Exploration and Use of Outer Space for Peaceful

Purposes, the two leaders agreed to a new initiative to expand civil space cooperation by exchanging flight opportunities for scientific instruments to fly on each other's spacecraft, and by exchanging results of independent national studies of future unmanned solar system exploration missions as a means of assessing prospects for further U.S.-Soviet cooperation on such missions. They also agreed to expand exchanges of space science data and of scientists, to enhance the scientific benefit that can be derived from the two countries' space research missions. They noted scientific missions to the Moon and Mars as areas of possible bilateral and international cooperation.[33]

Fig. 22.1 President Ronald Reagan and Soviet General Secretary Mikhail Gorbachev in Red Square during a break in their May 1988 Moscow summit. Earlier in the meeting, Gorbachev had proposed to Reagan that the United States and the Soviet Union undertake a joint human mission to Mars. (Photograph courtesy of Reagan Presidential Library)

Conclusion

By the end of the Reagan administration, the partnership in space station *Freedom* had been finalized and modest U.S.-Soviet cooperation in space science had begun. These achievements laid the foundation for a multidecade space station partnership and the dramatic expansion of U.S.-Russian space cooperation after the collapse of the Soviet Union, including Russia becoming a partner in what became known as the International Space Station.

The United States pursued space cooperation, among other objectives, as a means of demonstrating to other spacefaring countries its leadership position in space capability and achievement. But in the post-*Challenger* period, that leadership was increasingly being questioned. According to a 1987 report of a Task Force on International Relations in Space of the NASA Advisory Council, "unless a sense of purpose, vitality, and long-term vision is restored to the U.S. civil space program so that other nations will find the United States the most attractive partner for association and cooperation, they will be less interested in accepting U.S. leadership." The Task Force recommended that "the President should personally exercise an active leadership role and publicly recognize that space is an area of activity with particularly strong links to U.S. aspirations and standing in the world."[34] It would be up to Ronald Reagan in his final years in the White House to rise to this challenge.

Notes

1. Richard J. Smith, *Negotiating Environment and Science: An Insider's View of International Agreements, from Driftnets to the Space Station* (Washington, DC: Resources for the Future, 2012), 94–96.
2. Letter from James Fletcher to John Poindexter, November 14, 1986, Box 14, Outer Space Files, RRL. Craig Covault, "U.S., Europe Deadlock on Station Participation," *AWST*, November 24, 1986, 18; "U.S., Europeans Hold Further Talks on Space Station," *AWST*, December 1, 1986, 33.
3. Smith, *Negotiating Environment and Science*, 96–97.
4. Memorandum from Cap Weinberger to Acting Assistant to the President for National Security Affairs, "Space Station Negotiations," December 1, 1986, and memorandum from Fred Ikle to Acting Assistant to the President for National Security Affairs, "Space Station Negotiations," December 4, 1986, CIA-RDP92B00181R001901730023-5, CREST.
5. National Security Decision Directive 257, "Guidance to the U.S. Delegation for Negotiations with Western Europe, Japan and Canada on the Space Station," February 3, 1987, https://www.reaganlibrary.gov/sites/default/files/archives/reference/scanned-nsdds/nsdd257.pdf.
6. Smith, *Negotiating Environment and Science*, 99.
7. Ibid., 94–100.
8. Department of State, "Space Station Negotiations: February 25–27 Talks with ESA Members States," undated, CIA-RDP92B00181R001901730003-7, CREST.

9. Memorandum from Ralph Braibanti, OES, to Participants, Space Station BackstoppingGroup,"March31,1987,CIA-RDP92B00181R001901720002-9, CREST.
10. Smith, *Negotiating Environment and Science*, 101; Donald Fink, "Space Station Skirmishes," *AWST*, April 20, 1987, 11. Letter from Cap [Weinberger] to George [Shultz,] April 7, 1987, and handwritten note from "CP" (Colin Powell) to Frank (Carlucci), April 10, 1987, Folder 16765, NHRC; Theresa Foley, "International Negotiations Stalled by NASA/Military Station Dispute," and Jeffrey Lenorovitz, "Europe Maintains Cautious Attitude on U.S. Station Cooperation," *AWST*, April 20, 1987, 18, 20, 23.
11. Letter from James Fletcher to Frank Carlucci, with attached paper "Space Station Negotiations: NASA Views," April 13, 1987, Outer Space Files, RRL.
12. Smith, *Negotiating Environment and Science*, 100–102; Theresa Foley, "Defense Department Backs Down on International Station Demands," *AWST*, April 27, 1987, 42.
13. This account of the negotiations is drawn from Smith, *Negotiating Environment and Science*, 102–107. See also Theresa Foley, "Station Talks Conclude Without Firm Commitments," *AWST*, October 26, 1987.
14. Theresa Foley, "Space Station Partners to Sign Pact Starting 30-Year Agreement," *AWST*, September 12, 1988, 30.
15. Office of the Press Secretary, The White House, "Statement by the President," October 30, 1984, RAC 14, Papers of George Keyworth, RRL.
16. Memorandum from John Zimmerman to Mark Palmer, "Future of US-Soviet Space Cooperation: Joint Planetary Research?" January 16, 1985; memorandum from Thomas Simons, Jr. to Richard Burt, "Expanded US-Soviet Space Cooperation," February 5, 1985; handwritten note from "Gil" (Rye) to Jack Matlock, February 28, 1985; memorandum from Nicholas Platt to Robert McFarlane, "US-Soviet Space Cooperation," March 22, 1985; memorandum from Gil Rye to Robert McFarlane, "U.S./Soviet Space Cooperation," March 25, 1985, Box 35, USSR II Files, Papers of Jack Matlock, RRL.
17. Memorandum from Robert McFarlane to George Shultz, "U.S./Soviet Space Cooperation," with attached Non-Paper, July 29, 1985, Box 34, USSR II Files, Papers of Jack Matlock, RRL.
18. "Impromptu Dinner Toasts, Geneva (Reagan-Gorbachev) Summit," November 19, 1985, Papers of Jack Matlock, RRL.
19. Richard Barnes, Memorandum for the Record, "US/Soviet Space Cooperation Agreement," July 15, 1986, Box 6, Papers of James Fletcher, NARA.
20. For Sagdeev's reflections on his career, see Roald Sagdeev, *The Making of a Soviet Scientist: My Adventures in Nuclear Fusion and Space from Stalin to Star Wars* (New York: John Wiley & Sons, 1994). The account of Sagdeev's initial interactions with Gorbachev is on 268–271.
21. Memorandum from Lew Allen to Administrator, "Expert Talks with Soviets," September 15, 1986, Box 6, Papers of James Fletcher, NARA; Craig Covault, "White House, Kremlin May Revive Cooperative Space Programs," *AWST*, October 6, 1986, 23.
22. "Reagan, Gorbachev Endorse Civil Space Cooperation," *AWST*, October 27, 1986, 20. A copy of the State Department transcript of the summit discussions can be found at http://www.thereaganfiles.com/reykjavik-summit-transcript.pdf.

23. Interview with John Negroponte, November 20, 2017; Craig Covault, "Soviet Negotiators Agree to New Space Cooperation Pact," *AWST*, November 10, 1986, 27; letter from Cap Weinberger to George Schultz, November 13, 1986, Box 6, Papers of James Fletcher, NARA.
24. Memorandum from Deputy Director of International Relations, NASA, to Michael Michaud, Department of State, "Proposed Civil Space Initiative for Summit," October 6, 1987, Box 6, Papers of James Fletcher, NARA. Ronald Reagan: "Joint Statement on the Soviet-United States Summit Meeting," December 10, 1987. Online by Gerhard Peters and John T. Woolley, APP, http://www.presidency.ucsb.edu/ws/?pid=33803.
25. A copy of "The Mars Declaration" was provided to the author by Lou Friedman, former executive director of The Planetary Society. Also, The Planetary Society News Release, "Planetary Society Finds Broad Support for the Mars Declaration," November 23, 1987, http://digitalcollections.library.cmu.edu/awweb/awarchive?type=file&item=678192.
26. "To Mars, Via Moscow," *NYT*, December 24, 1987.
27. Transcript of interview with Mikhail Gorbachev, *WP*, May 22, 1988, https://www.washingtonpost.com/archive/politics/1988/05/22/it-is-the-fate-of-our-two-countries-to-live-together/a846faa9-8130-45b9-b1e2-60a52df5191d/?utm_term=.63c4eb2920f7.
28. Craig Covault, "Moscow Summit to Expand U.S./Soviet Space Ventures," *AWST*, May 30, 1988, 16–17.
29. Sagdeev, *Making of a Soviet Scientist*, 310–311.
30. Personal communication from Lou Friedman.
31. The White House, Memorandum of Conversation, "The President's First One-on-One Meeting with General Secretary Gorbachev," May 29, 1988, https://nsarchive2.gwu.edu//NSAEBB/NSAEBB251/15.pdf.
32. Sagdeev, *Making of a Soviet Scientist*, 311.
33. Ronald Reagan: "Joint Statement Following the Soviet-United States Summit Meeting in Moscow," June 1, 1988. Online by Peters and Woolley, APP, http://www.presidency.ucsb.edu/ws/?pid=35902.
34. Task Force on International Relations in Space, NASA Advisory Council, *International Space Policy for the 1990s and Beyond*, October 12, 1987, 32, 40.

CHAPTER 23

The Quest for Leadership

Keeping the United States the global leader in space capabilities and achievements was a constant theme throughout Ronald Reagan's time in the White House. A "basic goal" of the initial Reagan National Space Policy, set out on July 4, 1982, was maintaining "United States space leadership." As the administration entered its final two years, there was increasing attention to ensuring the Reagan legacy, in space as well as in other areas. Ensuring that the United States would maintain its leadership in the face of a very active Soviet space effort and increasing competition from Europe and Japan was a major focus of space policy-making in the Reagan administration during 1987–1988. One aspect of that effort was to associate other countries with the U.S. space effort as a means of demonstrating the superior U.S. status. But that strategy presumed that the United States indeed had a space program worthy of the leading position. This was not assured.

Indeed, there was a good reason for concern about the standing of the U.S. civilian space program. The October 5, 1987, cover of *Time* magazine displayed an image of the May 1987 launch of the Soviet *Energia* heavy-lift booster with the headline "Russia Takes the Lead"; the magazine's story was titled "Surging Ahead: The Soviets Overtake the U.S. as the No. 1 Spacefaring Nation." An assessment of European and Japanese space efforts prepared by the Central Intelligence Agency (CIA) suggested that the claim to U.S. leadership was based "on previous technology programs" and "past accomplishments." Because "civil space programs abroad" were growing in capability and ambition, said the CIA, U.S. space leadership was "lost in certain areas" and "at risk in others."[1]

A NASA Leadership Initiative

National Aeronautics and Space Administration (NASA) Administrator James Fletcher had become convinced during the post-*Challenger* debate in 1986 that NASA needed to set challenging goals for the post-1995 period in order

J. M. Logsdon, *Ronald Reagan and the Space Frontier*,
Palgrave Studies in the History of Science and Technology,
https://doi.org/10.1007/978-3-319-98962-4_23

to justify its shorter-term science and engineering efforts, in particular the space station. The head of NASA's Jet Propulsion Laboratory, Lew Allen, in early 1987 observed that Fletcher "appears to have concluded that he cannot muster needed support for NASA unless he selects a vision for the future (next century) that will give purpose and focus to current efforts." Fletcher wrote to Ronald Reagan on January 30, 1987, suggesting that the country needed "in the next few months to make some decisions on the long-term needs and goals of the civil space program." He observed a "growing consensus ... that the long term goal of the civil space program should be a manned base on Mars." Ronald Reagan in a message to NASA employees on the first anniversary of the *Challenger* accident had said: "The space station will be our gateway to the universe, our foothold in outer space, the keystone of our space program. With it as our base camp, we will be able to reach the planets and, perhaps one day, to the stars." Fletcher was hoping to take advantage of these sentiments.[2]

NASA in late 1986 and early 1987 had carried out an internal assessment of its future objectives and had settled on a future goal of "expanding human presence beyond the Earth into the Solar System"; the agency hoped to get White House endorsement of that goal. A March 1987 report on NASA goals produced by a task force of the NASA Advisory Council chaired by Apollo 11 astronaut Michael Collins argued that "the United States must be preeminent in space, and must be so perceived by its citizens, industrial nations, and the third world." The report proclaimed that "recognized leadership absolutely requires the expansion of human life beyond the Earth, since human exploration is one of the most challenging and compelling displays of our spacefaring activities." It added: "We should make exploring and prospecting on Mars our primary goal, and state so publicly." *The New York Times* reported that "momentum is building in the space agency and among science and aerospace leaders to make Mars the next major goal of the American civil space program" and that "the Mars 'initiative' is expected to be the central element in policy recommendations NASA will present to President Reagan in the next few months." One scientist quoted by the *Times* suggested that Reagan "might be in a mood to seize the opportunity to end his Presidency on a positive, forward-looking note" and thus "could dignify his Presidency with this one act."[3]

Ride Report

There were thus converging sources of advice to the NASA leadership that the appropriate long-range goal for the space agency was resuming human exploration beyond Earth orbit, with Mars as a particular goal. In addition to relying on those sources, Fletcher chartered an ad hoc internal study led by Sally Ride. She had been a member of the Rogers Commission investigating the *Challenger* accident; as that group completed its work, Ride had decided not to seek a third space flight and had come to NASA headquarters seeking a next way to help the agency face the future. Fletcher gave Ride the title special assistant to the administrator for strategic planning. Her study effort would be on a fresh look at NASA's future programs and goals. It would be carried out by a ten-person group. Between August 1986 and mid-1987, that group consulted with a wide range of people inside and outside of NASA, including in particular younger NASA employees with a stake in the agency's future.

The study group came up with four potential leadership initiatives for consideration:

1. ***Mission to Planet Earth***: a program that would use the perspective afforded from space to study and characterize our home planet on a global scale.
2. ***Exploration of the Solar System***: a program to retain U.S. leadership in [robotic] exploration of the outer solar system, and regain U.S. leadership in exploration of comets, asteroids, and Mars.
3. ***Outpost on the Moon***: a program that would build on and extend the legacy of the Apollo Program, returning Americans to the Moon to continue exploration, to establish a permanent scientific outpost, and to begin prospecting the Moon's resources.
4. ***Humans to Mars***: a program to send astronauts on a series of round trips to land on the surface of Mars, leading to the eventual establishment of a permanent base.

These initiatives were included in the group's August 1987 final report, titled *Leadership and America's Future in Space*. Reflecting that title, the report gave extended attention to defining the term "leadership," saying:

> Leadership cannot simply be proclaimed—it must be earned…
>
> Leadership does not require that the U.S. be preeminent in all areas and disciplines of space enterprise … Being an effective leader does mandate, however, that this country have capabilities which enable it to act independently and impressively when and where it chooses, and that its goals be capable of inspiring others—at home and abroad—to support them. It is essential for this country to move promptly to determine its priorities and to make conscious choices to pursue a set of objectives which will restore its leadership status.

Ride's report rejected a preference for a focus on human missions to Mars, saying that "settling Mars should be our eventual goal, but it should not be our next [human exploration] goal." That goal, the report suggested, should be establishing a lunar outpost in order to "push back frontiers, not to achieve a blaze of glory." The report also suggested that NASA establish an Office of Exploration, a step that was taken even before the report was publicly released.

There was no press conference as the report was released, and Ride, ever protective of her privacy, did not brief the White House or congressional committees on its contents. This lack of follow-up may have limited the report's influence on subsequent policy decisions, even though its concepts were widely known and discussed. *Time* commented that "the Administration evidently filed Ride next to Paine," that is, ignored the Ride report as it had so far ignored the 1986 report of the National Commission on Space. As she had planned when she came to NASA headquarters a year earlier, Ride retired from NASA as the report was released.[4]

The recommendations of the Ride report were in fact an excellent complement to those of the 1986 recommendations of the National Commission on Space, which had urged the United States to take the lead in "exploring, prospecting, and settling the solar system."[5] The Ride report's focus was on steps NASA could take in the short-to-medium term to support U.S. space leadership; the National Commission report, with its 50-year perspective, suggested the results of taking those steps.

Soviet Competition

In addition to trying to present Ronald Reagan and his associates with attractive prospects for a long-range space initiative, part of Fletcher's strategy for gaining White House attention to space issues was to make sure that Reagan was personally aware of the fact that the Soviet Union was taking significant strides in increasing its civilian space capabilities. To this end, he worked with the Intelligence Community to put together a briefing for the president on Soviet space activities, including space station *Mir*, in orbit since 1986, the May 1987 launch of the very heavy-lift booster *Energia*, and the imminent launch of a Soviet space shuttle. That briefing was scheduled for June 30, 1987.

White House Science Adviser William Graham objected, however. He wrote to Chief of Staff Howard Baker on June 24, saying that "despite my explicit requests to the NASA leadership, I have not been consulted on the policy recommendations likely to arise in this discussion … I have serious reservations as to the message of the briefing as I understand it through informal channels." What worried Graham was NASA having the chance to solicit presidential support for a NASA response to the Soviet space buildup without Graham or other NASA skeptics in the room. Graham suggested that "broader space policy issues … should also be brought to the President's attention, but only with the full participation of NASA, DOD, DOT and other affected agencies … Let me suggest that the subject briefing be postponed until a balanced, comprehensive set of views can be put forward."

National Security Adviser Frank Carlucci disagreed with Graham. He wrote to Chief of Staff Baker, saying that he "remained convinced the briefing should go forward as scheduled." He added: "When you, I and Jim Miller reviewed it on June 1, we agreed that it contained information that the President would find useful. Given recent Soviet launch tests and public announcements of far-reaching space initiatives, we believe that it is important to provide the President with the perspective this briefing contains." Carlucci did not succeed in his objection; based on Graham's concerns, the briefing was postponed. The postponement "angered NASA and further strained relations between space agency officials and Graham." Fletcher met with Baker on July 27 to "underline the importance of getting the space intelligence briefing back on the President's calendar as soon as possible."[6]

The briefing was rescheduled for August 7. The president spent almost an hour listening to a discussion of the Soviet space program and what should be the U.S. reaction to claims that the Soviet Union was now the leading nation in space.

Fletcher as Focus of Criticism

At this point in time, James Fletcher was not operating from a strong base in the space community. While he was striving to convince the White House of the need to set long-range space goals, his efforts were not always receiving positive reviews. A July 27, 1987, editorial in the widely read trade weekly *Aviation Week & Space Technology* titled "Space Leadership Void" commented that "reports circulating within NASA, throughout the aerospace/defense industry and in other government agencies, indicate that the NASA administrator has failed to either inspire or lead NASA or surround himself with a vigorous upper management that can. Morale is at a low ebb throughout NASA, programs and projects are in disarray and many capable managers have either left the agency or are contemplating departures." Noting that Fletcher in 1986 had protested that he did not want to return to NASA, the editorial suggested: "If that attitude continues to inhibit his performance as administrator, then he should in good conscience resign." But "a far better course of action" would be "for Fletcher to seize the initiative and demand the means to carry out the mandate he so reluctantly accepted."

The editorial also spread some of the blame on Ronald Reagan, "who gave the impression he would undertake a bold and visionary space effort when he became President." If Reagan did not support Fletcher in his efforts to exert space leadership, "history will judge him as the President who allowed the U.S. to become a second-class spacefaring nation."[7]

These were strong words, not entirely justified. Fletcher was trying to convince the White House to embrace a major space initiative, and the jury was still out on whether the president and his associates would support what he might propose. But the editorial did reflect the reality that many of the first-term space accomplishments of the Reagan administration were in mid-1987 seen to be in jeopardy.

SIG (Space) Reviews National Space Policy

As he sought support for the NASA leadership initiative, Fletcher was also still complaining about the involvement in making decisions on space issues of people from Cabinet departments such as Commerce and Transportation. He met with Cabinet Secretary Nancy Risque in late July; Risque commented rather sarcastically that Fletcher, "having exhausted all avenues … to get into the White House and to the President," was finally "giving us—Cabinet Affairs—

an opportunity to show the right stuff." The meeting between Fletcher and Risque included "a rather frank conversation about the President's use of cabinet government." After the meeting, Risque commented: "Dr. Fletcher still clearly believes any interagency process is/would be an intrusion into NASA's activities (and policy development)."[8]

However much Fletcher resented interagency engagement in setting civilian space policy, he could not escape it. By mid-1987, the fate of the NASA space leadership initiative became entangled with an Office of Science and Technology Policy (OSTP) and National Security Council (NSC)-initiated review of the July 4, 1982, National Space Policy. This review would be carried out under the direction of the Senior Interagency (Space) [SIG (Space)]. It would also be taking place at the same time as the ongoing examination of commercial space initiatives by the Space Commercialization Working Group of the Economic Policy Council (EPC) described in Chap. 21. Many of the same people participated in both reviews, but organizationally they remained separate. Meeting with Congressman Bill Nelson on July 30 to discuss the SIG (Space) review, National Security Adviser Frank Carlucci told Nelson that it was "not premised on an assumption that existing policy contains major flaws, or that the U.S. is becoming a second-rate space power ... We do not anticipate any fundamental changes in direction, but we aim to achieve a more coherent and integrated statement of policy."[9]

The principal rationale for the space policy review was the number of significant changes in the space arena since the 1982 policy was issued. They were spelled out in the terms of reference for the review: "establishment of SDI; commitment to a U.S. space station; STS and ELV failures, resulting in a policy of maintaining a more balanced mix of space launch vehicles; commercialization of ELVs; organizational initiatives in several governmental agencies; National Commission on Space report; increased foreign competition in space launch services, remote sensing, etc.; constraints on ASAT testing; continued progress of Soviet space program in both military and scientific areas." The schedule for the review called for it to be completed by the end of October, in time for any budget-related decisions to be incorporated into the Fiscal Year 1989 administration budget proposal. The terms of reference also noted that "the U.S. budget situation, the Administration's commitment to deficit reduction, and the Gramm/Rudman/Hollings legislation require that the budget implications of all proposed policy revisions be given careful scrutiny." This admonition did not bode well for an expensive space leadership initiative.

The lead NSC staff person for the policy review was Air Force Colonel Roger DeKok, who had replaced Jerry May in May 1987. DeKok came to the NSC from the Air Force Space Command. Unlike May, who had not been a success in his position as NSC director for space programs, DeKok would come to be highly regarded by those he worked with; one called him a "true gentleman" who was able to reconcile differing points of view and avoid personal and bureaucratic conflicts. DeKok would chair the Interagency Group (Space) [IG (Space)] during the policy review. Most of the review effort would be carried out by working groups reporting to IG (Space); there would be limited engagement in the review by the senior-level officials who were members of SIG (Space).[10]

NASA Moves Forward

As the space policy review got underway, with an initial meeting of IG (Space) on July 31, *Aviation Week* reported that "the White House and NASA have begun to recognize that the U.S. has lost its international leadership in space" and that "President Ronald Reagan, Vice President George Bush and key White House staff members are becoming involved in the effort to regain space supremacy." The report added that "the Administration has been prompted into action by pressure from NASA Administrator James Fletcher ... Fletcher earlier could not attract White House attention on vital space issues." As noted earlier, Fletcher on August 7 participated in a briefing to Reagan on the Soviet space program, and he had met with Bush on August 4 and again on August 10. According to one NASA official, prior to Fletcher's pressure "influential White House staff members" did not "have a good understanding of what is already underway or the future potentials of the civil space program." *Aviation Week's* editor, Donald Fink, suggested that the increased White House attention to space was "heartening and raises a glimmer of hope that President Reagan will finally be persuaded to take an active part in revitalizing what was once the world's preeminent space program. He has a great deal to do and little time to do it before his term of office ends next year."[11]

By mid-September Fletcher had also met with Attorney General Ed Meese, White House Chief of Staff Howard Baker, Office of Management and Budget (OMB) Director James Miller, Domestic Policy Adviser Ken Cribb, Cabinet Secretary Nancy Risque, and Science Adviser Bill Graham to alert each of them to the forthcoming NASA proposed initiative. The reception to his message was described as "generally positive with no evidence of major opposition to a new long-range initiative, but concern about cost."

Fletcher was advised that "we are passing through a turbulent period vis-à-vis the White House and Congress, but, on balance, the overall thrust of events seems moderately favorable for a bold, new initiative." Fletcher's advisors were uncertain "whether the Administration would place a leadership initiative sufficiently high on its agenda, and retains the time, persuasiveness, and political strength to promote and 'sell' the initiative to the Congress and the American people." But they suggested that whether or not this was the case, "the President deserve the chance to assert and ensure his stated policy of U.S. civil space leadership."[12]

Expanding Human Presence

As the space policy review got underway, its initial step was to compare U.S. national security and civilian and commercial space activities with those of both the Soviet Union and U.S. allies. That comparison, as noted in the introduction to this chapter, suggested that U.S. space leadership was increasingly under challenge. It also concluded that rhetoric suggesting that the United States was in danger of becoming a second-tier space power was overblown. Another activity was to develop a "policy baseline" that consolidated all formal space policy decisions embodied in the many National Security Decision

Directives since 1982 into a single document as the starting point for a new statement of National Space Policy. This baseline would be used to determine "whether deletions, additions, or modifications" of the 1982 policy statement were necessary.

One suggestion for an addition to the 1982 policy soon emerged, and would continue to be debated until the final policy statement was approved in December 1987. That suggestion was "to add an explicit endorsement for expansion of human presence in space beyond the earth into the solar system." This idea had emerged from NASA's internal review of the agency's strategic goals earlier in 1987. NASA saw it as a way of incorporating into national policy the results of the National Commission on Space and Ride reports, the NASA Advisory Council study, and NASA's other space leadership initiative efforts without requiring a decision on a specific destination or a specific schedule. By this time, it had become evident that the financial implications of proposing a program to send humans to the Moon or Mars were too great to have any chance of gaining administration approval; the human expansion statement was seen by NASA as the maximum commitment possible.

NASA suggested that adding a statement endorsing human expansion would be "21st century-oriented and generally endorses a much needed overall direction for manned civil space activity." Arguments offered in support of the proposed addition included the following:

- The solar system is the logical "next frontier" for expanding mankind's understanding and use of the universe.
- Expansion can open up new options for the human species, outside the limits of planet Earth. These include the expansion of the human economy through access to vast new resources of materials and energy, and the establishment of new economic enterprises. In the longer term, humans may be able to establish permanent settlements on some solar system bodies.
- The new language projects an outward-looking, strongly competitive America, reaping the many benefits that experience has shown will redound to Americans and all mankind. It rejects an inward-looking America that sees future change largely in terms of threat, cost, and risk.
- Opposition to this language is based in part on its allegedly overly general nature. This is a "catch-22" procedure which, if followed in recent years, would have delayed and possibly killed many of the president's historic initiatives.

Rebuttals to these arguments were:

- Inclusion of language endorsing expansion of human presence into the solar system creates raised expectations within the involved constituencies. Such expectations inevitably lead to near-term pressure to commit to specific programs—whether or not these programs serve the long-term needs of the overall national space program.

- Inclusion of language will be viewed as just another "pie in the sky" goal with no concrete way to achieve it. As Sally Ride has pointed out, "Leadership cannot simply be proclaimed—it must be earned." This nebulous statement simply proclaims.

Leading the opposition to the proposed statement, not surprisingly, was OMB, which argued that the statement would commit the nation "to huge expenditures for space exploration." The Department of Defense joined OMB in opposing the addition. The Department of State joined NASA in support of the human expansion thrust.[13]

Although the NSC had hoped to finish the policy review by the end of October, this turned out to be impossible. IG (Space) and then SIG (Space) met in November to hear reports from the various working groups; it turned out that there were continuing disputes over the language of the new policy statement. Another IG (Space) meeting was scheduled for December 11, to be followed by a SIG (Space) meeting on December 17. As those meetings approached, *Aviation Week* reported that the review was "deadlocked on whether or not U.S. space policy should call for a goal of preeminence in manned orbital flight and a second goal to expand man's presence into deep space beyond Earth orbit." Agreement on including these two goals and on other issues in dispute was reached by the time of the IG (Space) meeting; the December 17 SIG (Space) meeting, chaired by Deputy National Security Advisor John Negroponte, unanimously approved a new statement of National Space Policy and sent it to Ronald Reagan for his signature.[14]

A State of the Union Proposal?

Even before the policy statement was finally approved for transmittal to the president, NASA's Fletcher suggested that Ronald Reagan in his final State of the Union Address "should make a strong leadership statement on the future of the civil space program endorsing the proposed policy initiative on expansion of human presence and activity into the solar system." Fletcher noted that "members of the White House staff" had recently discussed with NASA "the possibility that the President's State of the Union message include a statement on the space program." He suggested that Reagan "has the opportunity to set forth a positive vision that will shape the space agenda for decades to come without an unacceptable near-term budget impact." This would "leave a legacy that ensures technological, economic and quality of life benefits for generations of Americans." Fletcher emphasized that NASA was not asking the president "to endorse a specific program of manned space exploration at this time. It is premature to decide now whether Mars, the Moon, or even another body in the solar system represents the appropriate pathway for future exploration."[15]

Reagan Approves Policy; Its Release Delayed

Ronald Reagan approved the new National Space Policy on January 5, 1988, as National Security Decision Directive-293. It was not immediately clear whether Reagan would announce the policy's highlights during his January 25 State of the Union Address or whether it would be made public through some other means. Even though the policy statement was classified at a very high level since it discussed military and intelligence space activities, much of its NASA-related content was soon leaked. *Aviation Week & Space Technology* in its January 18 issue reported that "President Reagan has signed a new National Space Policy aimed at providing direction to the U.S.'s faltering civil space program and setting NASA on a course for eventual development of a manned lunar base and manned flights to Mars." The story added that "NASA managers believe the policy represents a major victory for the embattled space agency, which fought hard against opposition from the Office of Management and Budget and other agencies, especially on key points regarding manned flight policy." An editorial in the same issue observed that "President Reagan has pumped badly needed steam into NASA's boilers … Our response … is a hearty 'right on!'" But, the editorial added, "pressure must be kept on those in the political process who choose to be nearsighted, faint of heart or preoccupied with counting beans." It asked "what better forum" for the president to announce the new policy "than next week's State of the Union address?"

The magazine's "scoop" with respect to the new policy undercut this suggestion. The story was quickly picked up by other media outlets, so there was no possibility of a dramatic presidential announcement akin to John Kennedy's May 1961 call for sending Americans to the Moon "before this decade is out." The following week, *Aviation Week* reported that its "exclusive story … changed White House planning," leaving the president's staff "puzzling how to announce something that had already been announced." As it turned out, there was no mention of the space program in Reagan's January 25, 1988, State of the Union speech.[16]

Instead, the White House decided to delay the formal release of an unclassified "fact sheet" summarizing the new policy until it could be made public at the same time as the results of the EPC review of space commercialization initiatives discussed in Chap. 21. That review in early January was deadlocked over the issue of how to go about supporting a commercially developed facility for microgravity research. The deadlock was not broken until a January 14 EPC meeting; the president approved the EPC-recommended set of initiatives only on February 9. A White House press conference to announce both the new National Space Policy and the 15 commercialization initiatives, including the commercially developed space facility, took place on February 11.

A New National Space Policy

An 11-page "fact sheet" summarizing the new policy was distributed at the press conference, as was a separate listing of the commercial space initiatives. With respect to the civilian and commercial elements of the new national space policy, most of which were unclassified, the language in the fact sheet was very close to that in the actual policy directive. The new policy stated that "a fundamental objective guiding United States space activity has been, and continues to be, space leadership." Echoing the formulation that had appeared in the Ride report, it noted that "leadership in an increasingly competitive international environment does not require United States preeminence in all areas and disciplines of space enterprise. It does require United States preeminence in key areas of space activity critical to achieving our national security, scientific, technical, economic, and foreign policy goals."

The policy listed six "overall goals of United States space activities." Sixth among them, "as a long-range goal," was "to expand human presence and activity beyond Earth orbit into the solar system." At the press conference accompanying the release of the fact sheet, NASA's Fletcher noted that this was the first time such a goal had been included in national space policy and that "this is a goal of enormous significance with potentially historic future implications." The policy noted that to begin to implement the goal, NASA would carry out "the systematic development of technologies necessary to enable and support a range of future manned missions." Specific destinations such as the Moon or the Mars were not mentioned; rather the technology development effort, called Pathfinder, would be "oriented toward a [future] Presidential decision on a focused program of manned exploration of the solar system."

The new National Space Policy was comprehensive in character, dealing with the civil, national security, and commercial space sectors. Reacting to the conflicts between NASA and the advocates of space commercialization, it said "the United States government shall not preclude or deter the continuing development of a separate, non-governmental Commercial Space Sector." At the press conference, Secretary of Commerce William Verity, in a slap at NASA, said that the new policy "ensures that no government agency will erect roadblocks that will hamper the activities of the private space industry. In fact, it specifically forbids such action." The policy validated the continuing role of the SIG (Space) as the forum for discussing issues related to national space policy, while the Commercial Space Working Group of the EPC would provide a "high-level focus" for commercial space issues. Creating a National Space Council had been considered, and rejected, during the review.[17]

Reactions to the New Policy

Reactions to the new national space policy were mixed. The Cleveland *Plain Dealer* said that the policy "evoked the best memories of NASA's great accomplishments" and was "a long overdue course correction" that would continue NASA "at the vanguard of the human exploration of the universe." The Dallas *Morning News* called the policy "comprehensive and farsighted" and "thoughtful." In an editorial in its February 29 issue, *Aviation Week* suggested that "one had to be an astute observer of the Washington bureaucratic process to ferret out the true significance" of the new policy, adding that "while any action on the U.S. space front is a welcome change from the stagnation and confusion that prevailed in the wake of the shuttle Challenger accident, Washington's political and budgetary realities make it painfully clear the policy statement comes much too late and offers too little."

There was particular criticism of Ronald Reagan; the editorial said that it was "disconcerting that President Reagan chose not to personally identify himself with the national space policy statement. He included no mention of it in his last State of the Union message to Congress and he failed to give it even token support by participating in the White House briefing at which it was finally unveiled. Is this any way for the 'Great Communicator' to inspire Washington officialdom, much less a U.S. public … Is the state of the U.S. space program such an embarrassment to the Reagan Administration that it wants only to go through the motions of articulating a space policy?"[18]

This criticism was harsh, but not unfair. Since approving a new approach to space station development in April 1987, Reagan had not been personally involved to any significant degree in space matters. The SIG (Space) review of national space policy and the EPC debates over commercial space initiatives had taken place without significant senior-level White House involvement; the recommendations emerging from both reviews had come to the president as consensus documents after extensive bureaucratic conflict. It is not clear whether Reagan was aware of the controversies that had preceded his approval of the review's conclusions. Fletcher at NASA had become increasingly frustrated at the president's lack of engagement with NASA's concerns. With the issuance of the new space policy, congressional resistance to the administration's space plans, and the imminent return of the space shuttle to flight, this situation was soon to change.

Ronald Reagan Reengages

The president made his first space-related speech in over a year on April 19, 1988. Speaking to a dinner organized by the Electronics Industry Association, Reagan said: "Tonight I'd like to talk to you about our nation's commitment to leadership and imagination … I'm talking about our space program. I believe it is time to look ahead and envision breaking the bounds of earthbound imagination … I look to the time, before the end of the first decade of the next century, when we may have manned visits to other planets." He asked: "Can we afford to stop our exploration and wait for others to pass us?"[19]

In that speech Reagan also gave strong support to the space station program. That support was needed as the space station came under attack in the Senate in mid-1988. As noted in Chap. 21, Reagan signed two letters to key members of Congress in support of the station in June of that year. These actions did not satisfy Fletcher. He wrote to new White House chief of staff Ken Duberstein on July 8, saying that "it is important that the White House continue to keep the 'heat' on Congress" and that it was "essential for the President to be seen as visibly associated with the space program." Fletcher suggested, among other possibilities, "a Reagan visit to NASA's Johnson Space Center to ... visit with the next Space Shuttle crew."

The shuttle was at this point scheduled to return to flight in August or September. When his suggestion did not get an immediate response, Fletcher on August 18 again wrote to the White House, suggesting that Reagan "visit the Johnson Space Center near Houston, Texas, during mid-September for the purpose of meeting the crew of Space Shuttle Discovery." Such a visit, he proposed, would be "a symbolic affirmation that the space shuttle is nearing its resumption to flight, the Space Station Freedom program is on track, and the country is behind the effort."

Fletcher met with Cabinet secretary Nancy Risque on August 22 to discuss this proposal. In preparation for that meeting, one of Risque's assistants noted that Fletcher was "once again singing the blues over the lack of presidential involvement and identification with the space program" and suggested that Fletcher would probably complain that Democratic candidates Michael Dukakis and Lloyd Bentsen were "paying more attention to space than President Reagan and Vice President Bush."[20]

Space and Manifest Destiny

Fletcher's persistence paid off. Reagan was planning a campaign visit to Texas on September 22, and a visit to the Johnson Space Center was scheduled between political events. This was a week before the planned return-to-flight launch of space shuttle *Discovery* on September 29. Because this was within the quarantine period for the shuttle's five-man crew, Reagan underwent a medical examination aboard Air Force One on the way to Houston to make sure he was not carrying any germs that might infect the astronauts.

Ronald Reagan's speech at the Johnson Space Center (Fig. 23.1) clearly demonstrated that he had not lost his belief in the great promise and special character of the U.S. space program. Indeed, his remarks were perhaps the most optimistic and far-reaching of any he had made during his eight years as president. As he stood next to the *Discovery* crew, he said:

> We are a nation that can achieve great dreams. Somewhere in America, there is alive today a small child who one day may be the first man or woman to ever set foot on the planet Mars or to inhabit a permanent base on the Moon. Let every child dream that he or she will be that person, that he or she may one day plant the Stars and Stripes on a distant planet...
>
> And you and I know that we're the nation that must do it, because in the next century, leadership on Earth will come to the nation that shows the greatest leadership in space. It is mankind's manifest destiny to bring our humanity into space; to colonize this galaxy; and as a nation, we have the power to determine whether America will lead or will follow.
>
> I say that America must lead. The Nation that has achieved the greatest human freedom on Earth must be the Nation to create a humane future for mankind in space, and it can be none other. It is only in a universe without limits that we will find a canvas large enough for the vastness of the human imagination.

Fig. 23.1 Ronald Reagan exits the stage, followed by NASA Administrator James Fletcher, after addressing employees at NASA's Johnson Space Center on September 22, 1988. Reagan is carrying a flight jacket present by the astronauts who would be aboard the space shuttle as it returned to flight a week later, almost 32 months after the *Challenger* accident. (Photograph courtesy of Reagan Presidential Library)

Mankind's journey into space, like every great voyage of discovery, will become part of our unending journey of liberation. In the limitless reaches of space, we will find liberation from tyranny, from scarcity, from ignorance, and from war. We'll find the means to protect this Earth and to nurture every human life and to explore the universe. Let us go forward. This is our mission; this is our destiny.

Then Reagan recalled: "One cold January day in 1986, I read part of a poem to a nation in grief. I want to leave you today with the rest of that poem because it's a poem about joy."

> Oh! I have slipped the surly bonds of Earth And danced the skies on laughter-silvered wings;
> Sunward I've climbed, and joined the tumbling mirth of sun-split clouds and done a hundred things You have not dreamed of—wheeled and soared and swung High in the sunlit silence.
> Hov'ring there, I've chased the shouting wind along, and flung My eager craft through footless halls of air...
> Up, up the long, delirious, burning blue I've topped the wind-swept heights with easy grace, Where never lark, or even eagle, flew;
> And, while with silent, lifting wings, trod the high untrespassed sanctity of space, put out my hand, and touched the face of God.[21]

As he spoke in Houston, Ronald Reagan had less than four months remaining of his eight years as president. He had arrived in Washington in January 1981 enthusiastic about the link between space leadership and his vision of American exceptionalism. The belief in that link had persisted throughout his eight years in the White House. However, its translation into administration space policies, programs, and budgets had seldom matched Reagan's vision.

Notes

1. A copy of the 1982 policy, portions of which remain classified, can be found at https://www.reaganlibrary.gov/sites/default/files/archives/reference/scanned-nsdds/nsdd42.pdf. "Surging Ahead: The Soviets overtake the U.S. as the No. 1 spacefaring nation," *Time*, October 5, 1987; Central Intelligence Agency, "The World's Civil Space Programs: A Comparison," undated, OA 19825, Papers of Gerald Driggs, RRL.
2. Letter from Lew Allen to Mary Scranton, April 8, 1987, Box 10, and letter from James Fletcher to the President, January 30, 1987, Box 4, Outer Space Files, RRL. Ronald Reagan: "Remarks to Employees of the National Aeronautics and Space Administration on the First Anniversary of the Explosion of the Space Shuttle Challenger," January 28, 1987. Online by Gerhard Peters and John T. Woolley, APP, http://www.presidency.ucsb.edu/ws/?pid=34452.
3. Letter from Daniel Fink to James Fletcher, March 16, 1987, with attached final report of NASA Space Goals Task Force, Box 3, Papers of James Fletcher, NARA; John Noble Wilford, "The Allure of Mars Grows as U.S. Searches for New National Goal," *NYT*, March 24, 1987.
4. NASA, *Leadership and America's Future in Space*, August 1987, 21, 12, 54–55. Lynn Sherr, *Sally Ride: America's First Woman in Space* (New York: Simon & Schuster, 2014), Chap. 8. Interview with Alan Ladwig, May 6, 2016. Craig Covault, "Ride Panel Calls for Aggressive Action to Assert U.S. Leadership in Space," *AWST*, August 24, 1987, 26. "Surging Ahead," *Time*, October 5, 1987.

5. National Commission on Space, *Pioneering the Space Frontier* (New York: Bantam Books, 1986), 5.
6. Memorandum from William Graham to Howard Baker, "NASA Presentation to the President," June 24, 1987, and memorandum from Frank Carlucci to Howard Baker, "NASA Presentation to the President," June 26, 1987, https://www.reaganlibrary.gov/sites/default/files/digitallibrary/smof/cos/baker-howard/box-004/40-27-6912132-004-001-2017.pdf. Craig Covault, "White House, NASA Act to Regain U.S. Space Leadership Role," *AWST*, August 10, 1987, 18; letter from James Fletcher to Howard Baker, July 28, 1987, Box 15, Papers of Nancy Risque, RRL.
7. Donald Fink, "Space Leadership Void," *AWST*, July 27, 1987, 9.
8. Memorandum from Nancy Risque to Senator Baker, "Dr. Fletcher – NASA," July 27, 1987, Box 15, Papers of Nancy Risque, RRL.
9. Memorandum from Alison Fortier for Frank Carlucci, "Congressman Bill Nelson (D-FL) Meeting in Your Office on July 30, 1987 at 2:15 p.m.," with attached "Talking Points on Review of National Space Policy," July 29, 1987, Outer Space Files, RRL.
10. "Terms of Reference: Review of National Space Policy," July 28, 1987, CIA-RDP92B00181R001701640039-0, CREST. Interview with Courtney Stadd, May 16, 2016. Unfortunately, the Reagan Presidential Library as of this writing (May 2018) has not processed DeKok's papers; this account of his time at the National Security Council and of the 1987 space policy review is thus necessarily incomplete.
11. Craig Covault, "White House, NASA Act to Regain Space Leadership Role" and Donald Fink, "A Glimmer of Hope," *AWST*, August 10, 1987, 18–19, 11.
12. Memorandum from Phil Culbertson, Associate Administrator for Policy and Planning, to Administrator, "Long-Range Leadership Initiative," with attached "Proposed Strategy for Winning Presidential Approval of the Long-Range Leadership Initiative," September 11, 1987, Box 5, Papers of James Fletcher, NARA.
13. National Space Policy Review, "Issue Paper," October 30, 1987 Draft, CIA-RDP92B00181R001801700021-1, CREST.
14. "Washington Roundup," *AWST*, December 7, 1987, 17.
15. Letter from James Fletcher to Howard Baker, December 10, 1987, Box 15, Papers of Nancy Risque, RRL.
16. Craig Covault, "President Signs Space Policy Backing Lunar, Mars Course," and "New U.S. Space Policy," *AWST*, January 18, 1988, 14, 7; "Washington Roundup," *AWST*, January 25, 1988, 15.
17. Fact Sheet, "Presidential Directive on National Space Policy," February 11, 1988, 1–2, https://www.hq.nasa.gov/office/pao/History/policy88.html. Office of the Press Secretary, The White House, "Press Briefing by Administrator of NASA, James C. Fletcher, Commerce Secretary, C. William Verity, and Transportation Secretary James Burnley on Space Policy," February 11, 1988, CIA-RDP92B00181R001801700012-1, CREST. National Security Council, "Cleared Q&A on National Space Policy," January 22, 1988, CIA-RDP90M00005R000300060022-4, CREST.
18. "NASA's Course Correction," Cleveland *Plain* Dealer, February 15, 1988; "Reagan is Right to Stress Exploration, Commerce," The Dallas *Morning News*,

February 13, 1988, copies in Box 15, Papers of Nancy Risque, RRL.; Donald Fink, "Too Late with Too Little," *AWST*, February 29, 1988, 9.

19. Ronald Reagan: "Remarks at the Electronic Industries Association's Annual Government-Industry Dinner," April 19, 1988. Online by Gerhard Peters and John T. Woolley, APP, http://www.presidency.ucsb.edu/ws/?pid=35704.
20. Letter from James Fletcher to Kenneth Duberstein, July 8, 1988; letter from James Fletcher to Nancy Risque, August 18, 1988; memorandum from "Todd" to "Nancy," August 19, 1988, Box 17, Papers of Nancy Risque, RRL.
21. Ronald Reagan: "Remarks at the Johnson Space Center in Houston, Texas," September 22, 1988. Online by Peters and Woolley, APP, http://www.presidency.ucsb.edu/ws/?pid=34875.

CHAPTER 24

The Reagan Space Legacy

As noted in the first chapter of this study, aerospace historian Andrew Butrica in 2003 suggested that "Ronald Reagan's two terms as president saw the United States undertake more new and more large space initiatives than any previous administration since that of John Kennedy," and that the programs and policies initiated during the Reagan administration would "cause the 1980s to be remembered as a major turning point in space history." He concluded that "Reagan would pass on a lasting legacy to the nation's space program." By contrast, during the last year of the Reagan administration, journalist William Broad observed in *The New York Times* that Reagan's legacy in space was "more reach than grasp," and that Reagan, while "one of the biggest space enthusiasts to occupy the White House," would leave behind "a host of costly projects and innovative policies" but "little achievement in the heavens," with a space program characterized by "muddle and disarray."[1]

The preceding chapters provide detailed evidence with which to evaluate these apparently conflicting claims.[2] In reality, they are not actually mutually exclusive, and both seem to have some degree of validity. While there may have been few short-term payoffs from the projects and policies adopted by the Reagan administration, as Butrica wrote 15 years after Reagan left the White House positive results from Reagan's initiatives were beginning to appear. The space station by 2003 was permanently occupied by an astronaut and cosmonaut crew, and on its way to completion and to beginning to be the kind of productive research facility that had been anticipated as Ronald Reagan had approved its development 20 years earlier. The space station partnership with Europe, Canada, and Japan had endured during those 20 years as a visible example of nations working together for a common purpose, and had been amplified as Russia joined the program in the 1990s. Cooperation between the United States and Russia in human spaceflight had become a central feature of the space programs of both countries. Working within the policy and regulatory

J. M. Logsdon, *Ronald Reagan and the Space Frontier*,
Palgrave Studies in the History of Science and Technology,
https://doi.org/10.1007/978-3-319-98962-4_24

framework first set up by the Reagan administration, U.S. firms offering commercial launch services had had modest success, and two new entrepreneurial companies, Space Exploration Technologies (SpaceX) and Blue Origin, were just beginning to attempt to enter the space launch market. Several firms were developing commercial Earth observation businesses, and entrepreneurial firms were beginning to once again explore other opportunities for privately financed space ventures. After its return to flight in September 1988, with a revised context for its operation, the space shuttle had been successfully launched 110 times, mostly carrying out missions for the National Aeronautics and Space Administration (NASA), including deploying, repairing, and servicing the Hubble Space Telescope; launching several planetary probes and other robotic spacecraft; carrying the European Spacelab for in-orbit research; rendezvousing nine times with the Russian space station *Mir*; and beginning space station assembly. Thus, Butrica was somewhat justified in his positive assessment of the Reagan administration space legacy, even if just as his book appeared a second fatal shuttle accident destroyed shuttle orbiter *Columbia*, killing another seven astronauts.

There is also justification for Broad's more skeptical assessment. In the years after Reagan left the White House, there was continual criticism of the lack of a clear statement of purpose to guide the American space program; Reagan had not articulated such a purpose, despite repeated suggestions that he do so. Setting a long-term goal in his 1988 National Space Policy "to expand human presence and activity beyond Earth orbit into the solar system" had been too general to give the hoped-for direction to NASA's human spaceflight effort. In the aftermath of the 2003 *Columbia* accident, the *Columbia* Accident Investigation Board deplored "the lack, over the past three decades, of any national mandate providing NASA a compelling mission requiring human presence in space." The Reagan administration had not succeeded in providing such a mandate.[3]

The "disarray" with respect to the overall purpose of NASA's efforts that Broad had noted in 1988 had persisted. NASA was frequently accused of "drifting." Ronald Reagan in his space speeches had occasionally spoken of rapid movement into space by the U.S. private sector; that had not happened, as private sector access to space for research purposes was lost after the *Challenger* accident and as the investment community lost its enthusiasm regarding the potential for rapid and large profits from space-based activities. The space station had been repeatedly delayed and subject to frequent redesigns.

Ronald Reagan was a romantic in his attitude toward space achievement, seeing it as a positive part of human progress that was destined to go forward under U.S. leadership. He thus perpetuated in his public remarks an extremely positive view of the potential of space in both tangible and intangible dimensions, and of the central role of human spaceflight in achieving that potential. Those who believed that the focus of space activities should be on exploitation leading to Earth-oriented payoffs, rather than continued human exploration, were critical of Reagan's endorsing the notion that the space program was "the last frontier."[4]

Reagan's enthusiasm for the space program as a frontier activity was a genuine part of his personality and of his ability to lead. However, that enthusiasm was held in check both by Reagan's concern for government fiscal responsibility and by his "bottom up" decision-making approach, which gave space skeptics equal voice with fellow enthusiasts in preparing issues for presidential decision. Only occasionally, as in the case of the space station, did Reagan allow his personal preferences to overrule the counsel of his senior advisers or lead to space budget increases. This reality led to a persistent gap between Reagan's space-related rhetoric and the actual policies and programs of his administration.

Assessing the Reagan Space Legacy

This study has given particular attention to five areas of space decisions during the Reagan administration. Those decisions were: (1) related to the space shuttle; (2) related to a space station; (3) related to commercial space activities; (4) related to international space cooperation; and (5) related to the overall purpose and goals of the U.S. civilian space program. The Reagan administration legacy in each of these areas is assessed below.

Ultimately, a Realistic Shuttle Policy

During the Reagan administration, more policy attention was given to the space shuttle than to any other element of U.S. space activities. The administration inherited and initially accepted the belief, dating from the origins of the shuttle program, that space shuttle operations would become frequent, routine, and relatively inexpensive. It based its shuttle policy on that belief for several years, in the face of increasing evidence that it was basically a myth. NASA engineers and their industrial contractors had been during the 1970s able to develop a technologically advanced, highly capable shuttle, but not one that could be operated on anything like a frequent, routine, or low-cost basis.

In the early years of the Reagan administration, as the shuttle began flights, the policy framework for operating it was based on the inherited myth rather than the emerging technological reality. This was not totally surprising, since many in the NASA human spaceflight community were deeply vested in that myth. Given prior successes, they had begun to exhibit what Gary Brewer in a perceptive analysis written soon after the *Challenger* accident characterized as "self deception, introversion and a diminished curiosity" about information coming from outside that community. Given the centrality of the space shuttle to NASA's identity and indeed to the objectives of national space policy, it was not in the interest of those responsible for the shuttle to heed warnings that it was a risky system, difficult and expensive to operate safely, and that a policy of sole dependence on the shuttle for U.S. government access to space was deeply flawed. As Brewer observed, "If there is only one game in town, even the hint of failure cannot be tolerated." The message that NASA was carrying out a flawed policy was ignored by the NASA leadership. Even as every signal from those outside of NASA who were aware of the shuttle's problems and limitations "shouts 'change,'" suggests Brewer, the "institution plows resolutely along."[5] This was the case as James Beggs so strongly resisted Pete Aldridge's campaign to provide an expendable complement to the shuttle for national security launches, as advocates of commercializing expendable launch vehicles (ELVs) struggled to remove the shuttle from competing for commercial launches, and as NASA pushed to increase the shuttle flight rate to an unrealistically high level. It took a long time for the Reagan administration to recognize its policy for operating the shuttle was from its inception ill-conceived.

Brewer also observed that "it seems to require the shock of heavy cannon to loosen dissent in a perfect place."[6] Certainly the tragic loss of seven people in the *Challenger* accident qualifies as "the shock of heavy cannon," and ultimately did lead to a more realistic policy for space shuttle use. Even with the accident, NASA's new administrator, James Fletcher, insisted on the need for replacing Challenger with a new shuttle orbiter and resisted as long as possible the multiple pressures to remove the shuttle from its role as a launcher for fee-paying users. Fletcher was likely a particularly poor choice to adapt NASA's shuttle policy to a post-*Challenger* reality, given that he had been the agency's head as the shuttle myth was first perpetuated. He also strongly resented the involvement of other government agencies in what he viewed as NASA's prerogatives, including the guidelines for shuttle operations.

It is an interesting speculation to ask what might have happened had there been no shuttle accident. In principle, Ronald Reagan and his White House associates could have become aware of the shuttle's problems and at some point initiated a change toward a more realistic shuttle policy. Secretary of Defense Cap Weinberger wrote Reagan on January 23, 1984, two years before the Challenger accident, saying that the country had made a "serious mistake" in depending on the shuttle for sole access to space. Pete Aldridge waged a vigorous and visible campaign to end national security dependence of the shuttle. Advocates of ELV commercialization were vocal in their criticism of a "shuttle only" policy. But President Reagan had himself become strongly committed to viewing the shuttle's human spaceflight capabilities as an indication of American exceptionalism, and he was not likely easily to change that perspective.

The reality, of course, is that the accident did happen. Even though it took an intense bureaucratic struggle and what many saw as an excessively long time to come up with a realistic post-*Challenger* space transportation policy, the Reagan administration developed and promulgated such a policy. That policy, while employing ELVs for missions not requiring human presence, allowed NASA, once the shuttle returned to flight, to take full advantage of the shuttle's unique capabilities for its missions, and kept the space shuttle in operation for another 20-plus years as a visible symbol of U.S. space leadership. Thus, ultimately developing a realistic policy for space shuttle use must be counted as a positive Reagan space legacy, even if it was preceded by five years of operating the shuttle under false premises and a preventable accident that took seven lives and took the shuttle out of service for almost three years.

Taking the Logical Next Step

In his book *The Space Station*, Hans Mark tells of a December 3, 1983, telephone conversation with James Beggs during which the NASA administrator informed his deputy that President Reagan had approved going ahead with the space station. Mark's response to this message was that "this was a historic occasion" and that "only twice before, in May 1961 for Apollo and in January 1972 for the space shuttle, had a president set a major new direction for the American space program." He added that the decision "spoke well for the political vision and sense of history that President Reagan possesses."[7] Indeed, the decision to make a permanently occupied space station a centerpiece of American human spaceflight for what has by now been more than three decades was a very significant step.

It was not, however, "a major new direction." NASA in 1970 had deferred, not abandoned, space station development, recognizing that developing a space shuttle first would eventually lead to that development. Creating a human outpost in Earth orbit had been central to NASA's plans from the space agency's very earliest years. What the space station decision represented was, as Beggs was fond of saying, "the logical next step" in a long-existing vision of developing the space frontier. That vision predated the creation of NASA; some even argue that NASA was created to realize it. The vision is centered on a belief that at some point "humans would leave the Earth's surface and explore the universe, just as their ancestors had crossed oceans to explore foreign lands." An orbiting outpost was seen by most space visionaries as a necessary step in the process of human exploration of the solar system, first of all the Earth's Moon and Mars.[8]

James Beggs was a believer in that space vision; he was also an astute political operator. Beggs and Mark had agreed that getting Ronald Reagan to approve a space station would be a major objective during their time at NASA. Beggs had initially hoped that Reagan, with his positive sense of America's future, would quickly embrace the space vision and on that basis early on in his administration would give NASA the go-ahead on the space station. But by the second half of 1982, Beggs made the political judgment that, however personally enthusiastic about the promises of space Reagan might be, other administration officials were very unlikely to recommend that the president approve a space station as a necessary part of achieving an expansive—and expensive—space vision. The NASA leadership and its White House allies, in particular the National Security Council's Gil Rye and Cabinet affairs liaison Craig Fuller, were thus faced with developing a strategy for getting the space station before the president for decision without tying it to accepting a long-term vision for space, even as that vision was actually what made the station "the logical next step."

Their strategy was to make the station decision a test case of the space policy development process adopted early on in the Reagan administration, centered on the Senior Interagency Group for Space [SIG (Space)] as the forum for bringing space decisions to the president. In late 1982, NASA agreed to a year long SIG (Space) study to answer the question, "What policy issues must be identified and resolved in order to establish the basis for an Administration decision on whether or not to proceed with development of a permanently-based, manned space station?" The goal was to have that study completed and a recommendation before the president by November 1983. That schedule was in essence followed, even as strong opposition to the station developed during the year-long study effort.

Beggs and Mark were walking a tightwire in trying to get a positive recommendation on the station out of the SIG (Space) process without tying it to the primary rationale of its being a necessary step toward human exploration beyond Earth orbit. Presidential Science Adviser Jay Keyworth recognized what they were up to, and tried to get NASA to acknowledge the linkage. NASA was able to resist, and instead tried to build the case for the station on two main arguments. One was that the space station was a required or facilitating capability for a wide variety of desirable space missions, particularly including ones with potential economic payoffs; the other was that the station was a necessary step beyond the space shuttle to ensure U.S. space leadership in the 1990s and beyond. Lurking in the background was the reality that the Soviet Union had developed a first-generation space station and was on its way to replacing it with a more advanced facility.

The first of these arguments was basically a failure. NASA and its contractors could not come up with evidence that there were a variety of important space missions for which a station was required. Perceptive space veteran Jimmy Hill of the National Reconnaissance Office would comment in mid-1983 "the apparent method used by NASA to develop civil 'requirements' was to list any potential civil space activity (regardless of priority, requirements, cost or technical feasibility) as a civil sector mission for conduct on the space station." In addition, the national security community denied having any need for a station. Thus, the majority of SIG (Space) members were opposed to going ahead with the station program, on the grounds that it was not justified by any set of requirements and that there were better uses for the resources that the station would require.

It was the second, less tangible, "leadership" argument that carried the day with Ronald Reagan. As he made his case for going ahead with the space station to the president on December 1, 1983, Beggs said that a space station would be "a highly visible symbol of U.S. strength" and that "the stakes are enormous: leadership in space for the next 25 years." Reagan resonated to this claim; he in his many space-related speeches had identified U.S. space success with American exceptionalism, and thus was willing to give his approval to the space station even though most of his senior advisers opposed the program. This was unusual behavior for Reagan; on most policy issues he waited until there was some sort of consensus among his inner circle before giving his assent to a course of action. With the respect to the space station, the "leadership" argument indeed convinced Reagan to lead. In announcing his decision to approve station development, Reagan said, "America has always been greatest when we dared to be great. We can reach for greatness again. We can follow our dreams to distant stars, living and working in space."

There were limits to Reagan's support of the space station, however, mostly in terms of the administration's financial commitment to the station program. Another part of the NASA approach to getting Reagan administration approval had been what Howard McCurdy in his book *The Space Station Decision* calls an incremental strategy—NASA would develop a space station "by the yard," starting with a modest and relatively inexpensive design intended to be supplemented in the future. While the result of NASA's campaign to gain presidential endorsement of the space station was a positive decision, in practice it was not one that got the station program off to a strong start. Ronald Reagan's rhetoric in announcing his decision to approve the station was soaring, but the actual budget for the program in its early years was very modest. McCurdy describes the presidential decision (and that of the Congress in subsequent months) to move forward with the station program as "a very weak commitment," noting that "policy makers reserved the right to appropriate only enough money to advance the program year by year and to reconsider the purpose of the project as circumstances changed." This is indeed what happened, as the station went through a series of redesigns during the remaining years of the Reagan administration and there were continuing threats to its budget.[9]

It is valid to question whether the "weak" character of Ronald Reagan's space station decision was appropriate for a program intended to be central to U.S. space activities for several decades. NASA Apollo-era administrator James Webb once described to the author the space station decision as akin to providing an airplane just enough forward momentum to get it off a runway and flying, but continually near stall speed. This was not a recipe for program success. The hesitant character of the space station decision allowed for the instability that plagued the station program during its first decade, until inviting Russia to join the space station partnership gave the program a solid geopolitical rationale.

The space station decision was in some ways similar to that made by the Nixon administration with respect to the space shuttle program. Dale Myers, who had been in charge of the shuttle program at NASA headquarters in the 1970–1973 period as the shuttle program was getting started and who after *Challenger* had returned to NASA as deputy administrator, in 1987 asked Fletcher "whether we are entering the same situation that we entered with the shuttle in 1972." Then, reflected Myers, "we got starved for budget from the first year on, contributing to design decisions and operational stringencies that might have contributed to the *Challenger* accident." He and Fletcher wondered whether the station program was headed down a similar path.

While Ronald Reagan's decision to approve space station development did not reflect a commitment of the character required for stability in a long-term undertaking, it *was* a positive decision. Furthermore, when the space station came under congressional attack in 1987 and 1988, Reagan was willing to express his support for the program in communications to key members of Congress. Reagan's rhetoric in support of the space station identified it as central to his vision of the future in space; in January 1987, he declared: "The space station will be our gateway to the universe, our foothold in outer space, the keystone of our space program. With it as our base camp, we will be able to reach the planets and, perhaps one day, to the stars." It is thus appropriate to identify the space station as Ronald Reagan's most lasting space program legacy.

Laying the Foundation for Space Commercialization

Ronald Reagan and his associates came into office with an ideological conviction that American free enterprise system was a powerful force for ensuring both societal and individual well-being, and that the role of the government was to create a policy and regulatory climate that would allow American business to thrive. This belief intersected with a message coming from entrepreneurs, established businesses, and a NASA eager to justify its activities to the new administration, that there were manifold new business opportunities in space. This conjunction led to the Reagan administration devoting significant policy attention in its first term in office to providing, as stated in its 1982 National Space Policy, "a climate conducive to expanded private sector investment and involvement in civil space activities." That attention was much more driven by ideological belief than it was by either technological or market realities.

Elements of policy attention included White House attempts to privatize government's civilian Earth observation activities; to support the creation of a commercial space launch business based on the use both of the existing ELVs that were to be phased out as the shuttle became the sole launcher for government space missions and of new launch vehicles developed by the emerging entrepreneurial space sector; and to create a series of policy incentives to encourage private sector activity in orbit. On its part, NASA used the potential for high-value activities in orbit as a justification for developing a space station and claimed to be ready to partner with private sector actors in exploring profit-making space activities.

At least a few in the Reagan administration were skeptical of the appropriateness of these commercially oriented initiatives, seeing them, in the words of junior White House staffer Lehman Li, as quite similar to "Democratic proposals for an industrial policy. Government would be making policy targeted toward a specific industry. Industry would be proposing the government policy changes needed to improve the industry's competitiveness." But such skepticism was overridden by enthusiasm for what a former secretary of commerce described as "working factories whose commercial potential has been tested. Properly exploited they will help transform the nation's $100 billion investment in the space program into sustained world leadership in an area of technology with as yet unimaginable applications for commerce and manufacture." Forecasts of multibillion dollar annual sales of products manufactured in the microgravity environment of space grabbed White House attention.

Beginning in 1981, the White House encouraged efforts to find ways to privatize both meteorological and remote sensing Earth observation activities, ultimately leading to a 1985 transfer of ownership of U.S. Landsat remote sensing satellites to a private operator. By 1992, that initiative had failed, and the satellites were returned to government ownership and operation. In 1983, President Reagan approved a policy directive intended to facilitate commercialization of ELVs; an Office of Commercial Space Transportation was established in the Department of Transportation to implement that directive. In 1984 Reagan issued another directive setting out the steps government would take to help business move into space. NASA in 1984 created an Office of Commercial Programs; the same year, the 1958 Space Act was amended to direct NASA "to seek and encourage, to the maximum extent possible, the fullest commercial use of space."

There was a fundamental problem with all this activity. It was based on ideology and hope, not reality. Warnings that the market for Earth observation data would not support a private business without government subsidy were ignored in the push toward privatization. While steps to commercialize space launch were taken, they were paralleled by White House approval of marketing the space shuttle as a launcher for fee-paying customers, a step described as a "paradoxical policy that would stymie the creation of a domestic launch industry." There were few demonstrations of the technical feasibility of most of the new space products with purported high economic payoffs, and little, if any, independent analysis of whether those products could compete with Earth-bound equivalents, given the high costs of operating in space. Writing at the 1984 height of the enthusiasm for space commercialization, this author suggested "large, new, commercial payoffs are still 15–20 years away; before then, the cupboard will be almost bare." Even that analysis was optimistic; payoffs from most of the commercial space activities touted in the early 1980s are just beginning to appear, and an industry based on materials processing in the microgravity space environment has never appeared.

Any chance that there would be short-term breakthroughs leading to commercial payoffs were also a casualty of the 1986 *Challenger* accident and the subsequent cutoff of private sector access to space for commercially oriented research. In addition, the date for space station entry into service was by the final years of the Reagan administration rapidly receding. David Thompson, president of the then-new entrepreneurial space company Orbital Sciences, suggested in late 1986 that "we are on the verge of a major policy failure. Our visions of sugar plums have stayed just that." By the end of the Reagan administration, *Aviation Week* was reporting that space commerce had been "stripped of the allure and gold rush claims of the early 1980s" and that the adjusted perception of the long-term and high-risk character of commercial space activities had "slowed or eliminated institutional investment interest" in those activities.[10]

While actual economic payoffs from space commercialization are thus not part of the Reagan space legacy, that does not mean that there were no lasting impacts of Reagan administration attention to that issue. Although a commercial launch business was slow to develop, the combination of the 1983 decision to support ELV commercialization and the 1986 decision taking commercial and foreign satellites off of the shuttle after the *Challenger* accident did allow the industry to survive.

More significantly, the early attention to space commercialization created the expectation that there would at some point in time be opportunities for private sector activities in space, and that the government had an important role in facilitating those opportunities. Many of the policy incentives discussed in the 1980s, such as government-industry partnerships and government procurement of space services rather than space hardware, are now in use. The emergence of a dynamic private space sector in recent years can trace its heritage to the Reagan administration encouraging the first-generation space entrepreneurs of the 1980s. The development of a commercial space sector can appropriately be considered a Reagan space legacy.

Another lasting impact of the Reagan administration attention to space commercialization was legitimizing government agencies other than NASA (and the Department of Commerce's National Oceanic and Atmospheric Administration, NOAA) as participants in civilian space policy development. The Department of Commerce and the Department of Transportation have stayed active in promoting and regulating private sector space activities, and continue to participate in interagency space policy processes. In their roles as NASA antagonists during the Reagan administration, representatives of both departments arguably played a productive role in forcing the space agency to begin to adapt to post-*Apollo* reality, a process still underway almost a half century after Americans traveled to the Moon.

Leading Through Cooperation

Reagan administration promotion of international space cooperation is perhaps Ronald Reagan's most clear-cut positive space legacy. Despite almost paranoid concerns among some officials in the Defense and State Departments regarding the possibility of space cooperation leading to unwanted technology transfer, the administration made cooperation in the space station a high-profile activity, advocating it at successive summit gatherings of the leaders of the United States and its closest allies. The international partnership in the space station is today held out as a model of successful cooperation, not only for future space activities, but for other high-technology undertakings. The fact that the partnership existed was an important influence in protecting the space station from cancellation at several points in its history. Persistent administration efforts after 1984 to reestablish space cooperation with the Soviet Union, while producing only modest achievements while Reagan was president, served as the foundation for increasing collaboration with post-Soviet Russia, leading to the 1993 merger of each country's space station program.

By demonstrating U.S. leadership through space cooperation, the space program remained an important instrument of U.S. foreign policy. In the run up to the space station decision, NASA's Beggs noted that "the space station uniquely lends itself to international cooperation. If we can attract that international cooperation, then other nations will cooperating with us in the resources they spend, rather than competing with us." The Reagan administration recognized this reality, and acted on it.

While there have been a number of tangible benefits to the U.S. space program from cooperation with other spacefaring countries, a primary payoff from that cooperation has been its contribution to the perception of U.S. space leadership in a world of the increasingly capable space efforts of other countries. Assistant Secretary of State John Negroponte in 1987 told the NASA Advisory Council Task Force on International Relations in Space that U.S. space activities "remain flagships of the American scientific and technological enterprise. Our space achievements, and the perception that we have greater capability than any other nation to act in space, have been important elements of our symbolic and world leadership." Space leadership, he stressed, supports the perception that the United States is "the most attractive partner for association and cooperation." This situation persists today.[11]

The Search for Purpose

President-elect Reagan's space transition team in December 1980 recommended "that the purpose and direction of the U.S. space effort be defined." Eight years later, as a new president prepared to take office, the National Academies of Science and Engineering chartered an ad hoc Committee on Space Policy to answer the question: "What are the U.S. purposes in space?" Creating this committee was an indication that, at least in the judgment of the leaders of the U.S. technical elite, the Reagan administration had failed to follow the advice of its transition team, and that the nation's space program was still without clearly understood "purpose and direction."

In its report, the Committee on Space Policy (this author was a member of the committee) suggested that "a key problem in U.S. civil space policy is the lack of a widely understood purpose, direction, and time scale," in particular for the U.S. human spaceflight effort. The committee also observed that "the lack of long-term, widely-supported, attainable goals … has led the U.S. space program to promise greater performance than its resources enabled it to deliver."[12]

Was this a fair criticism of the overall Reagan space legacy—leaving behind as it left Washington a space program without a clearly understood purpose or direction, promising more than it could deliver? Certainly, over the eight years that Ronald Reagan was in the White House, there had been repeated calls for setting a dramatic space goal such as a return to the Moon or beginning to prepare for human voyages to Mars. There was intense criticism from space-interested members of Congress and the broader space community, particularly after the *Challenger* accident, of the lack of presidential leadership in defining a future direction for U.S. space efforts. One of the primary complaints regarding the decision to develop a space station that it, like the space shuttle, was an infrastructure capability, a means to achieve broader goals, not a goal in itself.

There were available for choice well-conceived proposals for the future that the Reagan administration could have embraced. Among the many reviews of the space program that had been carried out during the Reagan administration, both the Ride report, *Leadership and America's Future in Space*, and the National Commission on Space report, *Pioneering the Space Frontier*, had proposed a "purpose and direction" for the U.S. space effort based on human travel to the Moon and Mars.

The detailed account of space policy development provided in this study suggests that the question of whether to adopt an ambitious space goal beyond the space station was never presented to Ronald Reagan for decision. It is not clear how Reagan would have reacted to such a proposal; given his rhetoric, there is some possibility he might have approved it. Because Reagan seldom took the initiative in proposing policy or program actions (the Strategic Defense Initiative being a notable exception), this implies that there was a continuing judgment among Reagan's advisers, aware of the many proposals for such a goal, that such an initiative was neither in the country's interest nor affordable in the context of the administration's fiscal policies. As the Democrats took control of the Congress, it was also judged not politically feasible. While a lower-level White House official like Gil Rye at the National Security Council may have hoped to get senior-level approval of significant space budget increases as part of creating a National Space Strategy, that hope was from the start unrealistic, given the administration's other priorities and political realities. Science Adviser Jay Keyworth's call for setting out the ambitious goals that a space station was intended to serve also found few supporters.

An additional constraint on adopting an ambitious space goal was the unanticipated multibillion dollar cost of replacing *Challenger*, procuring ELVs for national security and occasional NASA use, and redesigning expensive national security satellites so that they could be launched on those ELVs. The National Academy Committee on Space Policy commented that in the final years of the Reagan administration, "faced with spending constraints for all scientific and technological endeavors, the resources required to correct past policy mistakes—especially sole reliance on the space shuttle for access to space—have been diverted from more forward-looking efforts."[13]

The actual Reagan legacy with respect to articulating a "purpose and direction" for the U.S. space program was far from ambitious. The final Reagan National Space Policy did set out as a "long-range goal" expanding "human presence and activity beyond Earth orbit into the solar system." NASA's Fletcher claimed that "this is a goal of enormous significance with potentially historic future implications." His enthusiasm for the statement was not widely shared; most observers judged the statement to have little practical impact on space program activities. The final Reagan administration budget provided only a very modest $100 million to begin developing the technology needed for future exploration.

The absence of a defined long-range goal, particularly in terms of identifying one or more destinations for future travel, may well have been politically realistic; it is not at all clear that beyond the space community there would have been public or political support for such goal. Here the experience of Reagan's successor as president, George H.W. Bush, is instructive. On July 20, 1989, the

20th anniversary of the first lunar landing and exactly six months after becoming president, Bush set forth the kind of space vision that advocates had hoped to see from Ronald Reagan. He declared: "We must commit ourselves anew to a sustained program of manned exploration of the solar system and, yes, the permanent settlement of space." Steps toward that goal, he said, would be "first, for the coming decade, for the 1990's: Space Station Freedom, our critical next step in all our space endeavors. And next, for the new century: back to the Moon; back to the future. And this time, back to stay. And then a journey into tomorrow, a journey to another planet: a manned mission to Mars."[14]

The Bush proposal became known as the Space Exploration Initiative. Even though the Bush White House, under the leadership of a reestablished National Space Council, struggled to gain traction for the initiative, it was essentially stillborn.[15] This suggests that the Reagan administration's not setting out a clear "purpose and direction" for the civilian space program was not a policy failure, but rather a realistic judgment of the likely fate of such a step. A space effort with ambitions beyond returning the shuttle to regular operation and developing space station *Freedom* was not to be a Reagan leadership legacy.

Ronald Reagan and the Space Frontier

The notion of space as a "final frontier" was pervasive in Ronald Reagan's presidential rhetoric. This study has included many examples of that rhetoric. To Reagan, the concept of America's role in exploring frontiers was part of his positive, optimistic view of both U.S. history and the country's future.

It must be noted that this view has in recent years become increasingly controversial. Historian Roger Launius, for example, comments that "the construct of the frontier as a positive image of national character and of the progress of democracy has been challenged on all quarters and virtually rejected as a useful ideal in American postmodern, multicultural society." Launius cites another historian, Patricia Limerick, as pointing out that "the frontier myth, used as a happy metaphor by many, should be seen as a pejorative reflection," denoting "conquest of place and peoples, exploitation without environmental concern, wastefulness, political corruption, executive misbehavior, shoddy construction, brutal labor relations, and financial inefficiency."[16]

The relevance of the frontier metaphor to depicting space development will not be discussed here. Rather, the point is that it was a central element in the way Ronald Reagan thought about U.S. space achievements, and shaped his positive view of the link between space leadership and American exceptionalism. One of Reagan's biographers notes that "Reagan wasn't acting when he spoke; his rhetorical power rested on his wholehearted belief in all the wonderful things he said about the United States and the American people, about their brave past and their brilliant future." Activities on the space frontier were part of that future.

Another constant theme in Ronald Reagan's space rhetoric was "leadership." That objective may in fact have been, in Reagan's view, a sufficient declaration of purpose to guide the U.S. space effort. As long as the United States could claim a leading position with respect to other spacefaring countries, Reagan did not appear to push for a more ambitious program. When he was advised that the Soviet Union was threatening to surpass the United States in a particular area of space activity, such as developing a space station, he was willing to approve a U.S. response. One phrase that appeared in many speeches was, "We are first; we are the best; and we are so because we're free."

Even so, it was not leadership at any price. The final Reagan National Space Policy declared that "a fundamental objective guiding United States space activities has been, and continues to be, space leadership. Leadership in an increasingly competitive international environment does not require United States preeminence in all areas and disciplines of space enterprise. It does require United States preeminence in key areas of space activity." There were limits on spending to achieve the leading position in space.

Those limits were evident in NASA's budget allocations. When the Reagan administration took office in January 1981, the NASA budget sat at 0.84 percent of overall government spending; as the administration left the White House in January 1989, it was fundamentally unchanged, at 0.85 percent. This level budget share reflects an underlying reality. As much as Ronald Reagan personally favored the U.S. space program, he and his associates were not willing to increase its priority vis-à-vis other contenders for government support. In this, the Reagan administration was following the dictates of what this author has labeled the "Nixon space doctrine." In his March 1970 statement on the post-*Apollo* U.S. space program, Nixon had said "our space expenditures must take their proper place within a rigorous system of national priorities. What we do in space from here on in must become a normal and regular part of our national life." Ronald Reagan in effect accepted this policy approach as a guide to his space budget decisions; a top Reagan priority was controlling the increase in, if not reducing, government spending.

Many in the space community had hoped, as Ronald Reagan became president, that he would adopt a Kennedy-like attitude toward the space program, being willing to approve bold new initiatives to recapture the spirit of the Apollo program that had dissipated during the 1970s. What they got instead was a president who matched John F. Kennedy in rhetoric but was more similar to Richard Nixon in practice. The Nixon administration had limited the NASA budget to a significantly lower level than at the peak of Apollo, and within that limited budget had approved a major new human spaceflight program, the space shuttle. The Reagan administration kept the NASA budget at the steady, low level where it had settled during the Nixon administration, and within that constrained budget had approved another major human spaceflight program, the space station.

There were, however, important differences between Richard Nixon and Ronald Reagan in their attitude toward the space program. Even as the United States during the time he was in the White House carried out nine flights to the Moon, six of them landing Americans on the lunar surface, Richard Nixon made few space-themed speeches and seldom portrayed the Apollo astronauts as American heroes. Ronald Reagan continually used the space program as an example of American greatness and consoled the nation when seven astronauts died in the *Challenger* accident. While many internal to the space community criticized Reagan for lack of policy leadership in the post-*Challenger* period, it was his eloquent speech on the evening of the accident that the American public most remembered. Nixon was remembered as the person who ended U.S. space exploration; Reagan, as a president who at the end of his presidency claimed that "in the next century, leadership on Earth will come to the nation that shows the greatest leadership in space."

The Reagan space legacy, then, is a combination of memorable rhetoric and largely pragmatic practice. It is an exaggeration to claim that the eight years of the Reagan administration represented "a major turning point in space history." Rather, it was a period during which the space shuttle went from the centerpiece of access to space to an extremely capable system carrying out cutting edge NASA missions and serving to help construct a permanent outpost in orbit; during which the program to develop that outpost was begun; during which the foundational policies to support private sector space activities were first developed; and during which lasting international space partnerships were forged. Perhaps Ronald Reagan expressed it best as he on January 11, 1989, bid farewell to the country and to his associates in the "Reagan revolution," saying, "We did it. We weren't just marking time. We made a difference … All in all, not bad, not bad at all."[17]

Notes

1. Andrew J. Butrica, *Single Stage to Orbit: Politics, Technology, and the Quest for Reusable Rocketry* (Baltimore, MD: The Johns Hopkins University Press, 2003), 13. William Broad, "Reagan's Legacy in Space: More Reach Than Grasp," *NYT*, May 8, 1988, IV-9.
2. When material from earlier chapters is quoted in this analysis, the citations to original sources will not be repeated.
3. *Columbia* Accident Investigation Board, *Report*, Volume I, August 2003, 209.
4. An example of such criticism is in Alex Roland, "Barnstorming in Space: The Rise and Fall of the Romantic Era of Spaceflight, 1957–1986," in Radford Byerly, ed., *Space Policy Reconsidered* (Boulder, CO: Westview Press, 1989).
5. Gary Brewer, "Perfect Places: NASA as an Idealized Institution" in Byerly, *Space Policy Reconsidered*, 159, 163, 166.
6. Ibid., 159.
7. Hans Mark, *The Space Station: A Personal Journey* (Durham, NC: Duke University Press, 1987), 190–191.
8. For a discussion of the influence of the space vision on U.S. space policy, see Howard McCurdy, *Space and the American Imagination* (Washington, DC: Smithsonian Institution Press, 1997). The quote is on 1.
9. Howard McCurdy, *The Space Station Decision: Incremental Politics and Technological Choice* (Baltimore, MD: Johns Hopkins Press, 1990). The quote is on 233.

10. "Washington Roundup," *AWST*, October 6, 1986, 17; "Space Commerce Chastened" and Nicholas Kernstock, "Most Investors Shun High Risk Space Ventures," *AWST*, December 19, 1988, 7, 45.
11. Negroponte is quoted in Task Force on International Relations in Space, NASA Advisory Council, *International Space Policy for the 1990s and Beyond*, October 12, 1987, 20.
12. Committee on Space Policy, National Academy of Sciences and National Academy of Engineering, *Toward a New Era in Space: Realigning Policies to New Realities* (Washington, DC: National Academy Press, 1988), 22, 2, 1.
13. Ibid., 2.
14. George Bush: "Remarks on the 20th Anniversary of the Apollo 11 Moon Landing," July 20, 1989. Online by Gerhard Peters and John T. Woolley, APP, http://www.presidency.ucsb.edu/ws/?pid=17321.
15. For an account of the fate of the Space Exploration Initiative, see Thor Hogan, *Mars Wars: The Rise and Fall of the Space Exploration Initiative* (CreateSpace Independent Publishing Platform, 2013).
16. Roger D. Launius, "The Declining Significance of the Frontier in Space History?" July 30, 2012, https://launiusr.wordpress.com/2012/07/30/the-declining-significance-of-the-frontier-in-space-history/.
17. Ronald Reagan: "Farewell Address to the Nation," January 11, 1989. Online by Peters and Woolley, APP, http://www.presidency.ucsb.edu/ws/?pid=29650.

Index

J. M. Logsdon, *Ronald Reagan and the Space Frontier*,
Palgrave Studies in the History of Science and Technology,
https://doi.org/10.1007/978-3-319-98962-4

H

I

O

Printed by Printforce, the Netherlands